DATE DUE

JAN 30 1978	DEC 11 1998		
JUN 24 1985			
NOV 6 1989			
FEB 28 1991			
MAR 22 1994			
APR 7 1994			

Analysis of Metallurgical Failures

WILEY SERIES ON THE SCIENCE AND TECHNOLOGY OF MATERIALS

Advisory Editors: **E. Burke, B. Chalmers, James A. Krumhansl**

ANALYSIS OF METALLURGICAL FAILURES
V. J. Colangelo and F. A. Heiser
THERMODYNAMICS OF SOLIDS, SECOND EDITION
Richard A. Swalin
GLASS SCIENCE
Robert H. Doremus
THE SUPERALLOYS
Chester T. Sims and William C. Hagel, editors
X-RAY DIFFRACTION METHODS IN POLYMER SCIENCE
L. E. Alexander
PHYSICAL PROPERTIES OF MOLECULAR CRYSTALS, LIQUIDS, AND GLASSES
A. Bondi
FRACTURE OF STRUCTURAL MATERIALS
A. S. Tetelman and A. J. McEvily, Jr.
ORGANIC SEMICONDUCTORS
F. Gutmann and L. E. Lyons
INTERMETALLIC COMPOUNDS
J. H. Westbrook, editor
THE PHYSICAL PRINCIPLES OF MAGNETISM
Allan H. Morrish
HANDBOOK OF ELECTRON BEAM WELDING
R. Bakish and S. S. White
PHYSICS OF MAGNETISM
Sōshin Chikazumi
PHYSICS OF III-V COMPOUNDS
Otfried Madelung (translation by D. Meyerhofer)
PRINCIPLES OF SOLIDIFICATION
Bruce Chalmers
THE MECHANICAL PROPERTIES OF MATTER
A. H. Cottrell
THE ART AND SCIENCE OF GROWING CRYSTALS
J. J. Gilman. editor
SELECTED VALUES OF THERMODYNAMIC PROPERTIES OF METALS AND ALLOYS
Ralph Hultgren, Raymond L. Orr, Philip D. Anderson and Kenneth K. Kelly
PROCESSES OF CREEP AND FATIGUE IN METALS
A. J. Kennedy
COLUMBIUM AND TANTALUM
Frank T. Sisco and Edward Epremian, editors
TRANSMISSION ELECTRON MICROSCOPY OF METALS
Gareth Thomas
PLASTICITY AND CREEP OF METALS
J. D. Lubahn and R. P. Felgar
INTRODUCTION TO CERAMICS
W. D. Kingery
PHYSICAL METALLURGY
Bruce Chalmers
ZONE MELTING, SECOND EDITION
William G. Pfann

Analysis of Metallurgical Failures

V. J. COLANGELO

Consulting Metallurgist
Benet Research and Engineering Laboratories
Watervliet Arsenal

F. A. HEISER

Chief, Advanced Engineering Division
Benet Research and Engineering Laboratories
Watervliet Arsenal

A WILEY-INTERSCIENCE PUBLICATION

JOHN WILEY & SONS, New York • London • Sydney • Toronto

Library of Congress Cataloging in Publication Data:

Colangelo, Vito J.
Analysis of metallurgical failures.

(Wiley series on the science and technology of materials)
"A Wiley-Interscience publication."
Includes bibliographical references.
1. Metals—Fatigue. 2. Metals—Fracture. 3. Metals—Testing. I. Heiser, Francis A., joint author. II. Title.

TA460.C62 1974 620.1'6'3 73-19773
ISBN 0-471-16450-X

Printed in the United States of America

10 9 8 7 6 5 4 3 2 1

Preface

The proper analysis of a component failure can provide valuable assistance in determining the validity of a product design. Errors in selection and design as well as materials defects and shortcomings in processing can frequently be revealed. These problems can often be detected in prototype testing or early in production, which results in substantial savings in time and money. However, such analysis is often neglected. In many small organizations sophisticated materials and process engineering groups are not available. Too often, the responsibility falls to an engineer who lacks the background to interpret the available information.

In the course of several years of failure analysis it has become evident that a text clearly presenting the techniques and approaches of this specialty would be extremely useful to the working engineer. A comprehenisve work of this type is presently unavailable. Although much of the information has been published, the material is scattered throughout many sources and does not form a coherent presentation. In addition, the present time appears appropriate because of the significant advances that have occurred in the past decade, particularly in the fields of electron microscopy, fracture mechanics, and solidification.

In this volume we present a coordinated approach to failure analysis. Those destructive and nondestructive evaluation techniques commonly available are described, as are suggestions regarding their advantages, limitations, application, and meaning. Typical problem areas are approached from the viewpoints of physical and mechanical metallurgy. An attempt is made to show the interrelation between the practical and the theoretical, so that failure analyses can best be resolved and their recurrence prevented.

V. J. COLANGELO

F. A. HEISER

Troy, New York
July 1973

Contents

1

Introduction

1.1 FUNCTION OF FAILURE ANALYSIS

The primary reasons for conducting an analysis of a metallurgical failure are to determine and describe the factors responsible for the failure of the component or structure. This determination may be motivated by either sound engineering practice or by legal considerations.

From an engineering standpoint, the proper application of failure analysis techniques can provide a valuable feedback to design problems and material limitations. The optimum design is one in which the requirements are slightly exceeded by the capabilities in all circumstances. This aim is seldom realized because of the obvious difficulty in recognizing or defining precisely the various demands that the system must be called upon to meet. This latter aspect of the design requirements is generally met by a sound engineering device, the application of safety factors. However, how much of a safety factor is appropriate? To grossly overdesign the component is economically extravagant and can inadvertently overload other parts of the structure. Underdesigning of the component leads to its premature failure, is economically wasteful, and, most important, could jeopardize life. The role failure analysis plays in the overall design and production of a component is shown in Fig. 1.1. It is in this role, as a design adjunct, that failure analysis can play a maximum part, since the most sophisticated simulation testing can never duplicate the varied and interacting conditions found in actual service.

The legal reasons for failure analysis are equally compelling. Recently, the emphasis in product liability laws has shifted from the status of the plaintiff to the nature of defectiveness in a product.[1] Under present day law, regardless of the legal theory that the plaintiff may choose to proceed upon, he must prove that a defect in design or materials exists in the product. Consequently, if the aim is the successful defense or proof in a

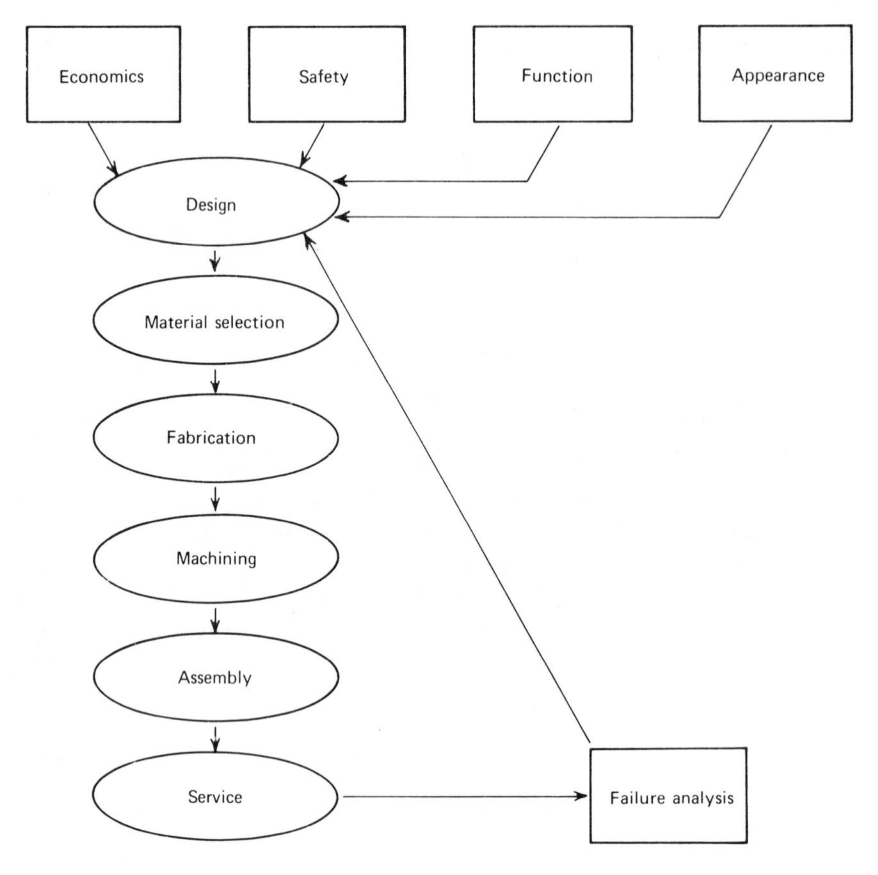

Fig. 1.1 The relationship of failure analysis to the design and production of a component.

product liability case, the question of whether a defect existed must be determined.

1.2 FACTORS RELATED TO FAILURE

Over the years the fundamental factors causally related to failure or shortening of service life have been identified:

1. Design.
2. Improper selection of material.
3. Heat treatment.
4. Fabrication.
5. Improper machining and assembly.

The actual failure may be due to any one of these factors acting independently or to the interaction of several of them. The exact cause is often not easy to ascertain and can only be resolved after an intensive investigation.

Upper management too often has little understanding of the factors and conditions leading to a failure. Frequently, after a field failure the component, perhaps greasy or encrusted with dirt and rust, is rushed to the engineering department where an immediate answer about the cause of failure is expected. This attitude is never helpful and can hinder an investigation by creating undue pressure on the investigating team. The only countermeasure is patient education.

1.3 INVESTIGATIVE PROCEDURE

When a metallurgical failure occurs and an investigation is required, it is reasonable to ask, "Where does one begin?" and to use the answer advanced by the King of Hearts in Lewis Carroll's *Alice in Wonderland*—"One begins at the beginning." For the investigating metallurgist, the beginning occurs when he is called in on the case; however, the component was conceived, designed, and fabricated during some previous period. In this section we show the significance of determining the history prior to the failure and outline the subsequent course of action.

1.3.1 Documentary Evidence

It is advantageous and even necessary to collect documentary evidence, such as test certifications by vendors, mechanical test data, and in-house evaluations and reports. This category of data also includes pertinent specifications and warranties as well as design drawings indicating dimensions and surface finish.

The examination of correspondence, too, can be illuminating, as letters between the producer and consumer or the technician and engineer. The importance of such "technical paper" cannot be overemphasized, since the questions of adherence to a procedure and compliance with a specification can become major points in the investigation, especially a legal one.

1.3.2 Service Conditions

Information about the actual operating or service conditions is extremely relevant. Data on the temperature level and range should be collected and compared with the design or intended service conditions to determine whether abnormal conditions were produced by improper operation, maintenance, temperatures, and so on. In a design where there has been no change in materials, design, or processing, the failure can often be attributed to a change in usage which created an abnormal service condition.

Equally significant are any data about the environmental conditions—composition of surrounding fluids, relative humidity, contamination, cleanliness conditions, and so on.

1.3.3 Materials Handling, Storage, and Identification

Failures can often be attributed not to conditions undergone in service but to deficiencies or errors in handling, identification, or storage. For example, tong marks received in handling can act as nuclei for quench cracks. Identification marks caused by stamping or etching can act as focal points for stress corrosion or fatigue.

A fatigue failure[2] which initiated from a numeral on the surface is shown in Fig. 1.2*a*. The electric etch produced temperatures high enough to austenitize the material, and the subsequent transformation resulted in brittle, untempered martensite. Abrasions such as those illustrated in Fig. 1.3 are also stress raisers and can create similar problems.

Storage conditions are also important. Welding electrodes used on hydrogen-sensitive materials must be kept dry. Failure to do so can result in hydrogen embrittlement of the weld.

Finished parts should have fingerprints removed and should be greased prior to storage to avoid corrosion.

Obviously, the failure to correctly identify materials can also result in materials problems. The error may be in the complete omission of any

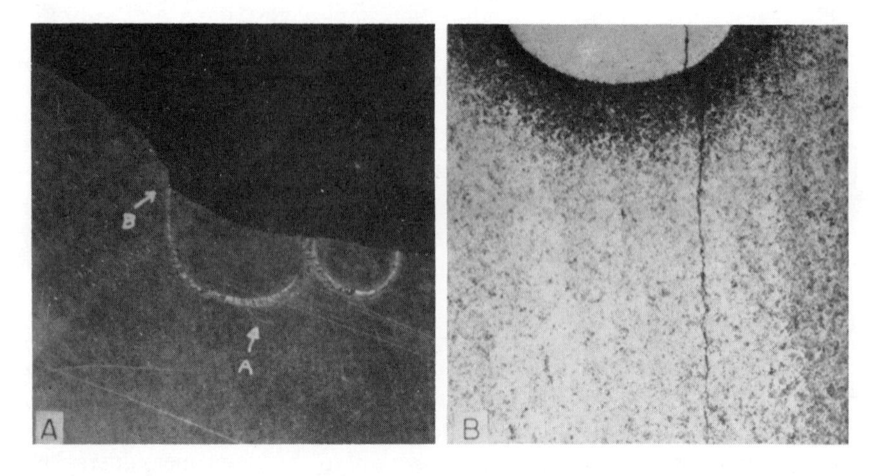

Fig. 1.2 Fatigue failure which initiated from electric etch marks. (*a*) Macroscopic view of fracture origin. (*b*) Metallographic section showing crack originating in etch mark.

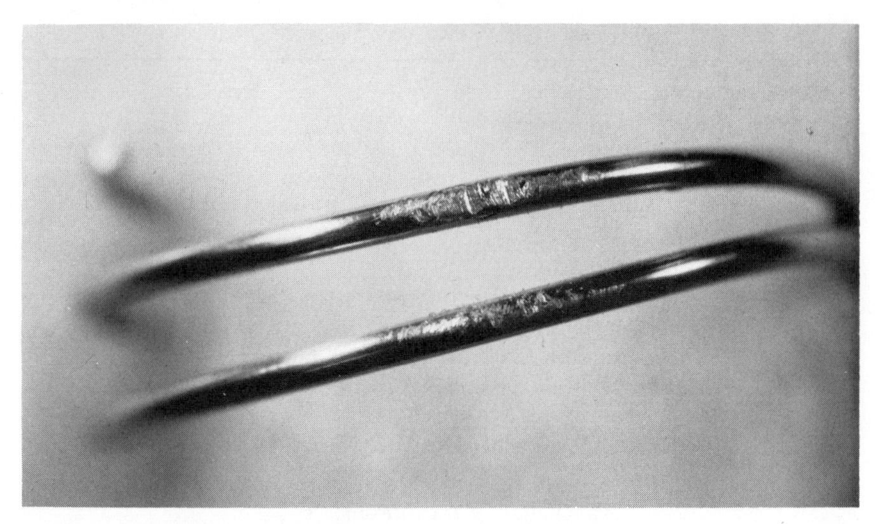

Fig. 1.3 Abrasions and gouges that can act as stress raisers.

identification or in the incorrect identification of the base metal or weld metal, or equally important, in the strength level.

1.3.4 Interviews

No investigation is complete without testimony from persons who have information about the failure—witnesses to the failure or personnel associated with the processing or testing. Such testimony, of course, is subject to bias, either unintentional or deliberate. Thus the failure may be directly or indirectly traced to a noncompliance with a prescribed procedure, and the witness will not care to implicate himself, (e.g., the operation of a furnace at too high a temperature or the use of a gauge known to be inaccurate).

This bias may be revealed by other testimony or by test data; or it may remain hidden, which is worse. The important point, however, is that the investigator should use the interview only as a tool; he should not place unreasonable emphasis on these data, but should analyze them judiciously.

1.3.5 Testing

The balance of the investigation should be concerned with various types of testing, both nondestructive and destructive. The test procedures are outlined in Table 1.1. The details of testing and the information obtainable are discussed in subsequent chapters.

Table 1.1 Procedural Sequence

I. Determine prior history
   A. Documentary evidence
      1. Test certificates
      2. Mechanical test data
      3. Pertinent specifications
      4. Correspondence
   B. Service parameters
      1. Design or intended operating parameters
      2. Actual service conditions
         a. Temperature data (magnitude and range)
         b. Environmental conditions
         c. Service stresses
   C. Details regarding failure as reported by field personnel

II. Nondestructive tests
   A. Macroscopic examination of fracture surface
      1. Presence of color or texture changes
         a. Temper colors
         b. Oxidation
         c. Corrosion products
      2. Presence of distinguishing surface features
         a. Shear lips
         b. Beach marks
         c. Chevron markings
         d. Gross plasticity
         e. Large voids or exogenous inclusions
         f. Secondary cracks
      3. Direction of propagation
      4. Fracture origin

1.4 EVALUATION OF DATA

The value of information gathered from macroscopic and microscopic examination and from the physical, chemical, and mechanical tests performed depends as much on the interpretation of the data as on the raw data themselves. Seldom are the raw data alone completely self-sustaining and sufficient to support the conclusions drawn. In the majority of failures, the true cause is revealed only by a systematic examination of *all* the facts related to the case. The investigator frequently must draw on prior experience in his background, which may be only casually related to the failure. Every item of data must be scrutinized and evaluated for its source, accuracy, and relevance to the entire investigation. The implication of each

Table 1.1—Continued

- B. Detection of surface and subsurface defects
  1. Magnaflux
  2. Dye penetrant
  3. Ultrasonics
- C. Hardness measurements
  1. Macroscopic
  2. Microscopic
- D. Chemical analysis
  1. Spectrographic
  2. Spot tests

III. Destructive
- A. Metallographic
  1. Macroscopic
  2. Microscopic
     - a. Structure
     - b. Grain size
     - c. Cleanliness
     - d. Microhardness
- B. Mechanical tests
  1. Tensile
  2. Impact
  3. Fracture toughness
  4. Special
- C. Corrosion tests
- D. Wet chemical analysis

item and the conclusion that it warrants must be logically analyzed and evaluated to establish whether it is reasonable and consistent with the balance of the information generated. Apparent conflicts created by incompatible data should be resolved by determining whether the datum was erroneous or true with the differences created by time, location of the test, or an abnormality. Seemingly incompatible data should not be arbitrarily rejected. These data may be indicative of some abnormality, and additional testing may be required to form a sound conclusion.

The unsuccessful performance of a structure or component can, in general, be traced to the following modes of failure: ductile or brittle fracture, fatigue, creep, corrosion, or wear. Failure may result via the independent action of any of these modes; however, the final failure is often caused by the simultaneous or sequential activity of several mechanisms.

One mechanism may create stress raisers while another may promote the initiation of a crack and its subsequent growth. Thus although there may be one primary mechanism, contributory mechanisms also exist. The role of each of these failure modes is examined in later chapters with the aim of developing a proficiency in the identification of the primary and secondary causes of failures.

REFERENCES

1. H. A. Wilson and R. J. Wampler, *The Law of Products Liability and the American Manufacturer*, private publication.
2. F. B. Stulen and W. C. Shulte, *Met. Eng. Quart.*, Vol. 5, No. 3, 30–38 (August 1965).

2

Mechanical Testing

A wide variety of material mechanical tests are available. In this chapter the more common ones are considered, such as tensile testing, impact testing, fracture toughness testing, and hardness testing.

2.1 TENSILE TESTING

A tensile test is performed by applying a uniaxial tensile load to a test bar. The load is gradually increased until the bar breaks. The increasing load is measured against the increasing elongation using an extensometer (Fig. 2.1). Usually, the tensile test bar has a round cross section, although flat plate specimens are also used. Figure 2.2 illustrates the various types of specimens.[1]

2.1.1 Engineering Stress-Strain

Tensile test data can be considered based on the engineering stress-strain or true stress-strain analyses. The main difference is that the former uses the original area and length, whereas the latter considers instantaneous values of area and length. Since the true stress-strain analysis is of little significance in a failure, it is not considered here. The engineering stress-strain analysis is used to determine the various mechanical properties cited in specifications. A typical stress-strain curve is shown in Fig. 2.3. Such curves are obtained from load-elongation diagrams by the following relations:

$$S = \frac{P}{A_0}$$

$$e = \frac{\Delta l}{l_0}$$

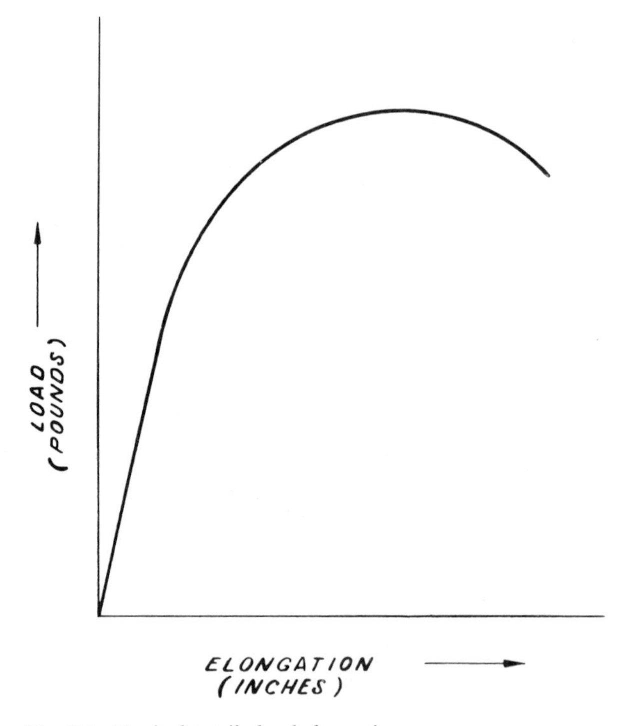

Fig. 2.1 Typical tensile load-elongation curve.

where

$$S = \text{stress}$$
$$P = \text{load}$$
$$A_0 = \text{original area}$$
$$e = \text{strain}$$
$$\Delta l = \text{elongation}$$
$$l_0 = \text{original gauge length}$$

In the elastic strain region, stress is proportional to strain according to Hooke's law:
where

$$S = Ee$$

E = Young's modulus, elastic modulus, or modulus of elasticity.

At the proportional limit, the stress-strain relation deviates from linearity. At a slightly higher stress, the elastic limit, the strain becomes

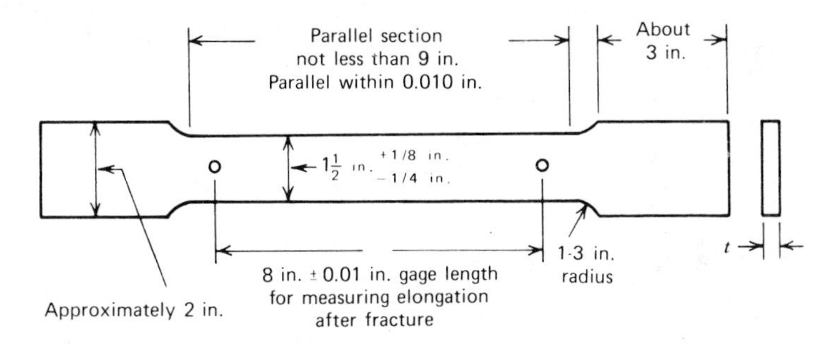

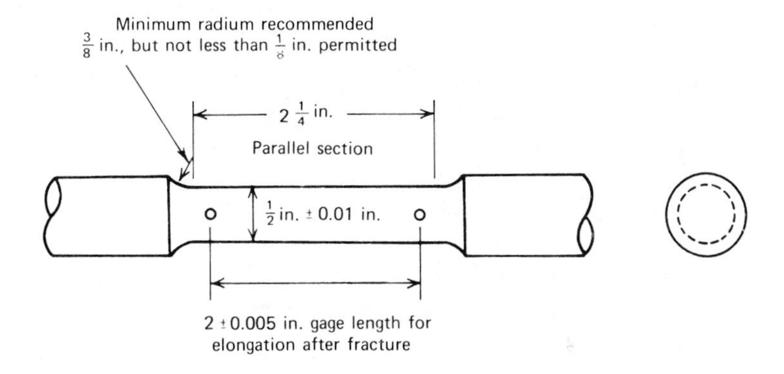

Fig. 2.2 Types of tensile specimens commonly used (Reproduced by permission, from Metals Handbook, American Society for Metals, 1948).

plastic. Proportional limit and elastic limit are of interest since it is usually undesirable to introduce plastic strain in service. However, these limits are very difficult to determine experimentally.

Yield strength (*YS*) is used as the limiting stress for elastic behavior. The yield strength is a defined quantity; it is a stress at which a certain value of strain occurs. There are a variety of methods for defining *YS*; the 0.2% offset method, shown in Fig. 2.4, is the most common. With this technique, 0.2% strain is measured on the horizontal axis, and a line parallel to the elastic modulus line is drawn. The stress at which this line intersects the curve is defined as the yield strength. It is always advisable to indicate the method used to calculate yield strength. In some materials, notably mild steel, a definite yield point is seen and it is unnecessary to use a defined yield strength.

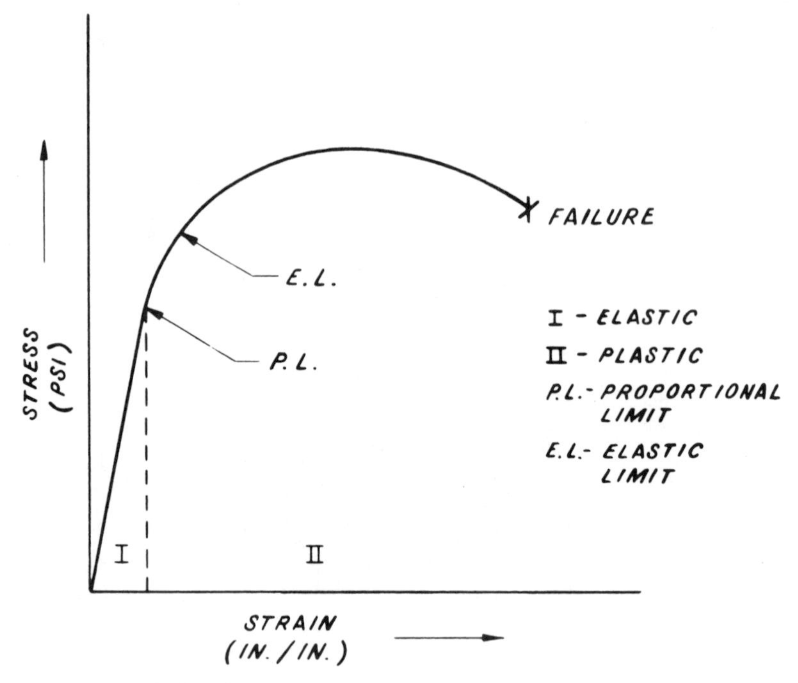

Fig. 2.3 Typical engineering stress-strain curve.

After the material becomes plastic, the load increases until it reaches a maximum. Unless the material is extremely brittle, the load then decreases until the specimen breaks. The maximum load corresponds to the ultimate tensile strength. The subsequent decrease in load implies a decrease in stress. However, stress increases until failure occurs since necking, that is, a localized decrease in cross-sectional area, occurs at the point of maximum stress.

Tensile ductility is measured after the completion of the test. Two measures of ductility are used, namely, elongation and reduction in area; both are expressed as percentages. Although both measure ductility, they are significantly different. Elongation (*El*) measures the uniform strain prior to necking, whereas reduction in area (*RA*) measures the localized strain that occurs in the necked region. In both cases, it is necessary to reassemble the broken test bar and then make the following calculations:

$$\%\ El = \frac{l_f - l_0}{l_0} \times 100$$

and

$$\% RA = \frac{A_0 - A_f}{A_0} \times 100$$

2.1.2 Instability in Tension (Necking)

Necking occurs at maximum load (not maximum stress). Prior to necking, the test bar is plastically deformed, which causes an interaction between the increase in stress (due to a decreasing cross-sectional area and increasing load) and the increasing strength (due to strain hardening). Instability occurs when the increase in stress is greater than the increase in strength.

When necking occurs the stress state changes from uniaxial to triaxial tension (Fig. 2.5), which restricts the tendency of the material to deform plastically.[2] Higher stresses are needed for additional plasticity.

Mohr's diagram, shown in Fig. 2.6, shows that changing from a uniaxial to a multiaxial stress system makes it more difficult to deform the material, since a multiaxial system lowers the shear stresses that cause plastic deformation. If the material is brittle, the tensile stresses cause it to fail before any noticeable plasticity occurs. Failure with little plastic deformation is often used as a criterion for brittle fracture.

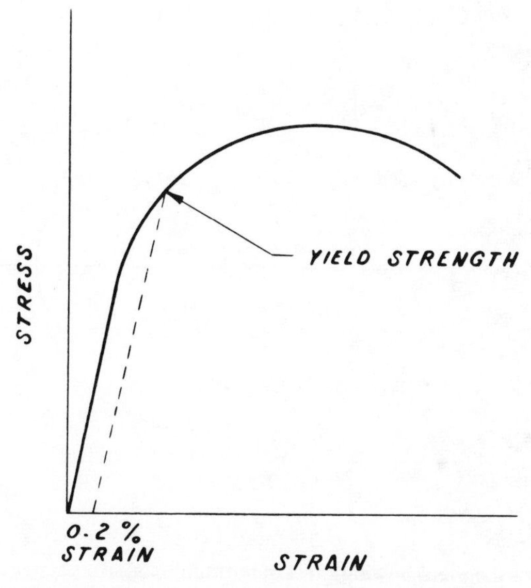

Fig. 2.4 Method of determining 0.2% *YS* from stress-strain curve.

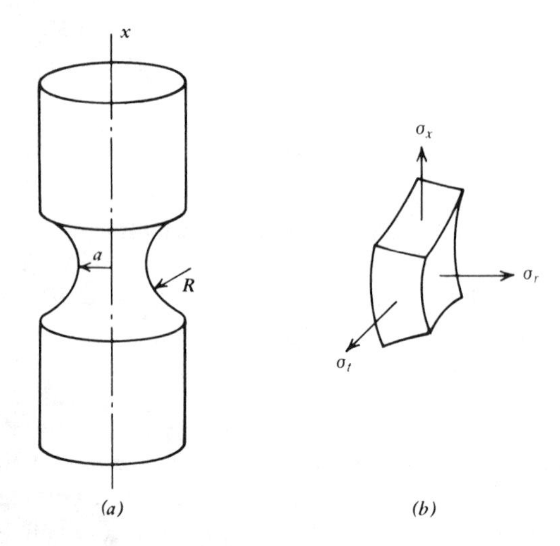

Fig. 2.5 Triaxial stress state with onset of necking (From Mechanical Metallurgy by G. Dieter. Copyright 1961, McGraw Hill. Used with permission of McGraw Hill Book Company).

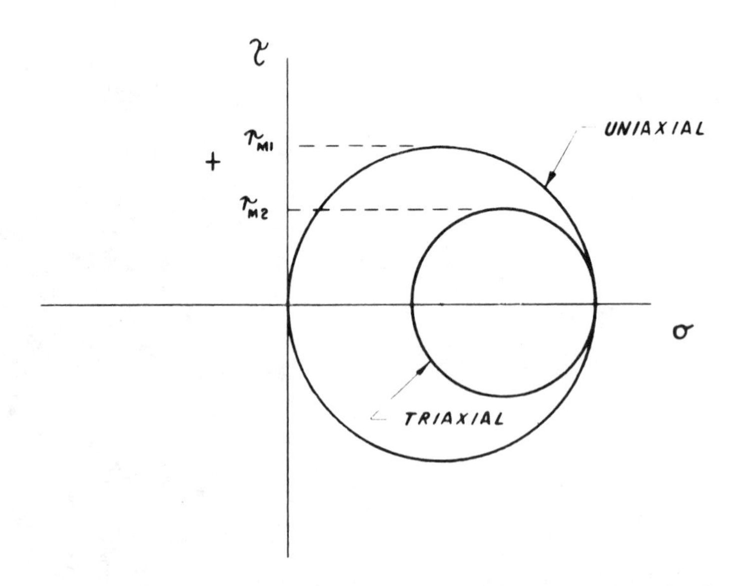

Fig. 2.6 Mohr's diagram showing effect of the multiaxial stress system on shearing stress.

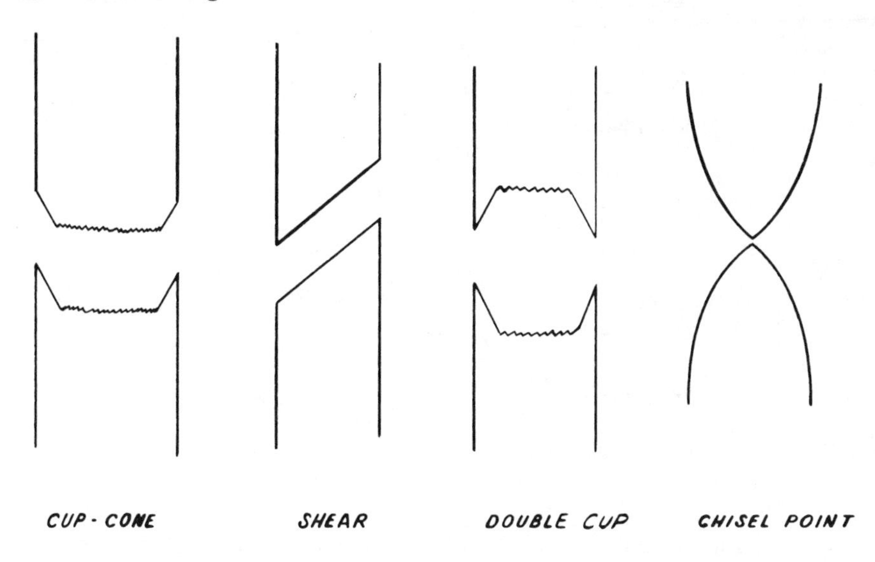

Fig. 2.7 Typical fracture profiles of tensile test bars.

When failure occurs, the necking leads to a variety of fracture surfaces, depending on the material's ductility (Fig. 2.7): cup-cone, double-cup, shear, and chisel point.

2.1.4 Notch Tensile Test

Tensile tests are conducted on smooth specimens. A modification, the notched tensile test, is employed to develop the notch sensitivity of the material.

The notch region is similar to the necked region of a smooth tensile bar. The information in the previous section, about the triaxial stress state, can also be applied to this test. The important parameters in a notch test are notch sharpness (a/r) (Fig. 2.8), and notch depth $[1 - (d/D)^2]$.

One property of prime interest is determined from a notch tensile test, that is notch strength, where

$$\text{notch strength} = \frac{\text{maximum load}}{\text{original area}}$$

A measure of the notch sensitivity is then determined by the ratio of notch strength to tensile strength (TS) to give

$$\text{notch sensitivity ratio } (NSR) = \frac{\text{notch strength}}{\text{tensile strength}}$$

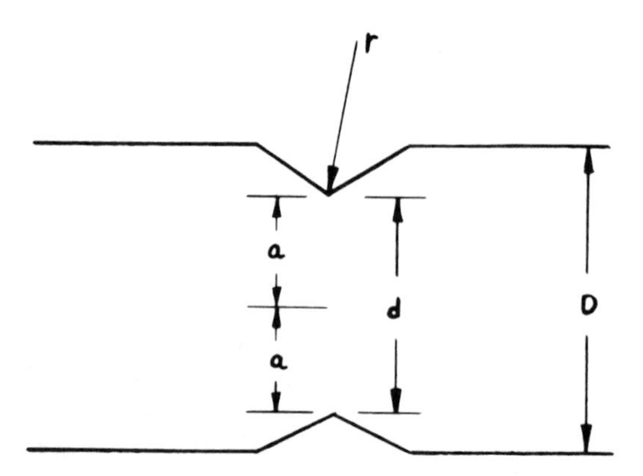

Fig. 2.8 Notched tensile bar showing important dimensions.

Several qualitative criteria are used to determine whether material is notch sensitive. The theoretical limit of NSR for ductile material is 3. If $NSR < 1$, the material is brittle. Any intermediate value indicates a differing extent of notch sensitivity.

The origin of fracture is a function of notch depth and material ductility. As the notch becomes deeper, $\sigma_{r\,\max}$ and therefore, the origin, moves closer to the center. However, as the material becomes more brittle, the origin moves closer to the root of the notch.

2.1.5 Torsion Test

Torsion testing is similar to tension testing since a load-deformation curve is also developed. In the torsion test, a solid or hollow cylindrical specimen is twisted. The deformation measured is the angle of twist. The measure of load is the twisting moment. A typical torsion curve resembles a tensile curve (Fig. 2.9).

Shear stress and shear strain can be calculated as follows:

$$\tau = \frac{16M}{\pi D^3} \text{ solid}$$

$$\tau = \frac{16M\ D_1}{\pi(D_1^4 - D_2^4)} \text{ tube}$$

where

τ = shear stress
M = moment
D = diameter
D_1 = inside diameter
D_2 = outside diameter

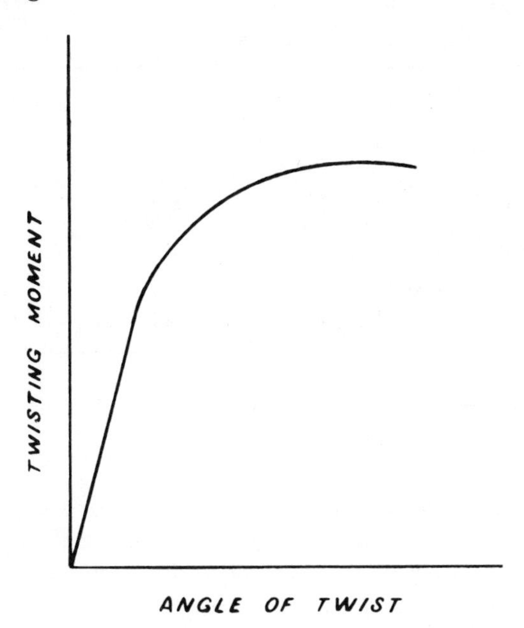

Fig. 2.9 Typical torsion test data showing moment versus angle of twist.

and

$$\gamma = \tan \phi = \frac{r\phi}{L}$$

where

$$\gamma = \text{shear strain}$$
$$\phi = \text{angle of twist}$$
$$r = \text{radius}$$
$$L = \text{length}$$

The torsional yield strength can be determined by the offset method used to measure yield strength. A modulus can be calculated from the elastic linear portion of the curve

$$G = \frac{\tau}{\gamma}$$

where

$$G = \text{modulus of rigidity}$$

The stress distribution in a torsion test is shown in Fig. 2.10.

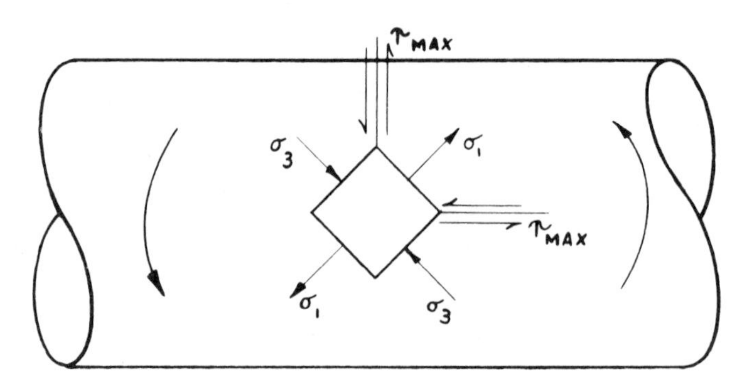

Fig. 2.10 Stress distribution in torsion test.

Tensile stresses lie normal to planes at 45° to the axis of the specimen; the shear stresses are parallel and normal to the axial direction. Ductile materials that fail because of shear stresses do so on planes normal to the axis, whereas brittle faliures due to tensile stresses separate along planes at 45° to the axis, for example, twisting a piece of chalk.

2.16. Factors Affecting Tensile Test Results

Tensile test results are affected by various conditions in the material, such as microstructure and grain size. Several factors associated with testing and location of test specimens are important to failure analysis studies. If they are not considered, erroneous data may result.

2.1.6.1 STRAIN RATE

Most materials are strain-rate sensitive at elevated temperatures, that is, as the strain-rate increases, the strength increases.[3] Although this sensitivity is more important at high temperatures, it is also seen in some materials, for example, mild steel, at room temperature (Fig. 2.11). In this case, the *TS* increases as the strain rate ($\dot{\epsilon}$) increases. Even though the strength at 1000°C is lower than that at room temperature, the increase, as $\dot{\epsilon}$ increases, is greater at 1000°C. Constant $\dot{\epsilon}$ tests are difficult to conduct and seldom used. Usually, constant cross-head speed is maintained.

2.1.6.2 TEMPERATURE

As temperature increases, strength decreases and ductility increases. Other phenomena may occur, for example, elimination of the yield point (Fig. 2.12).[4] Temperature effects may be compounded, depending on the physical metallurgy of the system. Precipitation, strain aging, and recrys-

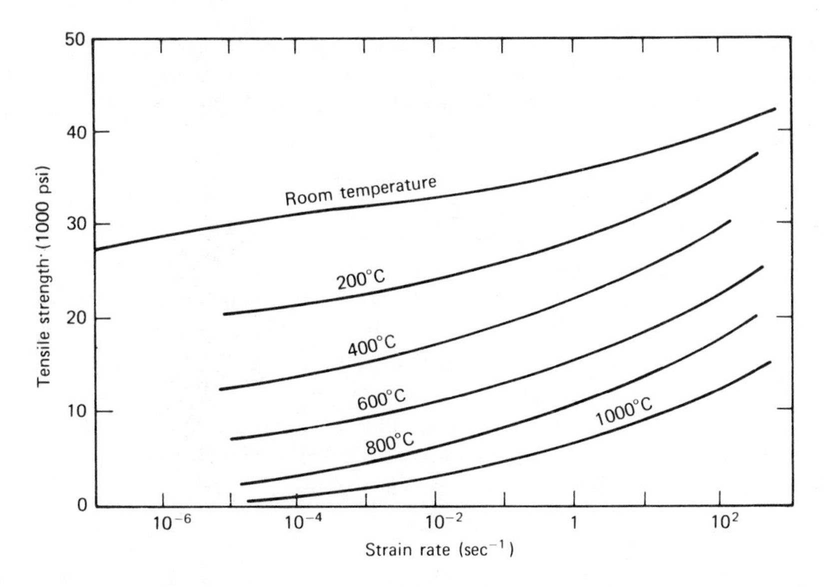

Fig. 2.11 Effect of strain rate on tensile strength for strain-rate sensitive material (From A. Nadai and J. Manjoine, J. Appl. Mech., Vol. 8, 1941. Used with permission of the American Society for Mechanical Engineers.)

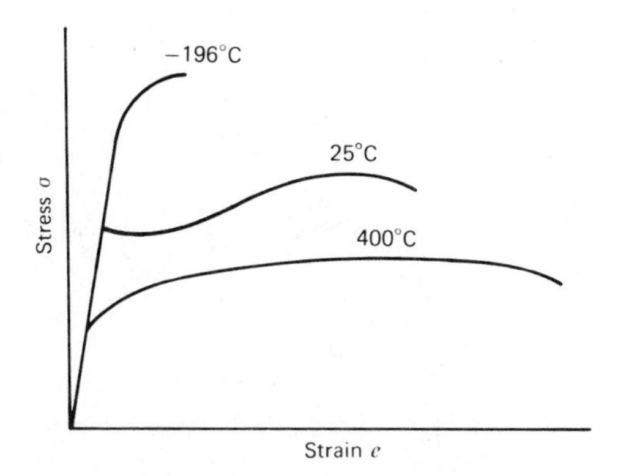

Fig. 2.12 Effect of temperature in elimination of yield point in mild steel (From Mechanical Metallurgy by G. Dieter. Copyright 1961, McGraw Hill. Used with permission of McGraw Hill Book Company).

tallization may occur. In certain cases, metallographic changes may cause an increase in strength at elevated temperature (secondary hardening) or a decrease in toughness (blue brittleness). These effects generally increase as exposure time increases.

2.1.6.3 ORIENTATION

Mechanical properties in wrought products are anisotropic, that is they vary with orientation of the test specimen in relation to the fabrication direction. This is due to the movement of metal during forming which aligns either crystal structure (crystallographic texturing) or microstructure and inclusions (mechanical fibering). Usually, the anisotropy is seen in ductility, particularly *RA*. For example, plotting *RA* versus rolling direction in sheet, as the orientation changes from parallel to the rolling direction (crack arrester) to normal to the rolling direction (short transverse), the *RA* decreases significantly (Fig. 2.13).[5]

Crack arrester	UTS (ksi)	YS (ksi)	RA (%)
P–F	84	54	56
P–F(H)	78	52	68
M(400)–F	175	110	15
M(800)–F	129	82	40
M(1000)–F	90	70	53
M(1000)–F(H)	128	117	56
Short transverse			
P–F	76	49	12
P–F(H)	73	49	30
M(400)–F	158	90	3
M(800)–F	102	72	9
M(1000)–F	99	72	13
M(1000)–F(H)	123	114	8

P–F = banded Pearlite-Ferrite
P–F(H) = homogenized Pearlite-Ferrite
M(X)–F = banded Martensite-Ferrite, tempered at X°F
M(X)–F(H) = homogenized Martensite-Ferrite, tempered at X°F

Fig. 2.13 Effect of orientation on *RA*.

2.2 IMPACT TESTING

Most failures start at stress concentrations, such as notches or cracks. The resistance to failure in the presence of a notch is controlled by the material toughness. A variety of tests have been developed to measure toughness; most are qualitative and comparative. Fracture toughness testing, which utilizes fracture mechanics concepts, is an exception to the rule. This test is potentially a design tool (see Section 2.3.3).

Many service conditions affect the toughness of a material and contribute to brittle failure. Among these are triaxial stress state, rapid loading rate, and low temperature.

To simulate these service conditions, tests that employ a notch and are broken over a range of temperatures under impact loading have been developed.

2.2.1 Charpy Impact Test

The Charpy and the Izod impact tests (Fig. 2.14) are the most widely used. The quantities measured in each test differ, but the general data are similar. For our purposes, only the Charpy test is considered.

Charpy tests can employ two different notches, namely, *V*-notch or key hole. Figure 2.15 shows the specimen loaded in an anvil and struck in impact on the side away from the notch, generating triaxial stresses at the notch root. Since the Charpy test is predominantly a comparative test, standard dimensions and close tolerances are maintained.

A variety of measurements can be made in a Charpy impact test: energy absorbed (ft-lb), percent contraction (necking), and percent fibrosity (macroscopic fracture surface).

A certain amount of energy is absorbed from the striking hammer when the specimen breaks. The absorbed energy is measured directly by using a calibrated scale.

During fracturing the test specimen bends and causes necking on the

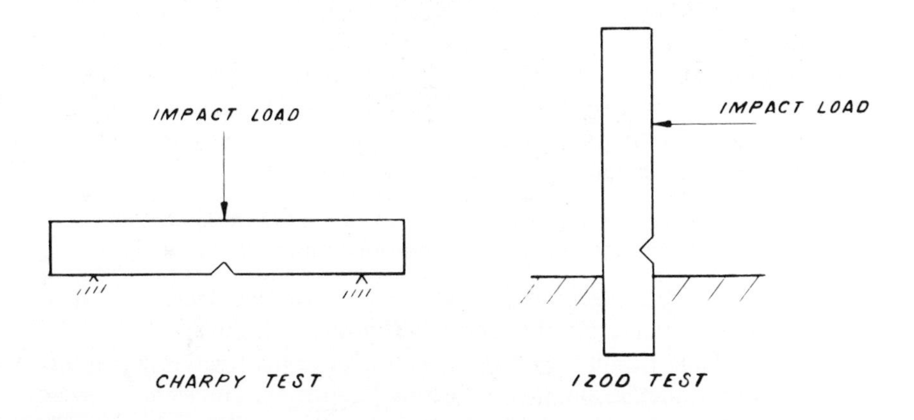

Fig. 2.14 Comparison of Charpy and Izod impact tests.

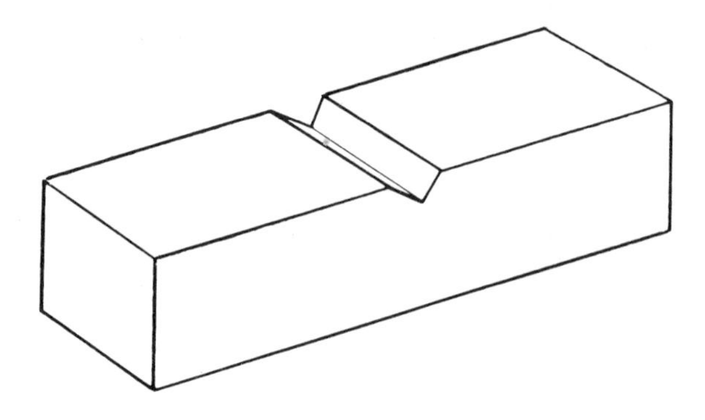

Fig. 2.15 Schematic of Charpy impact test specimen.

sides of the specimen, similar to that seen in a tensile test. After the test, the percent contraction is measured in the necked region by reassembling the test specimen halves. The amount of contraction is sometimes used to indicate toughness.

The fracture surface gives an indication of the toughness of the material. Tough materials show a fibrous or woody texture. Brittle materials show a smoother flat texture.

2.2.1.1 TRANSITION TEMPERATURES

Charpy bars can be broken at one temperature or over a range of temperatures. If a series of test bars is broken over a range of temperatures, a transition is seen, particularly in steels, regardless of the parameter measured (Fig. 2.16). Since the transition temperature indicates a change from high energy to low energy fracture with decreasing temperature, the lower the temperature, the better the material.

Theoretically, there are two transition temperatures, as shown in Fig. 2.17.[6] The significance of the fractures shown is as follows:

Ductility transition—temperature below which material is brittle.
Fracture transition—temperature above which material is ductile.

The curves usually seen in practice show only one transition. This might be expressed in terms of the double transition, as shown in Fig. 2.18. Cracks are difficult to initiate and propagate above the fracture transition, and behavior is ductile. Below the ductility transition, cracks are easy to initiate, and behavior is brittle. Between the two transition regions cracks are difficult to initiate, but once initiated they propagate relatively easily.[7]

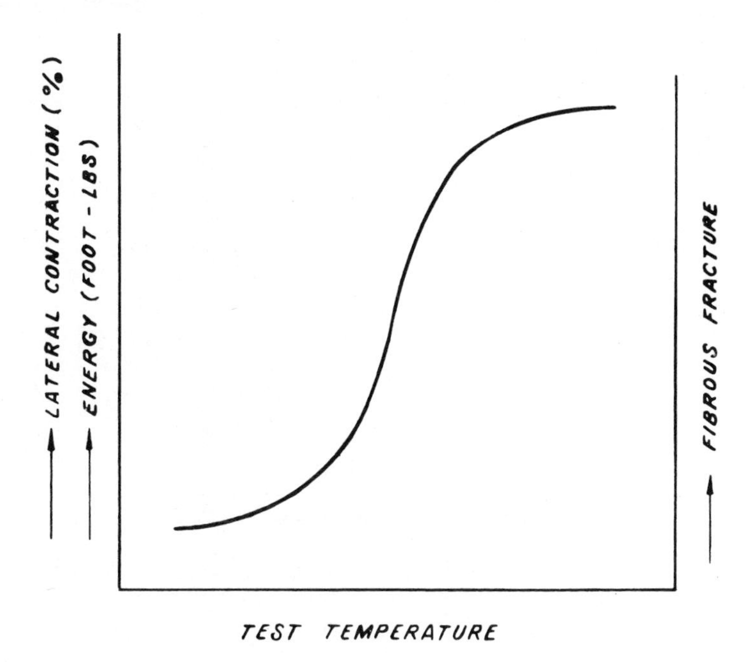

Fig. 2.16 Transition temperatures in Charpy impact testing.

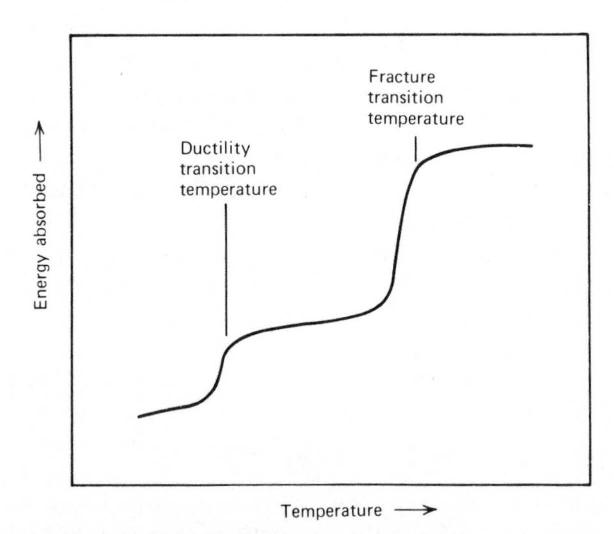

Fig. 2.17 Double transition temperature showing ductility transition and fracture transition (From Mechanical Metallurgy by G. Dieter. Copyright 1961, McGraw Hill. Used with permission of McGraw Hill Book Company).

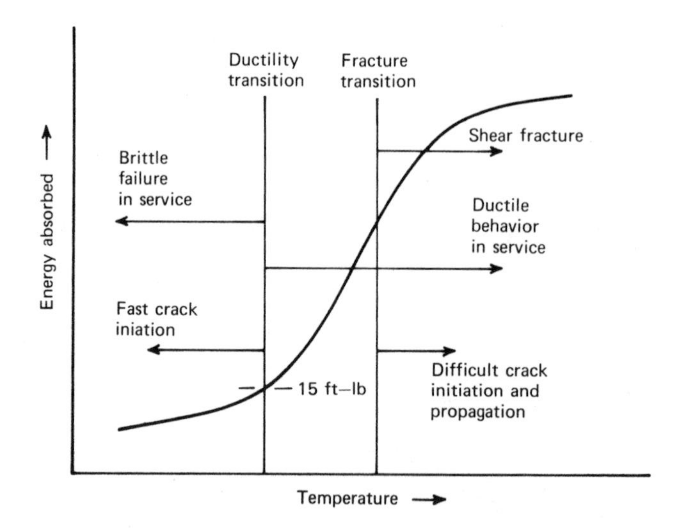

Fig. 2.18 Relating double transition concept in terms of behavior (From Mechanical Metallurgy by G. Dieter. Copyright 1961, McGraw Hill. Used with permission of McGraw Hill Book Company).

2.2.1.2 LIMITATIONS IN USE OF IMPACT TEST

The most serious drawback in the use of the Charpy test is that it is not quantitative and is not easily related to service conditions. Its primary use is as a material evaluator and separator. With sufficient empirical data, the values developed in a Charpy test do have meaning.

The test specimen is small. It is not representative of a whole structure, which is usually inhomogeneous. This accounts for the wide scatter usually seen in several Charpy tests from one structure. Therefore, a sufficiently large sample must be taken.

Charpy tests are not appropriate for some materials, notably high strength steel and nonferrous alloys. In such materials, the energy values are so low that the test is not sufficiently discriminatory.

To compare data from various sources, the machine must be properly calibrated when making Charpy tests. A machine out of calibration will usually give results that are too high.

2.2.2 Modifications to the Charpy Test

Modifications to the Charpy test are being evaluated to overcome some of its limitations, yet maintain its small size, relative simplicity, and general acceptability, particularly in the steel industry. Two of these modifications are the precracked Charpy test and the instrumented Charpy test.

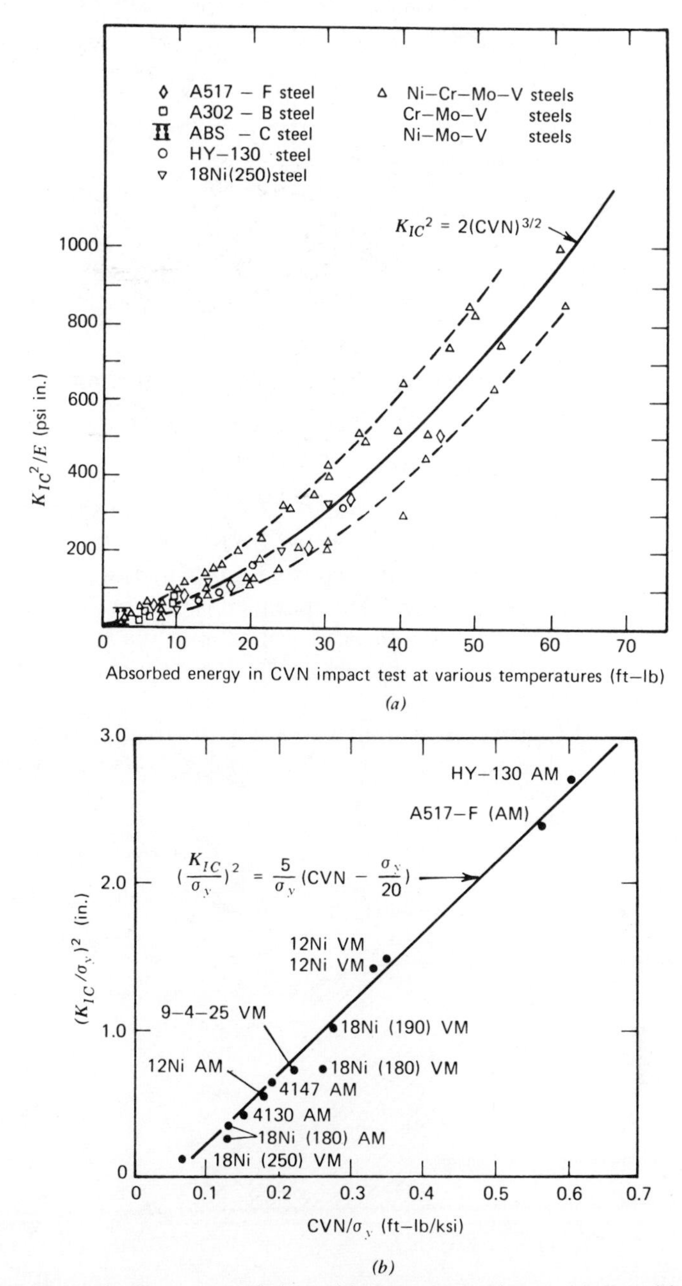

Fig. 2.19 Precrack Charpy correlation with fracture toughness. (*a*) Transition region. (*b*) Upper-energy shelf region. Tests were conducted at +80°F; VM—vacuum-melted; AM—air-melted. (From J. Barsom and S. Rolfe, ASTM STP 466. Used with permission of the American Society for Testing Materials.)

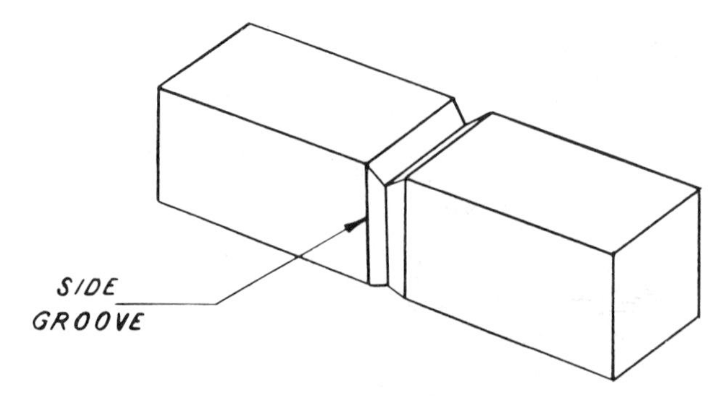

Fig. 2.20 Schematic showing side-grooves in precracked Charpy specimen.

2.2.2.1 PRECRACKED CHARPY TEST

Since one of the objections to the use of the Charpy test is the machined notch, the precracked Charpy test is being investigated. The specimen is the same as the Charpy specimen, except that a fatigue crack (precrack) is developed at the base of the notch. The specimen is then broken in impact. After fracture, the precrack depth is measured and the fracture area is calculated. This is then divided into the fracture energy to give in.-lb/in.^{2} (W/A).

The precracked Charpy test is similar to a fracture toughness test in that it utilizes a precracked specimen. The basic problem is its size. There is a

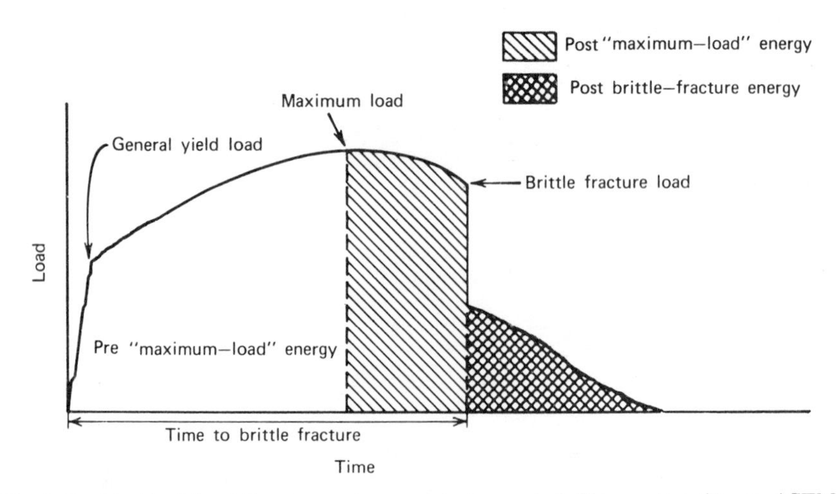

Fig. 2.21 Idealized load-time trace from an instrumented Charpy test (From ASTM STP 466. Used with permission of the American Society for Testing Materials).

stringent size criterion for a valid fracture toughness test, which is not ordinarily satisfied by a Charpy size specimen.

Despite its size drawback, there is still some interest in the precracked test. Attempts have been made to correlate the results with K_{IC} (fracture toughness). Figure 2.19 shows several suggested correlations for steel. However, neither has yet had sufficient usage to warrant direct application. Additional data are needed.

Several modifications to the original test are in use. Side grooves (Fig. 2.20) are often added to restrict the amount of necking. In addition to impact testing, precracked Charpy bars are also broken in slow bend.

2.2.2.2 INSTRUMENTED CHARPY TEST

A second test utilizing Charpy specimens is the instrumented Charpy test, which develops load-time information, in addition to fracture energy. An idealized load-time trace is shown in Fig. 2.21. It is possible to separate the total energy into energy required to initiate fracture, energy required to propagate brittle fracture, and energy associated with shear lip formation (necking).[9]

The precracked Charpy test is too small to determine valid fracture toughness. However, as the loading rate increases, the size for a valid test decreases. Therefore, the instrumented Charpy test might be used for dynamic stress intensity factors.

There are a number of problems with this test, primarily associated with accurate measurement of loads. It would be convenient to measure loads in the striker or anvil. However, responses that are erratic and difficult to interpret have resulted. Therefore, gauges are sometimes attached to the specimen, increasing the testing difficulties.

2.2.3 Alternative Impact Tests

Among the alternative impact tests, dynamic tear test (*DT*) (previously known as the drop weight tear test, *DWTT*), which is the most prevalent, is used on much larger specimens, particularly plate. The alternatives have a major advantage over the Charpy test size, since it is possible to test large plates rather than small specimens.

In the dynamic tear test,[10] a simple beam specimen with a minimum thickness of $\frac{5}{8}$ in. and a mechanically sharp notch is broken in impact by a guided weight with an energy of 250 to 1200 ft-lb. The specimens are not allowed to deflect more than a few tenths of an inch, which prevents gross plasticity. This test can be used to determine the nil-ductility transition temperature (*NDT*) of ferritic steels. This value is determined by testing a series of specimens over a range of temperatures. Thus *NDT* is the highest temperature at which a specimen breaks.

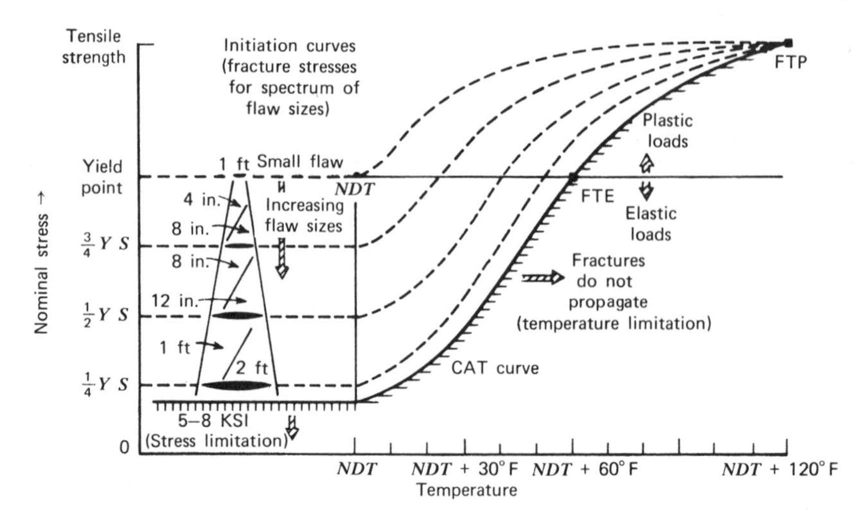

Fig. 2.22 Generalized fracture analysis diagram.

The *NDT* signifies that temperature at which the "small flaw" initiation curve (Fig. 2.22) falls to nominal yield strength stress levels with decreasing temperature. Figure 2.22 is a generalized flaw-size stress temperature diagram dervied from a wide variety of fracture initiation and fracture arrest tests and correlated with the *NDT* determined in a dynamic tear test.[12] In addition, the *NDT* may soon be used to determine dynamic stress intensity factors.[13]

However, the results are still qualitative and comparative. Data cannot be used in design; experience is necessary. Because of the relatively large size required, the tests are also less versatile than the Charpy test. They are confined in use primarily to plates.

2.2.4 Some Considerations in Impact Testing

Impact tests are qualitative. If sufficient history has been generated, the values can have considerable meaning. The classic example is the testing of ship plate during World War II. It was found that plate with C_v values of less than 15 ft-lb cracked.

If there is an option in specimen size, for example, plates, the largest possible specimen should be used. Test bars should be oriented so that the loads are in the same direction as the stresses in the specimen. In gun tubes, where the main stresses are tangential, Charpy specimens, such as those in Fig. 2.23, are taken. The stresses on the Charpy bar act in the same direction as the stresses in the tube.

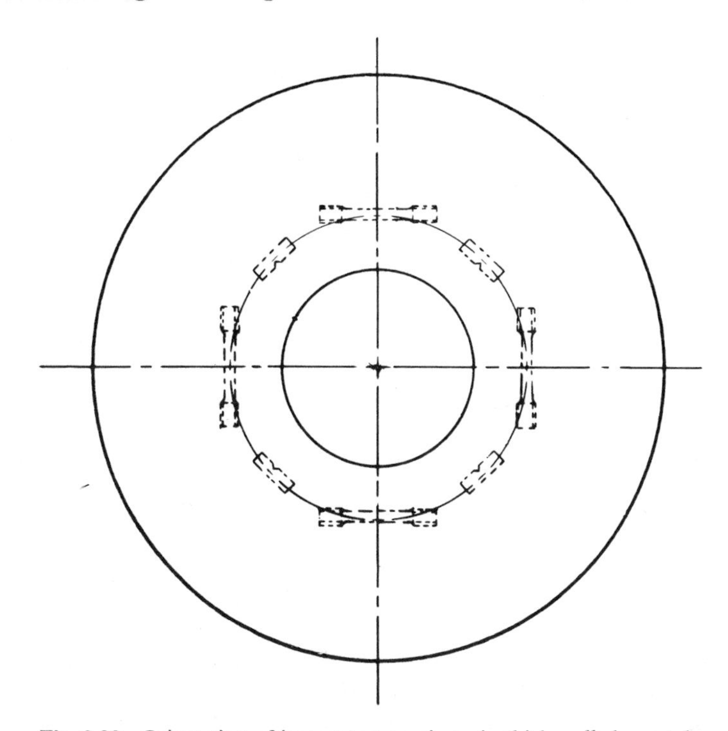

Fig. 2.23 Orientation of impact test specimen in thick walled gun tubes.

The newer modifications of the Charpy test may be useful. Since all these tests are qualitative, it is impossible to overemphasize the importance of experience.

The impact values can often be used to indicate a material or heat treatment problem. Several types of embrittlement can best be indicated by Charpy tests. A material with low yield strength usually has a high impact toughness. A combination of low yield strength and low toughness is indicative of inferior material.

2.3 FRACTURE TOUGHNESS TESTING

Most structures have built-in flaws; they may be mechanical, such as reentrant angles, or metallurgical, such as inclusions. At these flaws, stresses and strains are concentrated, and often lead to cracking. Premature fracture at nominal stresses considerably below the yield strength may ensue. In the presence of a flaw, the material often acts brittly.

Previously, several techniques for evaluating the brittle behavior of metals were described. They are qualitative tests primarily useful for comparing materials. These methods are often used in procurement specifications, with past experience as a basis.

To an engineer designing a part or to a metallurgist conducting a failure analysis, these techniques are not completely satisfactory. Fracture mechanics is more quantitatively useful, not only for the resolution of a failure but also, from the design standpoint, for preventing its occurrence or recurrence. In fracture mechanics, the parameter of interest is the fracture toughness.

2.3.1 Background

While studying the fracture strength of glass, Griffith noted that as the length of the rod increased, its strength decreased. This was attributed to the presence of surface flaws, since a greater probability of surface flaws existed as the length of the rod increased. As an out-growth of this work, Griffith developed probably the first analytical expression for determining the load carrying capacity of a material in the presence of a crack. His work is based on a conversion of strain energy to interfacial energy by crack extension and considers strain energy release rate, G, as a critical parameter. This is the amount of energy made available for crack extension per unit area of crack. Failure occurs at a critical value of G, or G_c.

Irwin and others have analyzed the same situation from a different viewpoint; they considered the stress state in the vicinity of a crack tip. Based on calculations of these stresses, they developed the theory of a stress intensity factor, K, which is a function of the applied stress, the crack size and geometry, and a size factor. Similar to the energy approach, failure occurs at a critical value of K, that is, K_c. The latter is predominantly used presently.

2.3.2 Stress Intensity factor

The relation of K (stress intensity factor) to σ (stress), a (crack length), and $f\ (a/w)$ (geometric factor) is called a K-calibration. These relations have been determined mathematically and/or experimentally and are available for a number of standard specimens.

When measuring the fracture toughness, three modes of fracture are considered (Fig. 2.24). In actual experience, the opening mode (Mode I) is the most important; thus it is this mode of fracture which is usually considered.

Finally, the concepts of plane strain and plane stress should be considered. In terms of the stresses and strains in a material, these mean that one of the principal strains or one of the principal stresses is zero, as shown in the following tables:

Plane Strain

$\sigma_x \neq 0$	$\epsilon_x \neq 0$
$\sigma_y \neq 0$	$\epsilon_y \neq 0$
$\sigma_z \neq 0$	$\epsilon_z = 0$

Plane Stress

$\sigma_x \neq 0$	$\epsilon_x \neq 0$
$\sigma_y \neq 0$	$\epsilon_y \neq 0$
$\sigma_z = 0$	$\epsilon_z \neq 0$

Because of the constraints imposed, plane strain conditions tend to apply on the interior of a specimen or material whereas, plane stress conditions apply at the surface because stresses normal to a free surface are zero.

2.3.3 Fracture Toughness Test Methods

Many types of tests have been developed. Usually, the available material influences the size of the specimen. If sheet material is involved, it would be proper to test a thin specimen taken from the sheet. Similarly, when conducting a failure analysis of a specific part, it would be desirable not only to take a specimen from the part itself, but also to orient the specimen

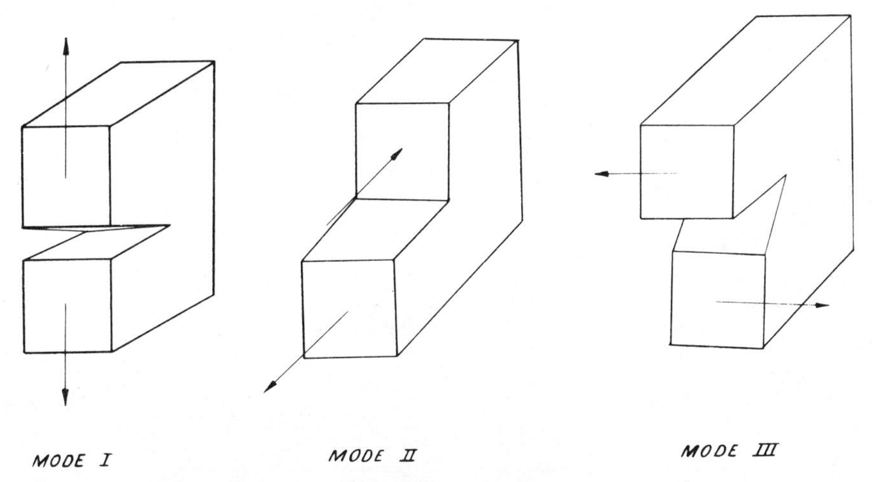

Fig. 2.24 Comparison of modes of fracture (From ASTM Spec. E208-66T. Used with permission of the American Society for Testing Materials).

so that it duplicated, as closely as possible, the same direction of loading as the part. Efforts should be made to evaluate the material close to the area that failed since it has been shown that toughness is not necessarily constant within a structure and, failures are often caused by local conditions, not average or bulk conditions.

Other factors influence the type and size of specimen used. A bending bar or a compact tension specimen usually require lower loads because of the mechanical advantage gained by the way the load is applied. The load necessary to determine K_c varies with the type of specimen because of the influence of geometric considerations in the fracture toughness equation, $K_c = \sigma\sqrt{a}\,f(a/w)$.

The geometric factor varies with the type of specimen, as shown graphically in Fig. 2.25. For the same relative crack depth (a/w), the multiplying

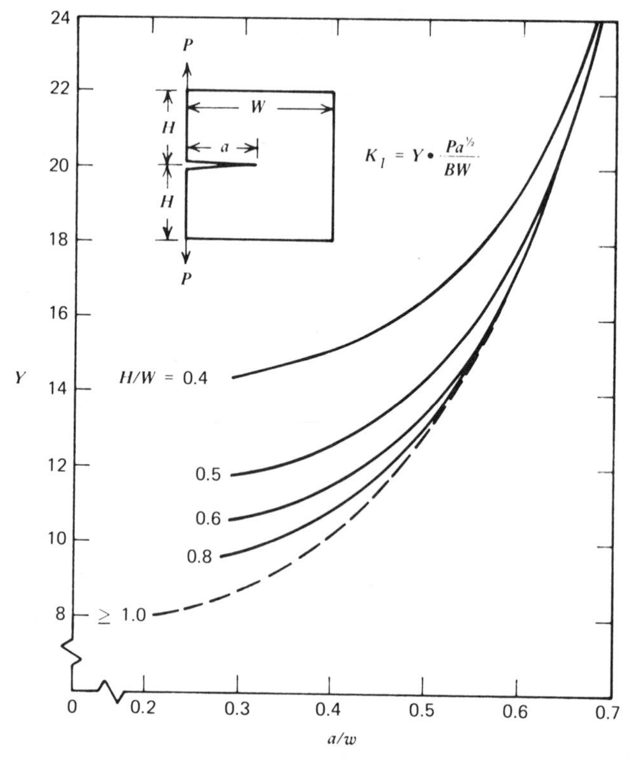

Fig. 2.25 Relation of $f(a/w)$ to relative crack length for different shape specimens (From W. Brown Jr. and J. Srawley, ASTM STP 410. Used with permission of the American Society for Testing Materials).

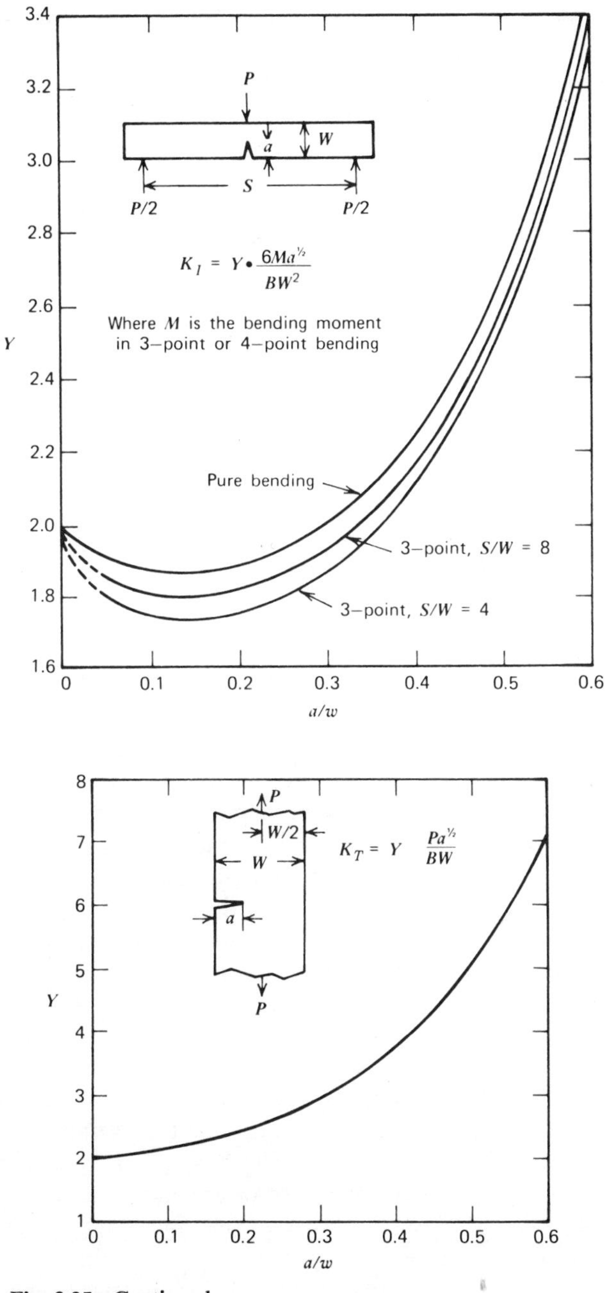

Fig. 2.25—Continued

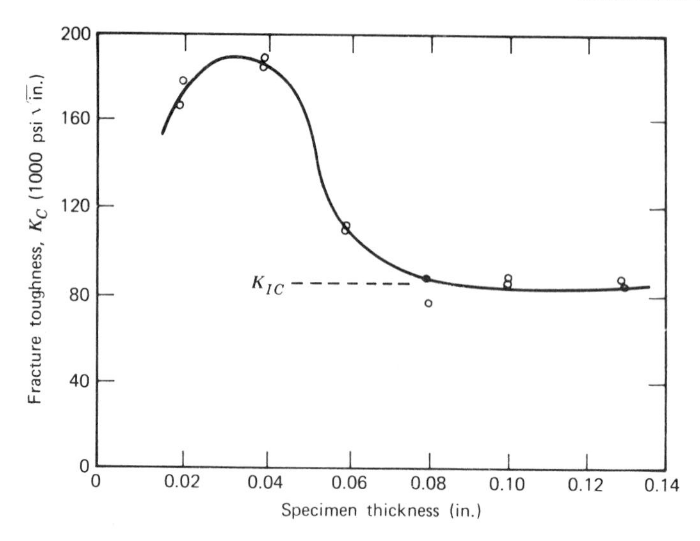

Fig. 2.26 Effect of specimen thickness on fracture toughness (From A. J. Brothers and S. Yukawa, Application of Fracture Toughness Parameters to Structured Metals. Used with permission of Gordon and Beach, Scientific Publishers).

factor is much higher for the edge notched bar than for the center notched bar, even though the load is applied similarly.[14]

When the term "fracture toughness" is used, the critical stress intensity factor under opening mode conditions is usually meant. Fracture toughness can be evaluated under other loading conditions, if this is desirable for a particular application. Generally, it is plane strain fracture toughness, K_{IC}, that is evaluated. It was previously noted that plane stress conditions apply at free surfaces whereas plane strain conditions tend to apply at the center. Also, the critical value of the stress intensity factor decreases as the specimen thickness increases (Fig. 2.26)[15] to a certain point, beyond which there is no further change. Beyond a certain point, the surface influence is unimportant and essentially plane strain conditions apply. This minimum value is K_{IC}.

A factor that is significant in fracture mechanics testing is the "plastic zone size," r_p. At the tip of a crack under load, a region of material that has been plastically deformed exists. This has an important influence since fracture toughness testing is a measure of cracking resistance, not plastic deformation resistance. As the plastic zone increases, the validity and significance of the test decrease.

As the yield strength decreases, the plastic zone size increases since plastic deformation is easier. Conversely, as the fracture toughness decreases, the plastic zone decreases, since the material shows an increasing propensity to crack rather than to deform. It has been shown that $r_p \alpha (K/YS)^2$, a factor that is important in determining necessary specimen sizes.

For a valid test, that is, a valid determination of K_{IC}, several conditions are necessary.

$$\text{Crack length prior to testing} \geq 2.5\left(\frac{K_{IC}}{YS}\right)^2$$

$$\text{Thickness of specimen} \geq 2.5\left(\frac{K_{IC}}{YS}\right)^2$$

$$\text{Ligament of specimen} \geq 2.5\left(\frac{K_{IC}}{YS}\right)^2$$

Although there are many specimen shapes that can be used, the American Society for Testing Materials (ASTM), which has been attempting to standardize test methods, currently suggests either a bend bar or a compact tension bar (Fig. 2.27). Even though the specific details, including direction of load application and K-calibration, are different, the general techniques of evaluation are similar. Consider the determination of K_{IC} using a bend bar, for which the K-calibration is

$$K = \frac{PS}{BW^{3/2}}\left[2.9\left(\frac{a}{w}\right)^{1/2} - 4.6\left(\frac{a}{w}\right)^{3/2} + 21.8\left(\frac{a}{w}\right)^{5/2} - 37.6\left(\frac{a}{w}\right)^{7/2} + 38.7\left(\frac{a}{w}\right)^{9/2}\right]$$

where

P = load (lb.)
B = thickness (in.)
S = span length (in.)
W = depth (in.)
a = crack length (in.)

Guidelines for determining specimen size are given above, but to use them, K_{IC} and YS must be known. The yield strength can be determined separately, but K_{IC} is the unknown parameter. Therefore, an estimate of K_{IC} is made, based on data from past experience or from the literature. The specimen size is determined with these values.

Unlike Charpy or notched tensile test bars that utilize a machined notch, fracture toughness tests utilize natural cracks. Usually, these are developed by machining a notch, then causing a crack by fatigue, that is, precracking.

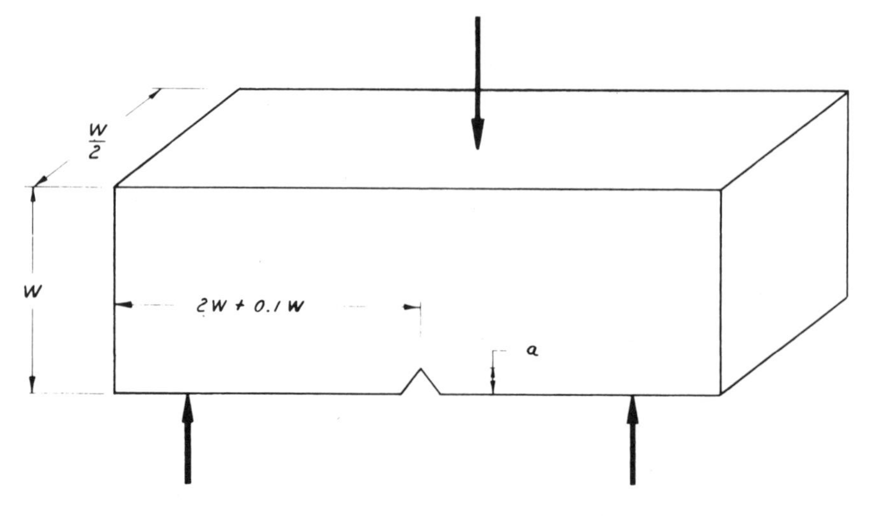

(a)

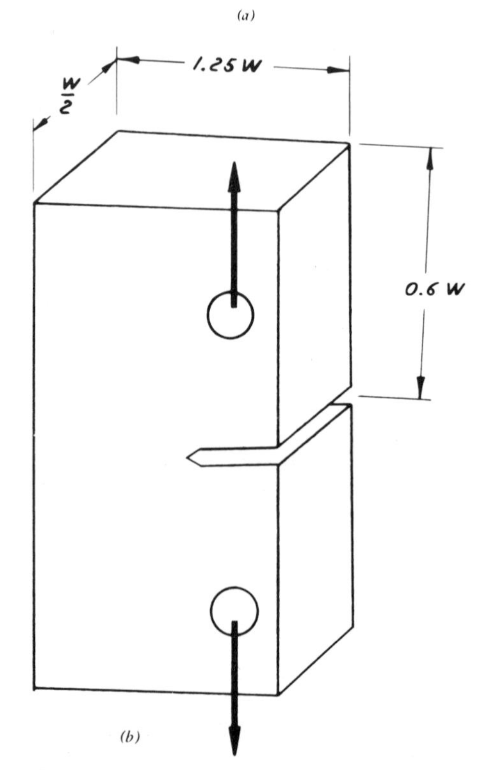

(b)

Fig. 2.27 Specimens for measuring K_{IC}, recommended by ASTM.

This is one of the critical factors in fracture toughness testing. The fatigue load must be evenly applied so that the crack has a flat front, parallel to the front and rear faces of the specimen. The load must be high enough that the crack can be generated in a reasonable length of time, but it cannot be too high or the plastic zone generated at the crack tip will too large and will significantly affect the results.

After precracking, the specimen is loaded on the testing machine and the load is applied. Usually, the rate of loading is slow, although there is some interest in dynamic fracture toughness, sometimes referred to as K_{ID}. The load is plotted against displacement,[16] for example, crack opening as measured by a compliance gauge (Fig. 2.28).

As the load increases, the material deforms elastically, and the load displacement plot is a straight line. As the load increases past a critical amount, several things can occur. In a brittle material the specimen fractures, and the load drops abruptly to zero. In calculating K_{IC}, the maximum load is used. In less brittle materials, a crack will suddenly occur at a critical point, the load will drop abruptly, but then will increase again. This may be repeated several times before failure. This phenomenon is called pop-in. When calculating K_{IC}, the load at pop-in is used. Finally, for more ductile materials, the load displacement plot deviates from linearity and may show no pop-in, even to failure. On these materials, interpretation of the curves is necessary to calculate K_{IC}. The ASTM has developed techniques for interpretation. With these types of test results, the most questionable values are obtained. If the deviation from linearity is too great, invalid results are obtained.

When the bar has been broken, the fracture specimen is examined and the average precrack depth is measured. Using the K-calibration and the

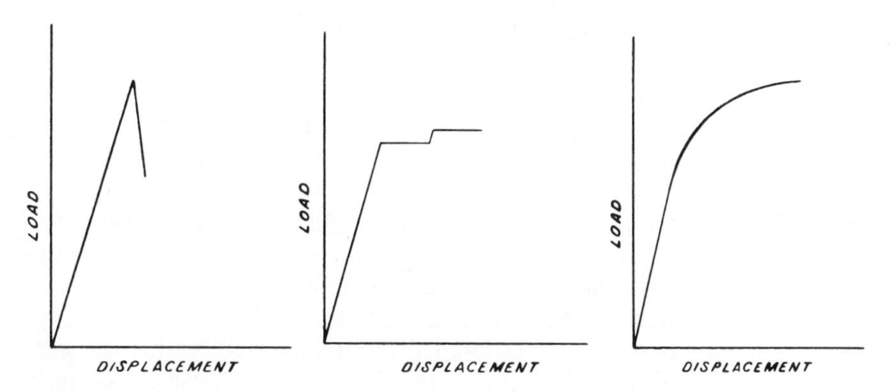

Fig. 2.28 Load versus displacement in fracture toughness test.

load, as described above, the critical K is calculated. At this stage, it is called K_Q, since it is not yet known whether it is valid. The various specimen parameters for valid K_{IC} are recalculated using K_Q and YS. If the sizes are adequate, $K_Q = K_{IC}$, if not, $K_Q \neq K_{IC}$. In the second case, additional testing would be required.

It is sometimes not important to know the precise K_{IC} in a failure analysis. It may only be necessary to know that the material is abnormally low. In those cases, standard size specimens could be used. If a precise evaluation of a failure is desired, for example, to determine if loads are higher than anticipated or if material requirements are inadequate, care must be taken in evaluating K_{IC}. Considerable testing may be required. However, the testing might then be minimized by using a specimen larger than necessary, since K_{IC} is not affected beyond a certain minimum size.

2.3.4 Using Fracture Toughness

As with any material parameter, there are general trends in the effects influencing K_{IC}. Generally these trends occur: (1) as strain rate ($\dot{\epsilon}$) increases, K_{IC} decreases, (2) as temperature (T) increases, K_{IC} increases, and (3) as yield strength (YS) increases, K_{IC} decreases. These effects are not really separable, since as $\dot{\epsilon}$ increases, YS increases and as T increases, YS decreases, so that there is a synergistic effect. Although these trends are generally observed, details for specific materials vary.

There are a number of drawbacks to the use of fracture mechanics. First, although K-calibrations are available for many configurations, they may not be available for the specific application under consideration. In this case, K_{IC} can only be used for comparison purposes. Second, the available material may not be adequate to provide a sufficiently large specimen; only comparative data can then be obtained. Finally, for very ductile materials, fracture toughness testing is simply not appropriate. For these materials, a tensile test may be more useful.

There are many advantages to a fracture toughness test. Its potential for acceptability is more widespread than for other tests for brittle material, primarily because of its background development in basic mechanics and because machined notches and inherent associated problems of machinability reproducibility are eliminated. It is potentially useful in design. In fatigue applications, it is potentially useful in predicting fatigue lives since the stress intensity factor is related to cyclic crack growth rates. Finally, when all the necessary information is available, it gives the failure analyst a tool to determine not only how the failure occurred but also under precisely what conditions. He can estimate the terminal crack depth, the actual loading conditions, and whether the specifications and design are correct. However, emphasis must be placed on "when all necessary information is available." If this is not available, fracture toughness then becomes a comparative test, not necessarily better than the other types of tests.

2.4 HARDNESS TESTING

The many hardness tests available fall into three general groupings: scratch, bounce, and indentation.

Only indentation methods have general application to metals; therefore, they are considered here.

The choice of a test method depends primarily on the size of the specimen available and the purpose for testing. Different tests are used to evaluate the bulk hardness of a large forging than those used to evaluate the surface hardness of a carburized piece of steel. Three types of tests are described: Brinell tests, Rockwell tests, and surface hardness tests.

2.4.1 Brinell Hardness Testing

The Brinell hardness test is used strictly for bulk hardness of relatively heavy sections. Of all the indentation methods, it requires the least surface preparation. It can be used for nonferrous material, but is more commonly applied to steel and iron.

To determine the Brinell hardness number (*BHN*), a 10 mm diameter steel ball is indented into the surface at a load of 3000 kg (500 kg for nonferrous metals) for 30 sec. The diameter of the indentation is then measured, using a low power microscope. Usually, two readings at 90° to each other are taken, and the average is used to calculate BHN, as follows:

$$\text{BHN} = \frac{P}{(\pi D/2)(D - \sqrt{D^2 - D_1^2})}$$

where

P = load (kg)
D = diameter of indenter (mm)
D_1 = diameter of indentation (mm)

The main source of error is in the measurement of the indentation. Plastic recovery occurs after the load is removed, but is negligible. However, two other phenomena can affect results.[17] These are "ridging" and "scribing in" (Fig. 2.29). The indentation diameter is measured as shown. When "scribing in" occurs, the actual diameter can sometimes be determined by covering the surface with a dull black coating.

Because of the depth of the indentation, the test specimen thickness should be at least 10× the depth of the indentation, and the distance from the center of the indentation should be not less than 2.5× the diameter. Neglecting either of these rules can lead to erroneous results.

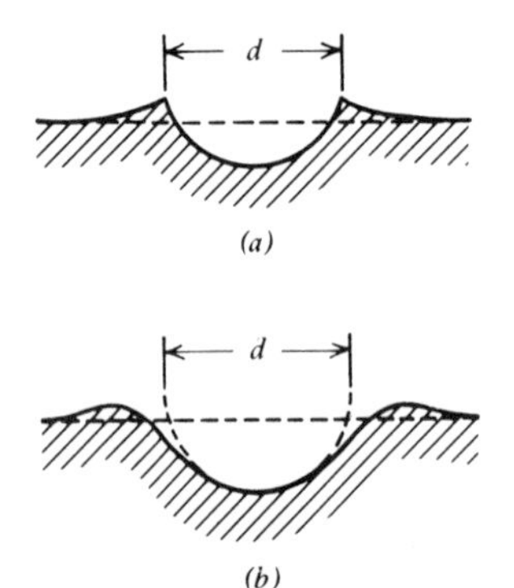

Fig. 2.29 (*a*) "Ridging" and (*b*) "scribing in" in Brinell hardness testing.

Attempts have been made to correlate BHN with tensile strength. Several rough estimates have been developed, for steel, BHN $\times$ 500 = *UTS* and for nonferrous materials, BHN $\times$ 300 = *UTS*.

One of the major advantages of Brinell testing is the size of the indentation. Since it is relatively large, compared to that of other methods, it actually tests a large amount of material. Therefore, it gives a better indication of overall hardness than the other methods.

However, it is still advantageous to take several readings to insure uniformity.

2.4.2 Rockwell Hardness Testing

The basic parameter measured in Rockwell hardness testing is depth of penetration. The sequence of operation is: (1) apply preload of 10 kg, (2) apply major load and (3) record depth of penetration automatically.

The Rockwell test is versatile; it can accommodate a variety of specimens. There are several indenters ranging from a $\frac{1}{16}$ in. diameter steel ball to a conical diamond penetrator used in combination with a variety of loads. The Rockwell hardness numbers (expressed for example as R_c 46, where *c* is the scale and 46 is the value) are arbitrary subdivisions of a scale. They have no quantitative physical significance. However, values of Rockwell hardness can be related to values of Brinell hardness.

Surface finish and flatness requirements are more stringent than those of the Brinell test. A ground finish is not needed, however, a reasonably smooth finish is necessary. The load should be applied normal to the surface of the specimen being tested. Any deviation will allow the indenter to slide, which causes an erroneous low reading.

The limitation on thickness mentioned for Brinell testing also applies. There are additional limitations. The specimen must be supported along the line of the axis of the load. It must be stable; any wobbling will cause erroneous readings. Overhang over the loading position should be minimized since this, too, causes false readings.

Since the indentation is smaller than BHN, Rockwell readings are more subject to inhomogeneities in the specimen. It is very important to take several readings to insure an adequate sample.

As with BHN, attempts have been made to relate Rockwell values to strength. No formulae have been derived; however relations have been developed for special cases. Yet, these are not rigorously derived and their usefulness is questionable. In some cases, the empirical relations are very useful. The most obvious is in the heat treatment of steel or, more significantly, in the strength of tempered steel. Fairly good relations have been developed between R_c values and strength. Often, in tempering steel, only hardness is specified. Despite this, the hardness test is not a substitute for a tensile test.

2.4.3 Superficial Hardness Testing

Sometimes it is desirable to measure the surface hardness of a component, for example, nitrided or carburized case. The loads of the methods previously described are too heavy. If used, they would undoubtedly indent not only the case but also the core. Therefore, they would be testing the core as well.

Several superficial hardness tests have been developed with the light loads and indentations. These techniques are also called microhardness tests and can be utilized to measure the hardness of a phase in a microstructure.

The most common techniques use a Knoop diamond indenter and a Tukon tester, which controls loads down to 25g. Like the Brinell test, this test requires measuring the size of the indentation. In this case, the indenter is a diamond with a 7–1 ratio of major to minor dimension. The hardnesss number is calculated from the formula

$$\mathrm{KHN} = \frac{P}{L^2 C}$$

where

KHN = Knoop hardness number
P = load (kg)
L = length of long diagonal
C = constant supplied by manufacture

Microhardness readings are relatively difficult. Caution must be taken in preparing surfaces since very fine ground surfaces are required. To measure the hardness of a case, it may be necessary to mount, polish, and etch the

Table 2.1 **Hardness Number Conversion Table**

Rockwell C-Scale Hardness Number	Brinell Hardness Number 10-mm Ball, 3000-kg Lead			Rockwell Hardness Number			Tensile Strength Approximate (1000 psi)
	Standard Ball	Hultgren Ball	Tungsten-Carbide Ball	A-Scale, 60-kg Lead Brale Penetrator	B-Scale, 100-kg Lead, 1/16 in. Diameter Ball	D-Scale, 100-kg Lead Brale Penetrator	
68	—	—	—	85.6	—	76.9	—
67	—	—	—	85.0	—	76.1	—
66	—	—	—	84.5	—	75.4	—
65	—	—	739	83.9	—	74.5	—
64	—	—	722	83.4	—	73.8	—
63	—	—	705	82.8	—	73.0	—
62	—	—	688	82.3	—	72.2	—
61	—	—	670	81.8	—	71.5	—
60	—	613	654	81.2	—	70.7	—
59	—	599	634	80.7	—	69.9	326
58	—	587	615	80.1	—	69.2	315
57	—	575	595	79.6	—	68.5	305
56	—	561	577	79.0	—	67.7	295
55	—	546	560	78.5	—	66.9	287
54	—	534	543	78.0	—	66.1	278
53	—	519	525	77.4	—	65.4	269
52	500	508	512	76.8	—	64.6	262
51	487	494	496	76.3	—	63.8	253
50	475	481	481	75.9	—	63.1	245
49	464	469	469	75.2	—	62.1	239
48	451	455	455	74.7	—	61.4	232
47	442	443	443	74.1	—	60.8	225
46	432	432	432	73.6	—	60.0	219
45	421	421	421	73.1	—	59.2	212
44	409	409	409	72.5	—	58.5	206
43	400	400	400	72.0	—	57.7	201
42	390	390	390	71.5	—	56.9	196
41	381	381	381	70.9	—	56.2	191
40	371	371	371	70.4	—	55.4	186
39	362	362	362	69.9	—	54.6	181
38	353	353	353	69.4	—	53.8	176
37	344	344	344	68.9	—	53.1	172
36	336	336	336	68.4	(109.0)	52.3	168
35	327	327	327	67.9	(108.5)	51.5	163

Table 2.1 Hardness Number Conversion Table—Continued

Rockwell C-Scale Hardness Number	Brinell Hardness Number 10-mm Ball, 3000-kg Lead			Rockwell Hardness Number			Tensile Strength Approximate (1000 psi)
	Standard Ball	Hultgren Ball	Tungsten-Carbide Ball	A-Scale, 60-kg Lead Brale Penetrator	B-Scale, 100-kg Lead, 1/16 in. Diameter Ball	D-Scale, 100-kg Lead Brale Penetrator	
34	319	319	319	67.4	(108.0)	50.8	159
33	311	311	311	66.8	(107.5)	50.0	154
32	301	301	301	66.3	(107.01)	49.2	150
31	294	294	294	65.8	(106.0)	48.4	146
30	386	286	286	65.3	(105.5)	47.7	142
29	279	279	279	64.7	(104.5)	47.0	138
28	271	271	271	64.3	(104.0)	46.1	134
27	264	264	264	63.8	(103.0)	45.2	131
26	258	258	258	63.3	(102.5)	44.6	127
25	253	253	253	62.8	(101.5)	43.8	124
24	247	247	247	62.4	(101.0)	43.1	121
23	243	243	243	62.0	100.0	42.1	118
22	237	237	237	61.5	99.0	41.6	115
21	231	231	231	61.0	98.5	40.9	113
20	226	226	226	60.5	97.8	40.1	110
(18)	219	219	219	—	96.7	—	106
(16)	212	212	212	—	95.5	—	102
(14)	203	203	203	—	93.9	—	98
(12)	194	194	194	—	92.3	—	94
(10)	187	187	187	—	90.7	—	90
(8)	179	179	179	—	89.5	—	87
(6)	171	171	171	—	87.1	—	84
(4)	165	165	165	—	85.5	—	80
(2)	158	158	158	—	83.5	—	77
(0)	152	152	152.	—	81.7	—	75

specimen. Flatness and squareness requirements are critical. The specimen must be flat and square (normal) to the axis of the load; even small dust particles can interfere.

These tests are not used for bulk measurements. They are used primarily for thin specimens, such as sheet, or for thin cases. They allow the measurement of certain types of specimens that could not be measured in any other way. For example, there is no way to even qualitatively approximate the strength of a case other than with superficial testing.

2.4.4 Hardness Test Conversions

Although each type of hardness test is appropriate for specific applications, it is often desirable to convert from one scale to another. A number of conversion tables and graphs are available, for example, Table 2.1, extracted from the 1959 SAE Handbook.[18] When utilizing a hardness conversion table, it is important to know the conditions governing its use. Because of differences in the tests, and testing conditions within each type of test, it is possible that different conversion values could be developed. For example, at high hardness, the relationship between Brinell and Rockwell hardness scales is affected by the type of ball used. Steel balls tend to flatten more than carbide balls, which results in large indentations and smaller BHN than with the carbide balls. Conversely, identical indentations made by steel and carbide balls correspond to different Rockwell values.

REFERENCES

1. *Metals Handbook*, 1948 Ed., ASM, p. 87.
2. G. Dieter, *Mechanical Metallurgy*, McGraw-Hill, New York, 1961, p. 251.
3. A. Nadai and J. Manjoine, *J. Appl. Mech.*, Vol. 8, A82, (1941).
4. G. Dieter, *op. cit.*, p. 256.
5. F. Heiser and R. Hertzberg, *J. Iron Steel Inst.*, December, 1971, p. 977.
6. G. Dieter, *op. cit.*, p. 379.
7. G. Dieter, *op. cit.*, p 380.
8. J. Barsom and S. Rolfe, *Impact Testing Metals*, ASTM STP 466, 1970, p . 281
9. R. Wullart, *Impact Testing of Metals*, ASTM STP 466, 1970, p. 148.
10. ASTM Specification E208–66T, 1966.
11. W. S. Pellini and P. P. Puzak, NRL Report 5920, March 15, 1963.
12. W. S. Pellini and P. P. Puzak, NRL Report 6030, November 5, 1963.
13. G. E. Nash and E. A. Lange, *J. Basic Eng.*, September 1969.
14. W. Brown, Jr. and J. Srawley, *Plane Strain Crack Toughness Testing of High Strength Metallic Materials*, ASTM STP 410, 1966, p. 12–14.
15. A. J. Brothers and S. Yukawa, *Application of Fracture Toughness Parameters to Structural Metals*, Gordon and Breach, Science Publishers, 1966, p. 55.
16. W. Brown, Jr. and J. Srawley, *op. cit.*, p. 41.
17. G. Dieter, *op. cit.*, p. 284.
18. *SAE Handbook*, 1959.

3

Nondestructive Testing

3.1 INTRODUCTION

Nondestructive testing encompasses all test methods that, when applied to a component, do not impair its subsequent utilization. This chapter is not intended to serve as a comprehensive treatise on nondestructive testing, but to familiarize the reader with various NDT methods, their limitations, and the expected results. These methods are particularly important in failure analysis. Structural materials are rarely perfect; usually they contain flaws or defects of varying size, although specifications are customarily applied to limit the severity of these variations. Since a failure often originates from a preexisting flaw or from one created from the combined action of the environment and the stress state, an important contribution of nondestructive testing is to determine whether the existent flaws or inhomogenities were related to the failure. Another important contribution is to characterize the flaws as to type, size, and location and to determine whether the failed component met the necessary specification requirements. Therefore, standardization of the procedures and the interpretation of the results are extremely important if valid comparisons are to be made.

A number of techniques are used in NDT; each is generally dependent on a different energy system. The techniques range from ordinary macroscopic examination with white light to the complex procedure of neutron radiography. As with testing in other fields, each method has an area in which it yields optimum performance though it can often be used successfully on marginal situations when the need arises.

3.2 LIQUID PENETRANT INSPECTION

Flaws or defects that intersect the surface, such as cracks, porosity seams, or laps, can be detected by using liquid penetrants. Subsurface defects are not detectable. Shrinkage, for example, is not usually detectable unless some prior machining operation has uncovered the defect. Liquid penetrant inspection may be applied to a variety of metals: aluminum, magnesium steels, stainless steels, brasses, and bronzes, as well as plastics, ceramics, and metallo-ceramics. These penetrants are compounded so that the viscosity is low, and wetting agents are usually added to allow the penetration and adsorption within the flaw interior. The penetrants also contain either a dye or a fluorescent material to increase the visibility of the defect.

The techniques require that the penetrant be applied to the surface to be tested and remain there for an appropriate length of time. The material is then flushed off using water or a solvent specific to the system. The surface is dried and a developer is then applied. The function of this developer is to pull the indicating material out of the defect and into the developer coat immediately over it. When dye penetrants are used, the contrast between the defect and background is strikingly apparent under ordinary light conditions. When fluorescent penetrants are used, the parts must be viewed under ultraviolet light to render the flaws visible. Cracks are indicated by a line indication, as shown in Fig. 3.1, and porosity or shrinkage appears as a grouping of point indications. The width of the flaw has a direct bearing on the ease of visibility, the tighter cracks being less prone to discovery. Therefore, parts are occasionally warmed slightly to facilitate inspection.

3.3 EDDY CURRENT INSPECTION

The eddy current technique and its variations are very useful in detecting defects at or near the surface of the component being tested. The technique is not confined to ferrous materials but can be used on any metal system. The method is based on the induction of eddy currents in the test specimen when a test coil bearing alternating current is placed on the metal surface. The magnitude of the response is dependent on the frequency and amplitude of the current, the magnetic permeability and electrical conductivity of the material being tested, and the size, shape, and orientation of the defects.

Since the procedure is sensitive to many variables, standardization of the test procedure is a prerequisite for any change in material or geometry. When large numbers of components similar in material or geometry must be tested, however, the procedure is ideal. Precisely because it is sensitive, this method can be utilized to determine the presence of a number of defective situations when comparative standards are available. Surface and subsurface cracks are easily detectable, but the procedure has also been used to determine the presence of intergranular corrosion,[1] the thickness of electroplatings[2] of various types, bonding of brazed jackets,[3] corrosion pits,

Fig. 3.1 Cracks as revealed by dye penetrant inspection.

variation in hardness of base materials, and weld defects.[4] It should be reemphasized that although these procedures are no doubt possible, their success in some cases depends on a considerable amount of development work in the area of probe and circuit design. They would, therefore, not lend themselves to the discovery of these defects in an isolated piece of hardware but would be more useful where large numbers of a component were being produced and comparative data could be generated.

In some instances the time and expense of special techniques may be warranted. For example, we investigated a failure involving extremely fine wire (0.007 in. in diameter) that cracked during a coiling operation. Longitudinal cracks in the wire prior to coiling were responsible for the condition. These could be detected by microscopic examination of metallographic sections, but obviously this method could not be used to screen hundreds of feet of wire. By using eddy current techniques, similar to those used by Myers and Renken,[5] the wire could easily be separated into good and bad material, thereby eliminating the problem.

3.4 MAGNETIC PARTICLE INSPECTION

Magnetic-particle inspection offers a means for the detection of surface and slightly subsurface discontinuities in ferromagnetic materials. The procedure is not applicable to nonferromagnetic materials, thereby excluding many structural metals, such as austenitic stainless steels, aluminum, magnesium, copper, brass, and titanium, from inspection by this method.

The test is conducted by creating a magnetic field in the test specimen. An indicating medium, either liquid or powder, containing high contrast magnetic particles is applied to the surface which is then examined for the presence of indications. These indications are caused by the concentration of magnetic particles in the area over the defect. Surface defects appear as sharp indications whereas subsurface defects yield indications that are broader and less defined. In general, the vagueness increases as the depth below the surface increases. Figure 3.2*a* shows magnetic particle indications caused by hydrogen cracks in a steel forging. Figure 3.2*b* shows the depth and extent of the cracking.

Defects that can be detected by this method possess a common characteristic—they have an interface between themselves and the bulk material. These include open defects such as cracks, seams, laps, porosity blow holes, and incomplete fusion of welds. Closed defects such as inadequate penetration, inclusions, and segregation that is so severe that a marked delineation occurs, can also be detected. Spurious indications caused by artifacts, too, occur occasionally. These may be due to overmagnetization, geometrical effects peculiar to the component, or magnetic

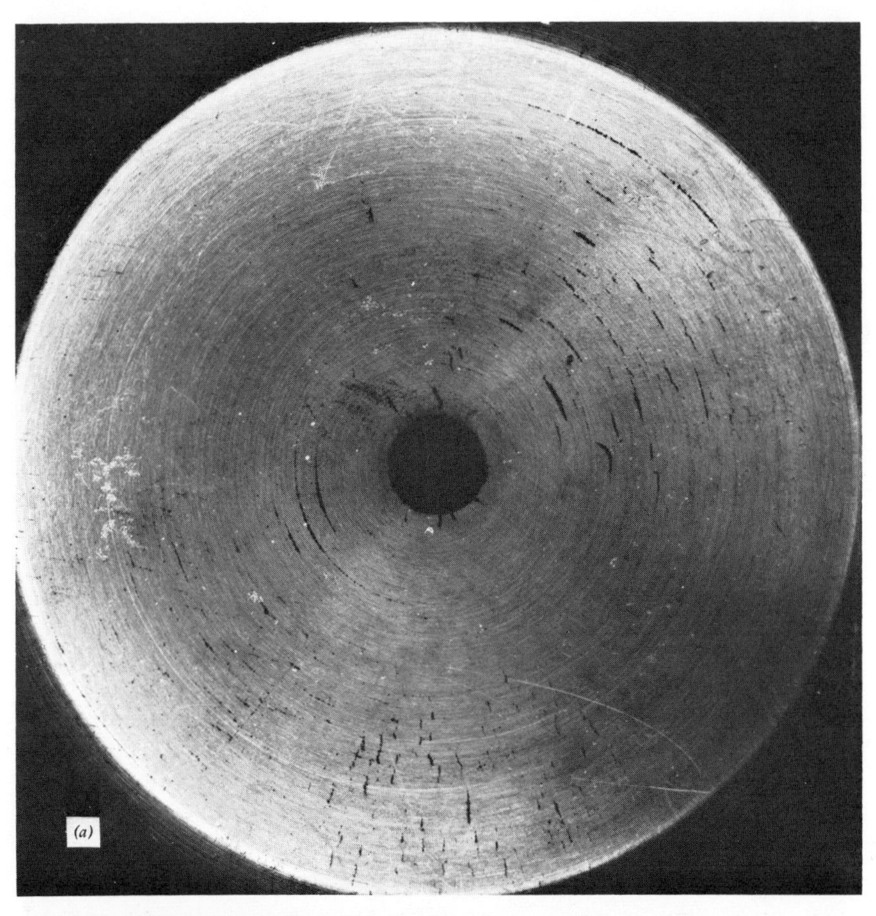

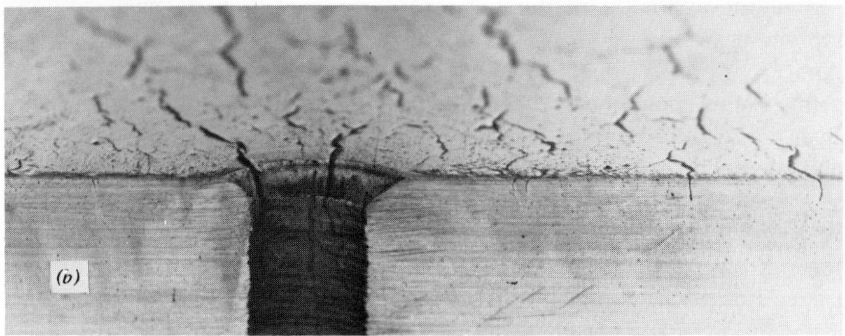

Fig. 3.2 (*a*) Magnetic particle indications. (*b*) Cross section showing extent and depth of the actual cracks causing the indications shown in (*a*).

writing caused by contact with an unmagnetized tool while the part is under test. When the authenticity of an indication is questionable, remagnetization is the best procedure.

3.5 RADIOLOGICAL EXAMINATION

The basis of radiological inspection is that X-rays and γ-rays can pass through materials that are optically opaque. The absorption of the initial X-ray by the material is a function of the thickness and nature of the material and the intensity of the initial radiation. The decrease may be shown by the classical absorption law

$$I = I_0 e^{-\mu t}$$

where

I = attentuated intensity
I_0 = initial intensity
μ = linear absorption coefficient
t = material thickness

The transmitted radiation is dependent on the thickness and nature of the material through which it passes; therefore, it can be used for the detection of internal flaws that may also have external manifestations. The primary methods for the detection of the radiation make use of a fluorescent screen (fluoroscopy) or a photographic image (radiography).

3.5.1 Radiography

A radiograph is the permanent record obtained when a sensitized film is exposed to X-rays or γ-rays passing through a test specimen. The degree of exposure is proportional to the intensity of the transmitted radiation at each point. Flaws such as cracks and voids, being hollow, will therefore show up as darker areas whereas refractory inclusions will appear as lighter areas, since they tend to absorb a greater amount of radiation. The absorption by a material depends mostly on its atomic number and to a lesser extent on its thickness. The ability of radiographic procedures to detect a flaw depends on the density of[6] the flaw, its size, geometry, and orientation. As a check on the adequacy of the radiographic technique, a standard test specimen or penetrameter is placed on the source side of the subject. This test device is usually of a simple geometric configuration, for example, a step block, a series of drilled holes, or a series of wires, made of a material radiographically similar to the subject.

The usual method of specifying sensitivity is

$$\text{sensitivity} = \frac{\text{thickness of smallest wire, hole, etc., whose image can be observed on radiograph} \times 100\%}{\text{thickness of subject}}$$

Thus a smaller numerical value implies that a smaller defect will be detectable and the radiograph will have better sensitivity. However, there is no simple relationship between penetrameter sensitivity and flaw sensitivity, and a 1% penetrameter sensitivity does not mean that all flaws 1% of the thickness in size will be detectable.

Some direct relationships do exist with three types of flaws, provided that they are air filled (the relationships do not hold for flaws filled with slag or sand):

1. *Pipe.* Pipe is a linear discontinuity of essentially circular cross section and as such corresponds to the sensitivity of a wire indicator, except that the effect on the image is reversed. Since the pipe is hollow, the thickness of the material has been decreased; with the wire it has been increased.
2. *Round holes or voids.* The ratio of the volume of a sphere to a cylinder of the same radius is 2:3. Therefore, the use of a stephole penetrameter gives a good approximation of the defect size if this correction is applied.
3. *Massive voids.* If the voids have a very large area, sensitivity is a function of contrast and corresponds to simple thickness—sensitivity as measured with a stepped-wedge.

3.5.2 Fluoroscopy

The impingement of transmitted radiation on a coated screen produces a fluorescence of varying intensity. This image may be examined for evidence of characteristic defects much in the same manner as on ordinary radiographs. One of the primary differences is that a radiograph is a negative image whereas fluoroscopy yields a positive image, that is, the brightness of the image is proportional to the intensity of the transmitted radiation. The advantages of this method over radiography are: lower costs, since film requirements are absent; speed of test, since no delay is incurred by processing; and scanning capability, since the part may be rotated about, thereby changing the viewing angle. The disadvantages are the lack of a permanent record, which makes subsequent destructive testing a problem, and an image quality that is, in general, inferior to that attainable by radiography.

3.5.3 Evaluation of Data

Radiological methods can be used to determine the presence of flaws that are grossly different in composition because they are void, contain gas, or are inclusions. In most cases, cracking is an unquestionable basis for rejection. However, other defects, particularly those found in castings such as gas porosity and shrinkage are almost always present, and rejection must be based on the degree of defectiveness. This requirement means that specifications must be considered and that interpretation is also a factor in the evaluation. It has been stated that if perfection were the requirement, then acceptability would be an easy matter. Since the degree of a defect is not easily quantifiable, various standards have been prepared for a number of defect types.

Table 3.1 Reference Specifications for Radiographic Practice[a]

Number	Title	Origin	Date of Latest Issue
B.S.2600: 1962 (being revised)	General recommendations for the radiographic examination of fusion welded joints in thickness of steel up to 2 in.	U.K.	1962
B.S.2737: 1956	Terminology of internal defects in castings as revealed by radiography	U.K.	1956
B.S.2910: 1962 (being revised)	General recommendations for the radiographic examination of fusion welded circumferential butt-joints in steel pipes.	U.K.	1965
B.S.3451: 1962	Testing fusion-welds in Al and Al-alloys.	U.K.	1962
B.S.3971: 1966	Image Quality Indicators (I.Q.I.) and recommendations for their use.	U.K.	1966
B.S.4080: 1966	Methods for NDT of steel castings.	U.K.	1966
B.S.4097: 1966	Gamma-ray exposure containers for industrial purposes.	U.K.	1966
B.S.499: 1965 Part 3	Terminology for fusion welding imperfections as revealed by radiography.	U.K.	1965
B.S.1500: 1958	Fusion welded pressure vessels for use in Chemical, Petroleum and allied industries.	U.K.	1958
B.S.2633: 1966	Class 1 metal arc welding of steel pipelines for carrying fluids.	U.K.	1966
B.S.3351: 1961	Piping systems for the petroleum industry.	U.K.	1961
B.S.4206: 1967	Methods of testing fusion welds in Cu and Cu-alloys.	U.K.	1967
MIL–STD–271A (Ships)	Military Standard Nondestructive Testing Requirements for Metals (Radiography, Magnetic Particles, Liquid Penetrant, Ultrasonic, Helium Leak, for determining the presence of surface and internal discontinuities in metal).	U.S.A.	1959

Table 3.1 **Reference Specifications for Radiographic Practice[a]—Continued**

Number	Title	Origin	Date of Latest Issue
MIL–I–6870A	Military Specification Inspection Requirements Nondestructive; for Aircraft Material and Parts (Magnetic Particle, Penetrant, X-ray, Ultrasonics).	U.S.A.	1953
API STD 1104	API Specifications for field welding of pipe lines.	U.S.A.	1965
	ASME Boiler and Pressure Vessel Code—Power Boilers (Section 1); Nuclear Vessels (Section 3).	U.S.A.	1969
	ASME Boiler and Pressure Vessel Code—Unfired Pressure Vessels (Section VIII).	U.S.A.	1969
AWS /D3.3–53	'Rules for Welding Piping in Marine Construction.' 'X-ray standard for high pressure—high temperature steam piping.' (Issued in conjunction with above in separate binding).	U.S.A.	1953
ASTM.3.71	Reference Radiographs for steel castings up to 2 in. thickness.	U.S.A.	1964
ASTM.E.155	Reference Radiographs for Inspection of Aluminum and Magnesium Castings.	U.S.A.	1964
ASTM.E.99	Reference Radiographs for Steel Welds.	U.S.A.	1963
ASTM.E.155 Series II	Reference Radiographs for Inspection of Aluminum and Magnesium Castings.	U.S.A.	1964
ASTM.E.94	Tentative Recommended Practice for Radiographic Testing.	U.S.A.	1962
MIL–R–11470 (ORD)	Military Specifications—Radiographic Inspection; Qualification of Equipment, Operators and Procedures.	U.S.A.	1953
ML–STD–453	Military Specifications—Radiographic Inspection.	U.S.A.	1963
MIL–I–6870B	Military Specification—Inspection, Radiographic.	U.S.A.	1965
MIL–R–11468	Military Specification—Radiographic Inspection, soundness requirements for arc and gas welds in steel.	U.S.A.	1951
MIL–R–11469 (ORD)	Military Specification—Radiographic Inspection.	U.S.A.	1953
NAVSHIPS No. 250–692–2	X-ray standards for production and repair welds.	U.S.A.	1961
NAVSHIPS No. 250–537	Radiographic standards for bronze castings.	U.S.A.	

Table 31. Reference Specifications for Radiographic Practice[a]—Continued

Number	Title	Origin	Date of Latest Issue
ASTM.E.142	Standard Method for controlling Quality of Radiographic testing.	U.S.A.	1964
ASTM.E.186	Tentative Reference radiographs for heavy walled (2 to $4\frac{1}{2}$ in.) steel castings.	U.S.A.	1965
ASTM.E.192	Standard Reference radiographs of investment steel castings for aerospace applications.	U.S.A.	1964
ASTM.E.242	Tentative reference radiographs for appearance of radiographic images as certain parameters are changed.	U.S.A.	1964
ASTM.E.272	Tentative reference radiographs for high strength Cu-base and Ni-Cu alloy castings.	U.S.A.	1965
ASTM.E.280	Tentative reference radiographs for heavy-walled ($4\frac{1}{2}$ to 12 in.) steel castings.	U.S.A.	1965
ASTM.E.310	Tentative reference radiographs for Tin–Bronze castings	U.S.A.	1966
AWS.D2.0.56	Standard specifications for welded highway and railway bridges.	U.S.A.	1956
AWS.3.3.53 D3.4.52	Rules for welding piping in marine construction.	U.S.A.	1953
MIL–R–45774(1)	Military Standard; soundness requirements for fusion welds in Al and Mg missile components.	U.S.A.	1963
MIL–STD–779	Reference radiographs for steel fusion welds 0.03 to 5.0 in.	U.S.A.	1968
NAVSHIPS 250–692–2	X-ray standards for production and repair welds.	U.S.A.	1961
AMS–2635B	Radiographic Inspection.	U.S.A.	—
AMS–2650	Fluoroscopic Inspection	U.S.A.	—
B.31.7	Code for pressure piping.	U.S.A.	1968
B.31.8	Gas transmission piping systems.	U.S.A.	1967
MIL–A–11356D	Cast steel armour.	U.S.A.	1962
57–0–10	Radiographic procedure.	U. S.A.	1949
NAVSHIPS 0900–000–1000	Fabrication, welding and inspection of ship hulls.	U.S.A.	1967
MIL–STD–746A	Radiographic testing requirements for cast explosives.	U.S.A.	1963
MIL–C–6021G	Military Specification. Castings Classification and Inspection for Aeronautical Applications.	U.S.A.	1967

Table 3.1 Reference Specifications for Radiographic Practice[a]—Continued

Number	Title	Origin	Date of Latest Issue
IIS/IIW-6-58 (being revised)	Recommended Practice for the radiographic inspection of fusion welded joints for steel plates up to 2 in. (50 mm) thick.	I.I.W.	1958
IIS/IIW-35-59 (V-115-59) (being revised)	Recommended Practice for the X-ray inspection of fusion welded joints on aluminium and its alloys and magnesium and its alloys up to 2 in. (50 mm) thick.	I.I.W.	1959
IIS/IIW-26-59 (V-113-59)	Recommended Practice for the radiographic inspection of circumferential fusion welded butt-joints in steel pipes up to 2 in. (50 mm) wall thickness.	I.I.W.	1959
IIS/IIW-62-60	Recommended practice concerning radiographic image quality indicators (I.Q.I.).	I.I.W.	1960
IIS/IIW-85-61	Radiography of welds of boilers and pressure vessles (Part 1: Principles).	I.I.W.	1961
IIS/IIW-269-67	Recommended practices for the radiographic inspection of fusion butt-welded joints in steel from 2 to 8 in. (50 to 200 mm) thickness.	I.I.W.	1967
IIS/IIW-275-67	Draft recommended practice for the radiographic examination of resistance spot welds on aluminium and its alloys.	I.I.W.	1967
IIS/IIW-316-68	Image quality values for steel.	I.I.W.	1968
	'Collection of reference radiographs of welds in steel' (with 3 supplements).	I.I.W.	1962
	'Collection of reference radiographs of welds in aluminium and aluminium alloys.'	I.I.W.	1963
DIN-54-109	Bestimmung der Bildgüte von Röntegen- und Gamma—Filmaufnahmen an metallischen Werkstoffen.	Germany	1962
DIN-54-111	Prufung von Schweissverbindungen metallischer Werkstoffe mit Roentgen- oder Gamma-stahlen Aufnahme von Durchstrahlungsbildern.	Germany	1969

[a] From Ref. 6.

Halmshaw[6] has compiled a list of international specifications and standards that are in common use. These are presented in Table 3.1. Some of these standards include sets of reference radiographs,[7,8] which can be used to establish the acceptable degrees of severity.

3.5.4 Types of Defects

Various types of defects can be identified. Perhaps the major use of radiological inspection is in the examination of castings, undoubtedly because they are very prone to internal defects. In castings, the procedure can be used to identify and assess such defects as cracks, cold shuts, shrinkage porosity, misruns, gas holes, inclusions, severe segregation, hot tears, and core shifts. Welding and forging defects that involve void space and inclusions obviously can also be inspected radiologically. Specific radiographic examples are given in later chapters where casting, forging, and welding defects are discussed in detail.

Radiography or fluoroscopy can also be utilized in the evaluation of total assemblies to determine whether an assembly is correct. Figure 3.3 illustrates a throttle cable that was not fully inserted iinto the sheath and thus led to a malfunction. The radiographic procedure of inspecting com-

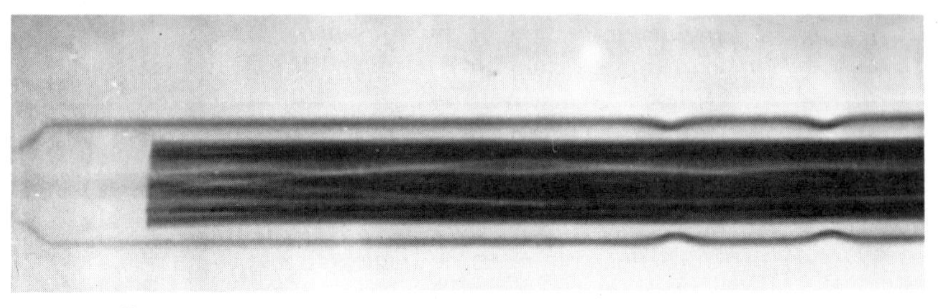

Correct

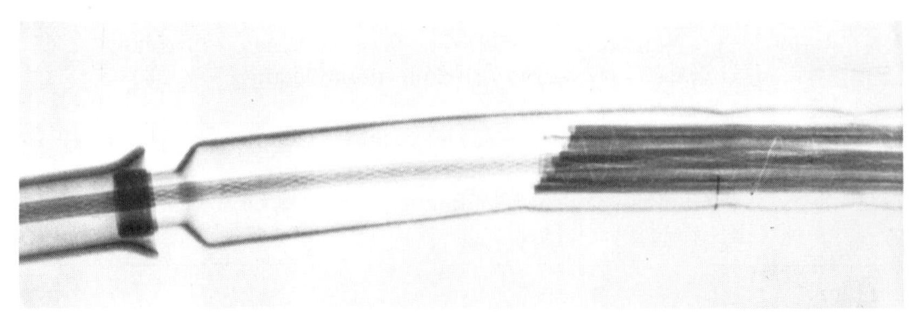

Incorrect

Fig. 3.3 Radiograph illustrating improper assembly of a throttle cable. (Courtesy of M. E. Barzelay).

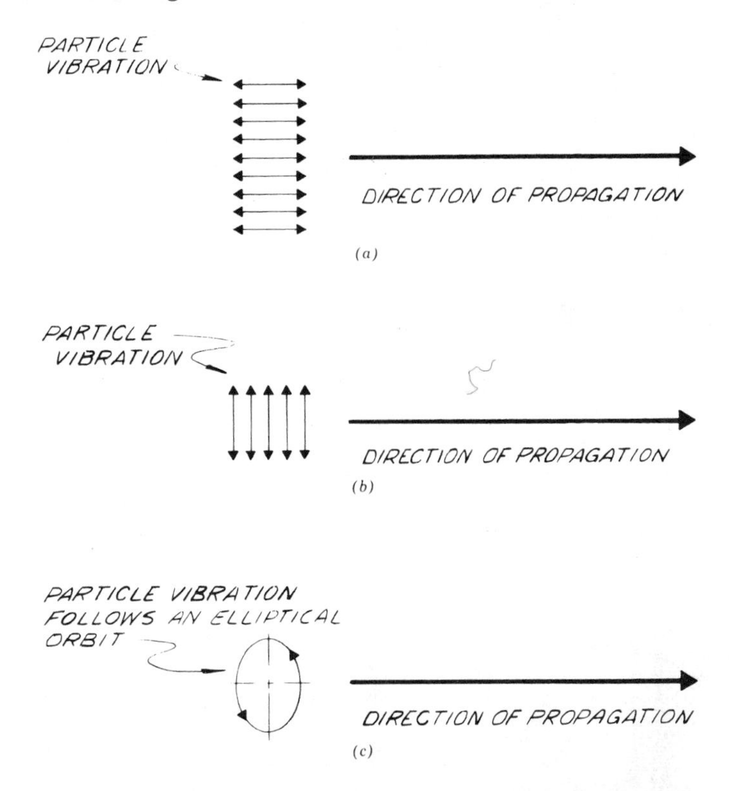

Fig. 3.5 Various modes of ultrasonic vibration. (*a*) Longitudinal waves, similar to audible sound waves, are generated by the thickness expansion of a piezoelectric plate. The particle displacement is in the direction of propagation. (*b*) Shear waves propagate with particle vibration transverse to the direction of wave travel. The velocity of shear waves is approximately half that of longitudinal waves. (*c*) Surface (Rayleigh) waves travel with little attenuation in the direction of propagation, but their energy decreases rapidly as the wave penetrates below the surface. The particle displacement of the wave motion follows an elliptical orbit.

Techniques employing longitudinal waves are usually used on objects of a rectilinear geometry whereas techniques using shear waves are most often used on curvilinear specimens such as pipe and tubing (as shown in Fig. 3.6) or where access to the total surface of the component is blocked.

Surface waves readily follow the free surfaces of parts being inspected and are reflected from surface cracks and discontinuities. Therefore, the technique is especially useful for the inspection of components in which the defects are usually associated with the surface, such as in springs, mill rolls, die blocks, and so forth. The depth to which the surface wave extends

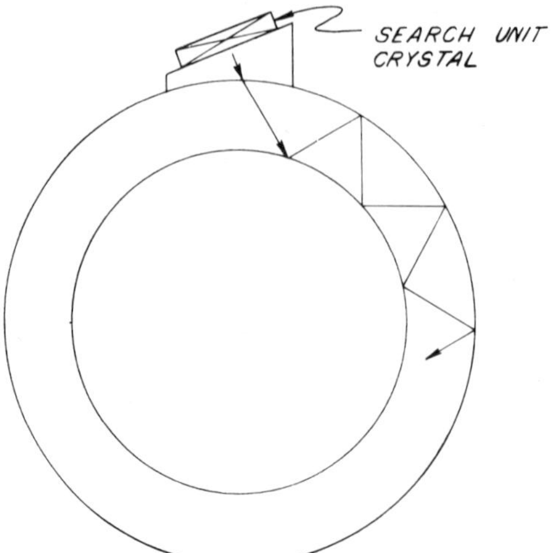

Fig. 3.6 Angle beam technique using a shear wave.

is generally reported as about one wavelength. It is precisely this concentration of energy at the surface and the ability to follow curved and convoluted surfaces that makes it an extremely sensitive technique for the determination of surface flaws.

3.7.2 Test Methods

There are several techniques that are in general use. Of these, the most widely known is the basic technique—pulse echo. In this technique the transducer emits a high frequency pulse, then stops to receive the echo of that pulse either from the opposite side of the specimen or from an intervening defect. These signals are usually displayed on a cathode ray tube after suitable conversion and amplification. The travel time of the reflected wave is accurately determined and appears as a pulse to the right of the initial pulse.

From its position, the location of a defect within the specimen can be established. In addition, an approximate concept of the size, shape, and orientation of the defect can be determined by checking the specimen from another location or surface.

In general, the higher the frequency is, the greater the sensitivity. However, at the higher frequencies the sensitivity can be so great that the wave is blocked by such things as grain boundaries, inclusions and other metallurgical variables. Therefore, one must choose between maximum sensitivity and practicality. The pulse echo technique has general applicability and may be used for either contact or immersion testing or with longitudinal, shear, or surface waves.

Other techniques are also used; for example, in highly attenuating materials the through-transmission method is quite useful. In this technique a receiving transducer is placed in a position to receive the wave emitted from a transmitting transducer. In defect-free materials, this energy beam is essentially undiminished in magnitude. When a flaw exists, the beam is partially blocked and the signal received is diminished proportionately. If the defect is sufficiently large, the signal will be totally extinguished. The obvious disadvantages of this procedure are that accessibility to two surfaces is required and that the procedure gives no information about the location of the defect. One advantage, however, is that it can be used in energy-limited situations, such as with large castings or with large grain size material.

Thickness measurements can also be determined from one side of a component by using the resonance method of ultrasonic inspection. A standing wave is induced in the material by varying the frequency of the input so that the incoming wave and the reflected wave are in phase. Since the resonant fundamental frequency is a function of the material thickness, this thickness can readily be determined once resonance is established by utilizing the following equation:

$$F = \frac{V}{\lambda} = \frac{V}{2T}$$

where

F = fundamental frequency (MHz)
V = velocity of sound (10^6 in./sec)
T = thickness (in.)
λ = wavelength (in.)

Changes in resonance are caused by variations in the wall thickness. The technique, therefore, can be used to detect variations in the walls of pressure vessels, tubing, and thin-wall piping, caused by process variables or by service conditions, for example, pitting or general corrosion.

The methods just described can be utilized to determine the presence of internal defects in forgings, castings, welds, and semifinished products, such as bars, rods, and tubing. An excellent presentation on the topic of flaw detection in metal structures has been made by Krautkramer.[12] The test is usually conducted through the smallest dimension of the component, that is, transverse to major axis. For our purposes, the aim is the detection of flaws or cracks that may have contributed to the failure. The types of defects that can be determined using ultrasonics are those in which a discontinuity or interface exists such as cracks caused by forging, quenching, fatigue and hydrogen flaking. Nonmetallic inclusions, those resulting from deoxidation and other chemical reactions, and those caused by the entrapment of foreign materials (i.e., slag, furnace lining, etc) can be detected using ultrasonics providing these inclusions are sufficiently large relative to the section size being tested.

The primary candidates for detection in castings are internal defects such as shrinkage, blow holes and porosity. Segregation is not usually detectable unless accompanied by grosser defects. The complexity and finish of the surface, however, can limit the use of ultrasonics for the inspection of castings. Weld defects such as slag inclusions, incomplete penetration, cracking, and porosity can be detected quite readily in most welded structures. It has even become common practice to ultrasonically field inspect welds made on certain construction projects, such as bridges.[13]

Ultrasonic techniques utilizing surface waves have been used to determine the degree of bonding in honeycomb structures and surface defects in valve springs, mill rolls, and other products, such as thin extrusions, where the flaws would probably occur near the surface.

3.8 METHODS OF IDENTIFICATION

Since a failure analysis investigation often revolves around the question of whether the component was fabricated from the specified material, the employment of a rapid nondestructive test method can often provide the answer. A number of tests have been utilized for this purpose, some requiring little equipment and others requiring the use of elaborate apparatus.

1. *Spot tests.* These are simple analytical tests based on chemical reactions of the compositional elements, with reagents applied to the surface. The presence or absence of the suspected elements is revealed by indicators added to the dissolved spot. By the application of sequential tests, the composition of an alloy may be verified. Tabulations of various spot tests[14] have been made, which facilitates the determination of a number of metals and alloys.

2. *Spectrographic analysis.* When a metal is subjected to an arc, the electrons of the various compositional elements are excited. When these electrons return to their respective ground states, light is emitted. This emitted light is passed through a diffraction grating, resulting in a spectrum. Examination of the spectrum can reveal how much of an element is present since each element results in a line at a definite wave length postion and the intensity of that line is proportional to the amount of the element.

3. *Magnetic response.* Carbon and alloy steels are strongly magnetic up to the Curie (Ci) temperature. Austenitic stainless steels, the 200 and 300 series, however, are nonmagnetic.

Ferritic stainless steels, 405, 430, 442, and 446, and martensitic stainless steels, 410, 431, and 440A, are also magnetic.

4. *Spark tests.* Although these tests are semidestructive in nature and may appear to be quite crude, they do offer a means to an approximate analysis, with little apparatus such as might be required in the field. Fairly complete listings of various alloy responses have been catalogued.[15,16]

REFERENCES

1. R. C. Robinson, ASTM, Special Technical Publication, 1958, p. 223.
2. G. D. Linsey and H. L. Libby, *ibid.*
3. W. J. McGonnagle, *ibid.*
4. W. J. Warren, *Corrosion*, Vol. 10, October, (1954), 318–323.
5. R. G. Meyers and C. J. Renken, *Electronics*, Vol. 31, No. 39, 72–73 (September 1958).
6. R. Halmshaw, *Industrial Radiographic Techniques*, Springer-Verlag, New York, 1971.
7. ASTM Reference Radiographs, Steel Casting, E186, E192, E71, ASTM, Philadelphia.
8. ASTM Reference Radiographs, Aluminum and Magnesium, E155, ASTM, Philadelphia.
9. W. J. McGonnagle, *Nondestructive Testing*, McGraw-Hill, New York, 1961.
10. J. F. Hinsley, *Nondestructive Testing*, MacDonald and Evans, London, 1959.
11. R. C. McMaster, *Nondestructive Testing Handbook*, Vol. II, Ronald Press, New York, 1963.
12. J. Krautkramer and H. Krautkramer, *Ultrasonic Testing of Materials*, Springer-Verlag, New York, 1969.
13. New York State, using American Welding Society Specification D10, #409.
14. F. Feigel, *Spot Tests, Inorganic Applications*, Vol. I, Elsevier Publishing Co., Amsterdam, 1954.
15. ASM, Metals Handbook, *Spark Testing*, 1948, pp. 397–399.
16. *Nickel Bull.*, Vol. 15, No. 7, 102–104 (July 1942).

4

Fractography

Fractures often leave characteristic markings. The ability to recognize them simplifies the resolution of a failure analysis.

Fractography can be considered on three scales: macroscopic, microscopic, and electron fractographic. In this chapter, typical examples of identifiable features are shown. Although there is a certain amount of overlapping with material presented in later chapters, unifying the examples in one chapter is beneficial to the reader.

4.1 MACROSCOPIC

A good general rule to apply when considering what scale of magnification should be used in examining a fracture is: the higher the magnification (a) the more expensive the examination, (b) the more skill necessary in the handling and preparation of materials, and (c) the more time consuming the examination.

A macroscopic examination requires minimum preparation. However, the rules cited earlier about precautions in handling should be applied, since further study may eventually be warranted. The macroscopic result often provides sufficient information to explain the cause of a problem or to guide further work by supplying a clue to its origin.

When conducting a fractographic examination, it is advantageous to approximately assemble the fragments, being careful to avoid touching the fracture surfaces. Contact between the surfaces can cause abrasion, which can interfere with subsequent examinations. Some of the factors to consider are: the distortion associated with the fracture, dislocation of the fracture surfaces, corrosion products, the number, size, and location of fragments, the roughness or smoothness of the fracture surfaces, and any relation of the fracture to external damage, such as nicks, or to design features, such as angles or radii.

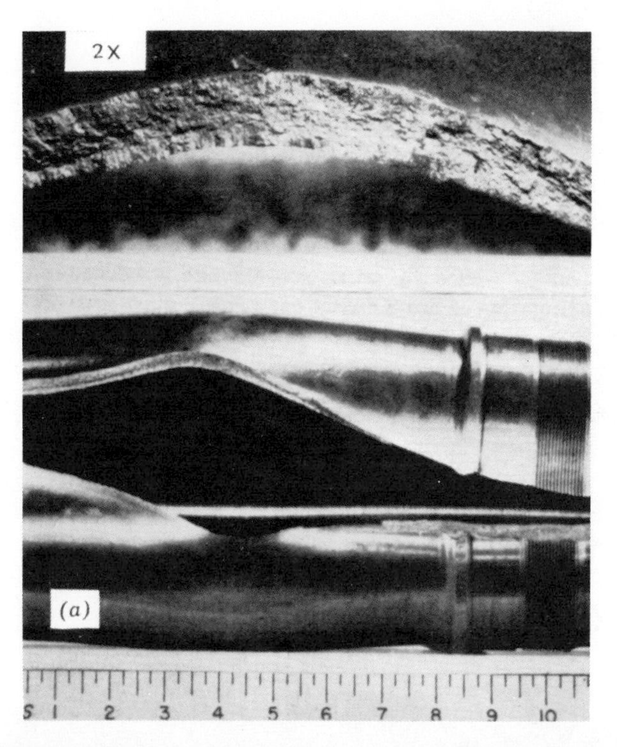

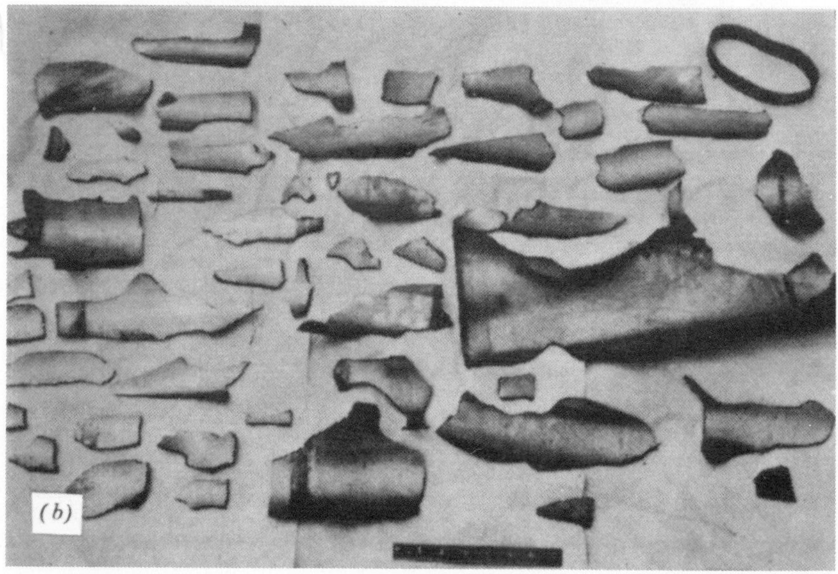

Fig. 4.1 Comparison of failure of (*a*) ductile and (*b*) brittle thick wall tubes.

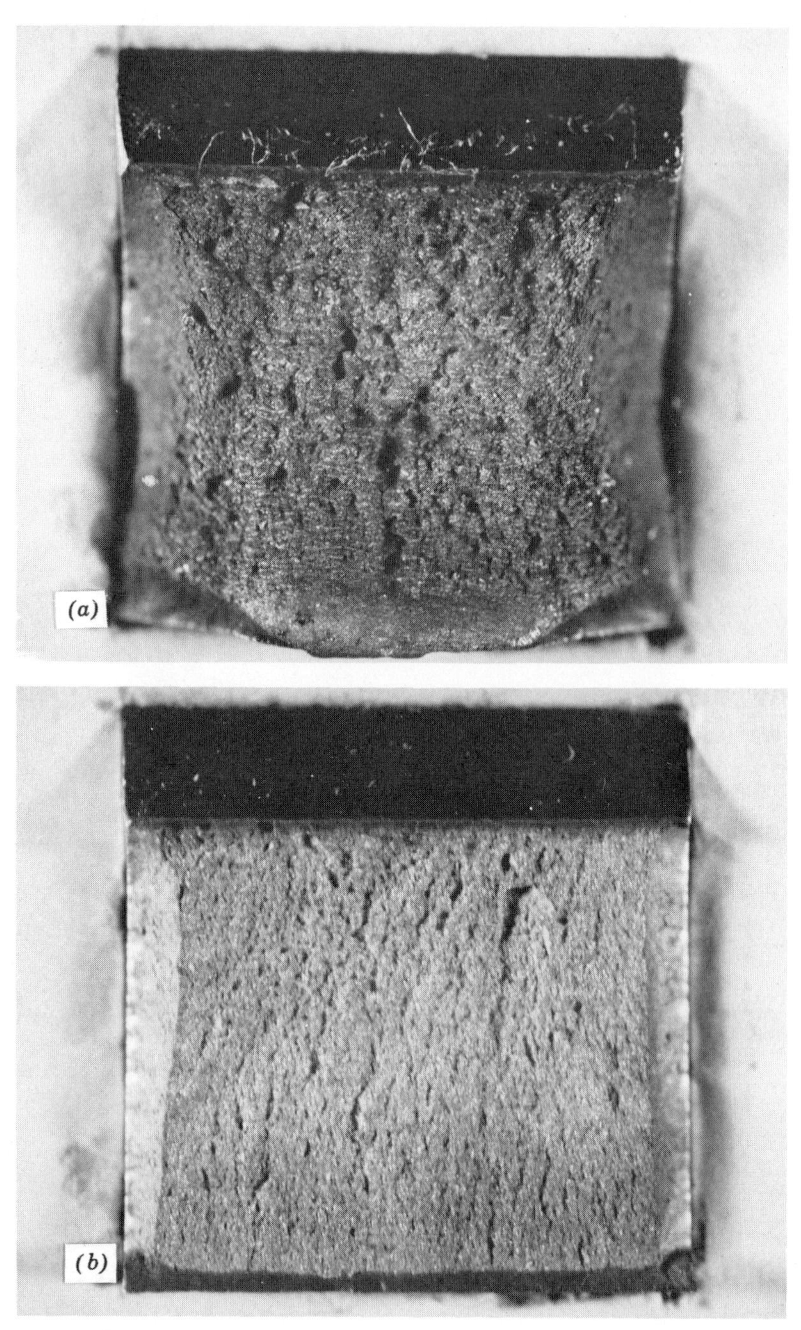

Fig. 4.2 Difference in texture on (*a*) ductile and (*b*) brittle fracture surfaces.

One of the most obvious macroscopic features of a failure is the amount of plastic deformation. Figure 4.1 shows the difference in failure pattern between a ductile and a brittle material.[1] In a ductile failure there is always plastic deformation, since a certain amount of energy is absorbed when the metal is deformed, before new surfaces are created (i.e., fracturing), whereas in brittle materials most of the energy goes into fracture. In brittle failures, the fractured pieces often fit together fairly well. In outward appearance, ductile fracture surfaces are rough because of the plastic deformation, whereas brittle failures are smooth (Fig. 4.2).

The fracture origin can be determined by analyzing the chevron pattern often seen (Fig. 4.3). These features point toward the origin of fracture.[2] The origin in a multifragment failure can often be located by assembling the fragments and tracing the chevron markings.

To explain a fracture, it is necessary to determine if the fracture was progressive, for example, a certain amount of crack growth per load application (fatigue) or if the failure occurred in one application of load (overload). Because a tensile test fails by overload, its typical fracture surfaces are representative of a general tensile overload failure. Tensile failure in a reasonably ductile material initiates internally at inclusions or voids (Fig. 4.4), and ultimately fails by shear on 45° planes at the surface, that is, forming shear lips, resulting in the typical cup-cone fracture.[3] Conversely,

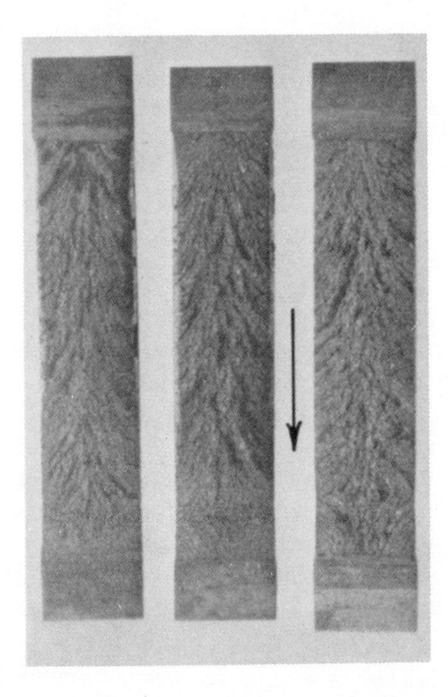

Fig. 4.3 Chevron pattern with arrow showing direction of crack propagation (Courtesy of Canadian Metallurgical Quarterly, Vol. 5, July-September, 1966, p. 207.).

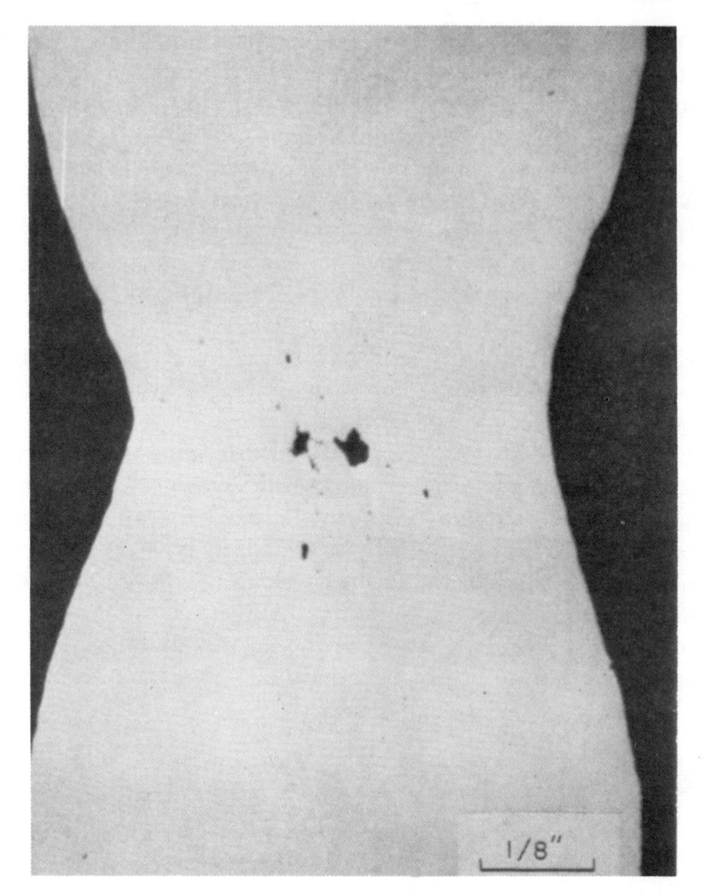

Fig. 4.4 Internal fissuring in cup portion of cup-cone tensile fracture (Courtesy of H. Rogers, Transactions, AIME, Vol. 218, 1960, p. 498.).

brittle tensile failures are smoother, without exhibiting the shear lips. As the test temperature is increased, the ductility increases, the strength decreases, and the fracture appearance changes (Fig. 4.5).[4]

The appearance of fatigue failures has often been described as brittle, because there is very little gross plastic deformation associated with them, and the fracture surfaces are fairly smooth. However, fatigue failures are usually easy to distinguish from brittle failures. Because they are progressive, they leave characteristic marks. On a macroscopic scale,[5] they are seen as beach marks (Fig. 4.6), which represent delays in the fatigue loading cycle. Just as chevron markings can be used to locate a fracture origin, so too, can beach marks. As crack length increases, beach marks increase in

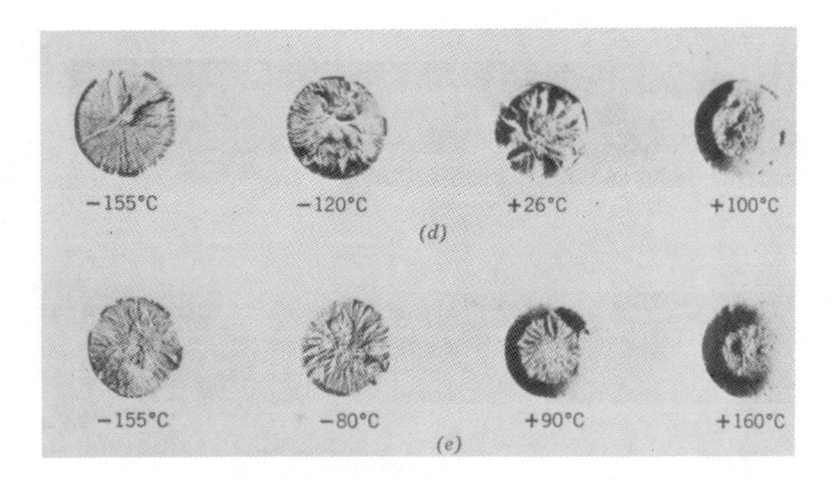

Fig. 4.5 Change in fracture appearance of a tensile bar with increasing test temperature (Reproduced by permission, from Transactions, ASM, American Society for Metals, 1962.).

Fig. 4.6 Beach marks on fatigue fracture surface (Reproduced by permission, from How Components Fail, American Society for Metals, 1966.).

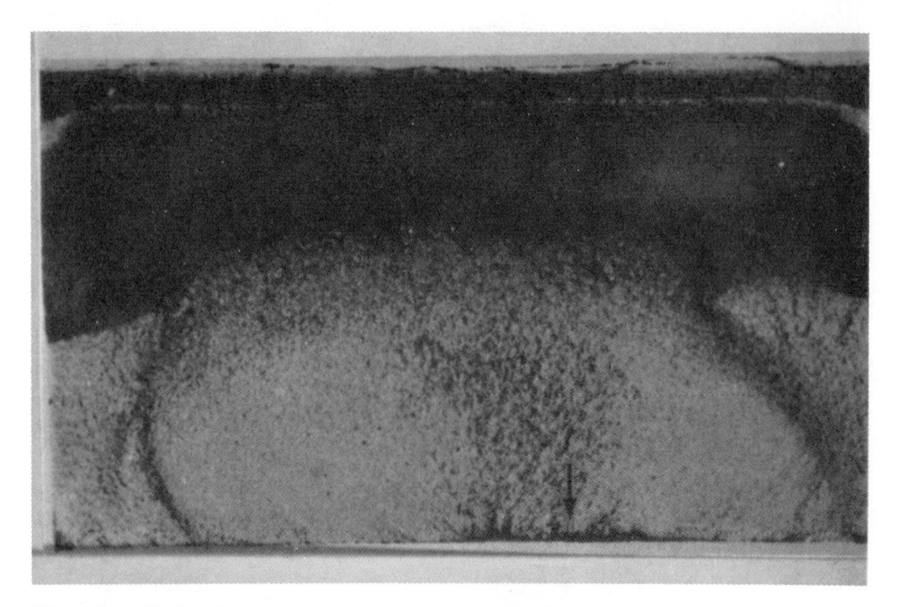

Fig. 4.7 Fatigue fracture without beach marks (Reproduced by permission, from How Components Fail, American Society for Metals, 1966.).

size and spacing, so that the location of the smallest beach mark indicates the origin of the failure. However, fatigue failures do not always show beach marks (Fig. 4.7). Therefore, their absence does not necessarily preclude fatigue.[6]

Although beach marks appear generally the same regardless of the type of load application, specific features vary.[6,7] A typical example of a tensile fatigue failure is shown in Fig. 4.8. Figure 4.9 is an example of torsional failure of a shaft. Note that the origin of the fracture can easily be traced. In gears, fatigue failures usually occur as a result of bending stresses by undermining of the teeth (Fig. 4.10). One fact common to all fatigue failures is that the fracture is caused by a tensile load generated by tension, rotation, or bending, and the fracture path is normal to the tensile load.

The stress state to which a component is subjected influences the fracture appearance. A tensile test is uniaxial; however, when necking occurs the stress pattern is triaxial. This causes internal fissuring which leads to failure. In some tensile fractures of plates the fracture often changes from essentially flat and normal to the tensile axis to a slant of 45° to the tensile axis (Fig. 4.11). This fracture mode transition is associated with a change in stress state from plane strain to plane stress.[8] The plane stress conditions at free surfaces account for the shear lips often seen on fracture surfaces,

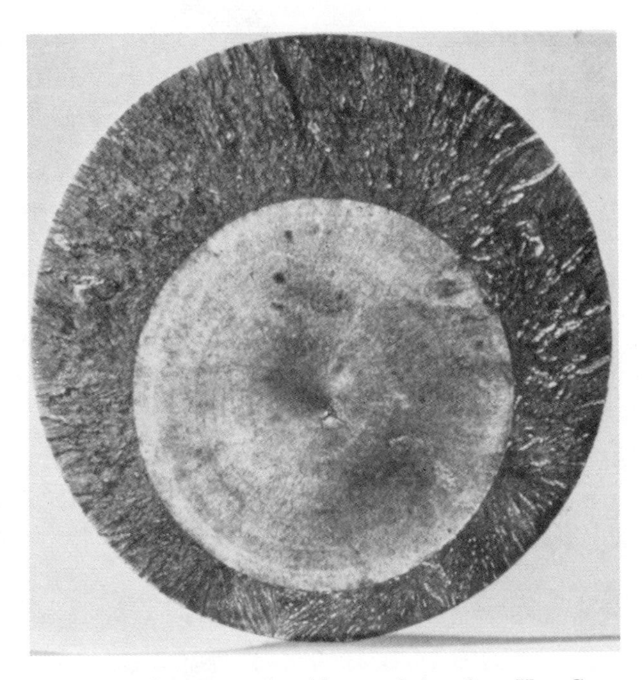

Fig. 4.8 Tensile fatigue failure (Reproduced by permission, from How Components Fail, American Society for Metals, 1966.).

Fig. 4.9 Fatigue failure of a shaft in torsion.

Fig. 4.10 Fatigue failure of a gear due to undermining of the teeth (Reproduced by permission from How Components Fail, American Society for Metals, 1966.).

Fig. 4.11 Transition from flat to slant fracture in a tensile test of steel plate.

which are actually the cone part of a cup-cone fracture. The absence of a shear lip is evidence of a brittle-type fracture; these lips can be found on both overload and fatigue fractures.

Material and processing defects also leave characteristic markings. In the molten state, a metal has a higher tolerance for gas than in the solid state. If precautions are not taken during solidification, the solid metals may have a greater amount of gas than is tolerable, that is, supersaturation. This is true for hydrogen in steel,[9] when, during forging, the combination of stress and gas can lead to hydrogen cracks or flakes (Fig. 4.12). These are pre-existing cracks, which weaken the structure.

During heat treatment of steel, it is often necessary to cool the metal rapidly to develop desired mechanical properties. Unfortunately, if the

Fig. 4.12 Flakes (hydrogen cracks) in heavy steel forging (Courtesy Bethlehem Steel Corp.).

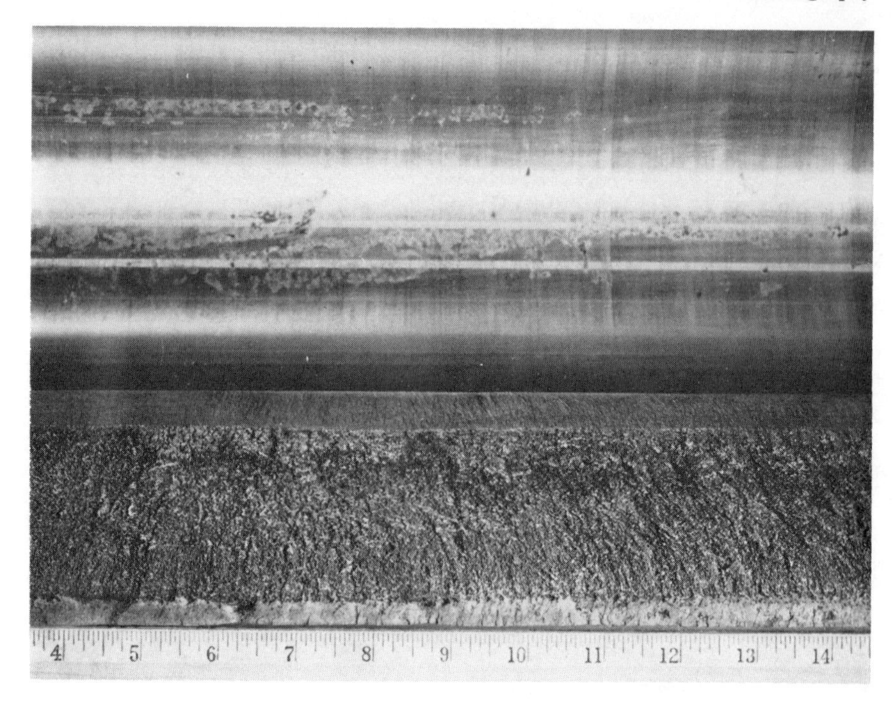

Fig. 4.13 Quench crack in heavy wall steel tubing.

cooling rate is too rapid or if it is improperly performed, quench cracks can result (Fig. 4.13). These, too, can lead to failures. In addition to the characteristic shape of the quench crack, a second indication is the discoloration of areas of the fracture surface. Tempering normally follows a quenching operation and the temperatures used leave a discoloration. The appearance of a discoloration on the fracture surface indicates that the surface was exposed to the furnace atmosphere; therefore, the component was cracked prior to tempering.

The prior existence of a crack can also be proven by the presence of corrosion products. If corrosion is present on the fracture surface, a crack had to exist for some time prior to failure. If the fractured parts were mistreated, however, this conclusion is not necessarily valid. Inadequate protection of the failure prior to examination can also result in corrosion.

Improper processing is often the cause of a failure. In a welded structure, insufficient care can result in arcing between the weld rod and the component. This not only causes a mechanical notch and subsequent stress concentration, but also results in metallurgical damage. In steel, for example, this may result in the formation of brittle untempered martensite.

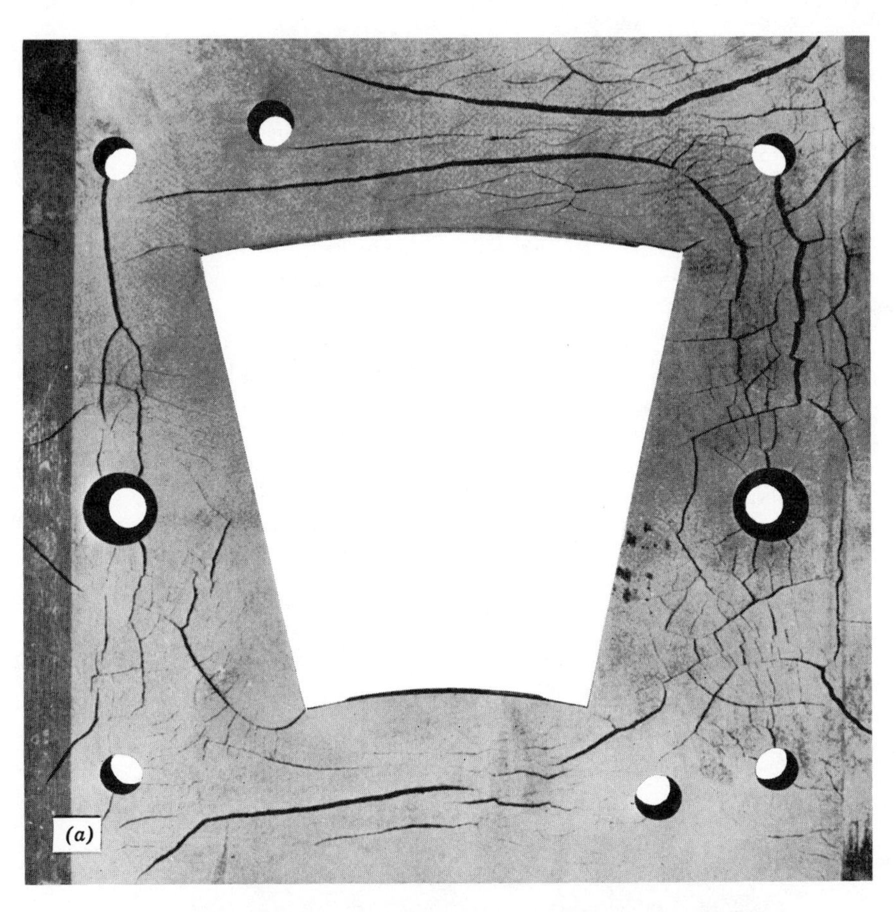

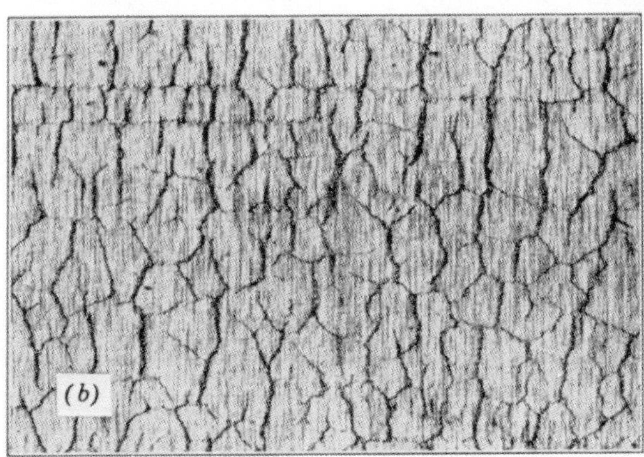

Fig. 4.14 Cracks due to improper grinding procedure (*a*-Courtesy Bethlehem Steel Corp.; *b*-Reproduced by permission, from Metals Handbook, American Society for Metals, 1958.).

Similarly, it is sometimes found that weld beads are laid in unspecified regions. This can also result in untempered martensite and a stress concentration region. Although both conditions can be seen macroscopically, it is often necessary to utilize microscopic examination.

The presence of nicks or gouges also indicates carelessness in handling. If a component is dropped or struck, a discontinuity may result. These mechanical notches also act as stress concentrators. If the normal loading places the notch in a high tensile stress region, failure may occur.

Any manufacturing process that results in high localized heating can cause cracking, which can occur even during machining (Fig. 4.14). Usually, such cracks are associated with grinding, and result when too great a cut is taken.[10,11] They are a type of quench crack, since the heavy cut causes excess heating followed by rapid cooling and cracking.

Although surface discontinuities are generally more detrimental, subsurface flaws can also result in failure. Figure 4.15 shows a fatigue failure that was traced to an inclusion,[12] and Fig. 5.10 shows an overload failure due

Fig. 4.15 Fatigue failure whose initiation site was an inclusion (From the Principles of Physical Metallurgy by G. Doan. Copyright 1953, McGraw Hill. Used with permission of McGraw Hill Book Company.).

to an inclusion. In both cases, the cause was easily identified, since it was due to the stress concentration at the inclusion.

Identifying the origin of a failure is not always so clear-cut. Other techniques must often be utilized to augment the visual analysis.

4.2 MICROSCOPIC

Microscopic examination, that is, up to 1000×, is seldom performed on fracture surfaces themselves. The lack of depth perception markedly reduces its usefulness. However, although microscopic examinations are often made only to augment other phases of investigation, they can sometimes be used to pinpoint a problem and to explain its cause.

There are two general reasons for conducting a microscopic examination: to examine the microstructure to determine if prior processing (e.g., heat treatment) was proper, or to examine the relation of crack path to the microstructure. When general microstructure is being studied, any material may be examined. As in all metallographic work, care must be taken in mounting, polishing, and etching. However, it is not necessary to insure that the edge is not rounded. When the relation of crack path to microstructure is of interest, however, a crack profile is examined, and additional precautions must be taken to preserve the edge. Therefore, a coating (e.g., nickel plating) is applied to the fracture surface prior to mounting the sample.

The properties of most engineering materials can be altered by heat treatment. Changes in microstructure are usually associated with the change in properties. When conducting an investigation of a failed part it is advisable to evaluate its general microstructure. Although it may not be possible to explain a failure by its general microstructure, occasionally, some phase of heat treatment will have been improperly performed and the microstructure will be obviously incorrect. For example, in a quench and tempered steel, the presence of ferrite indicates that the quenching practice was inadequate. Unfortunately, the changes in microstructure between good and bad performance are often so subtle as to be indistinguishable. This is especially true for low-allow high hardenability steel.

Metal movement during forming processes,[13,14] such as forging, results in the formation of flow lines (Fig. 4.16). They are due to chemical segregation from the original solidification of the ingot, and cause banding in mild steel (Fig. 4.17) and ingotism (Fig. 4.18). Flow lines are often meaningless in terms of performance. However, the macroscopic ductility across them is less than along them. In structures, it is best to orient the flow lines in the direction of the maximum stress, if possible.

Inclusion failures or inclusion-matrix fractures have been associated with

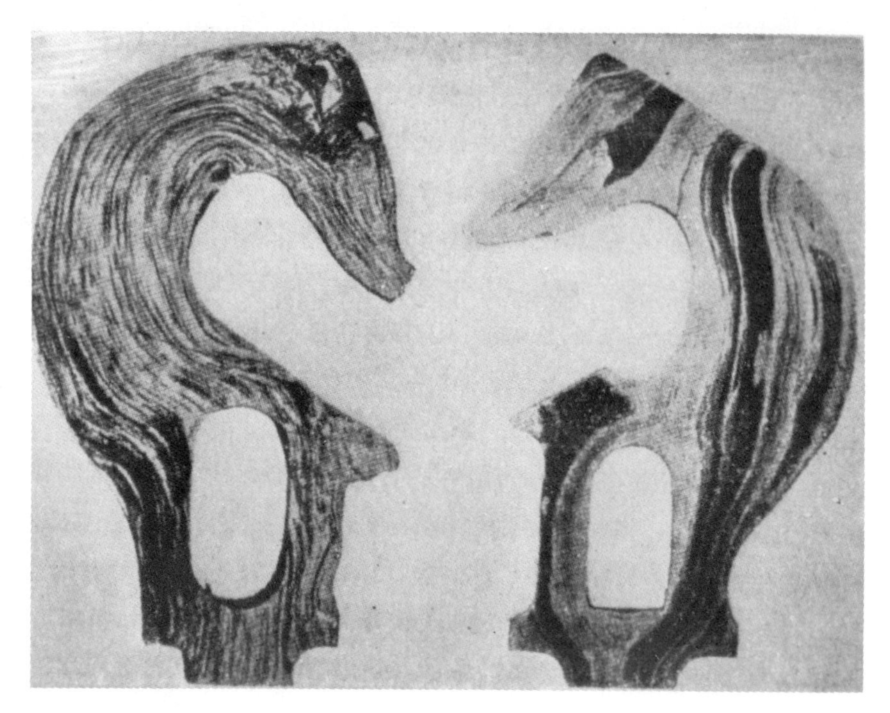

Fig. 4.16 Flow lines in steel (From The Principles of Physical Metallurgy by G. Doan. Copyright 1953, McGraw Hill. Used with permission of McGraw Hill Book Company.).

all types of fractures, static, dynamic, and fatigue. When a metal is deformed, not only is the matrix deformed but also the inclusions. As with flow lines, ductility transverse to inclusions is less than that parallel to them. Figure 4.19 illustrates fractured inclusions that caused failures.

Fractures have typical macroscopic and microscopic appearances. When only the fractured pieces are available for analysis, little information can be gained by microscopic examination. However, there are often secondary cracks associated with a fracture. These cracks do not totally penetrate the piece and are usually representative of the main fracture. They also show both sides of the fracture surfaces and their junctures. Failures sometimes wipe out important features. However, these cracks do not experience the same degree of abrasion, and are often intact.

Macroscopically, brittle fractures are often distinguishable from ductile fractures by the amount of plastic deformation. This same distinction can be seen microscopically. Figure 4.20 compares ductile and brittle fracture through ferrite in a low carbon steel. In this case, the brittle fracture was

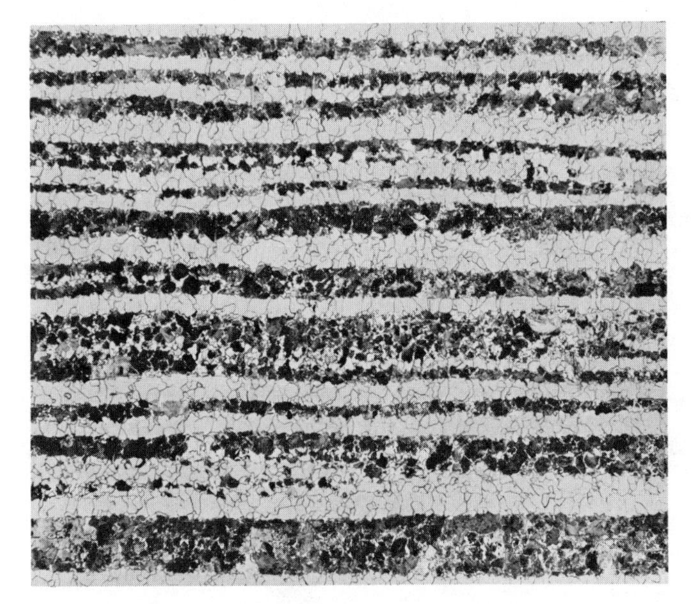

Fig. 4.17 Banding in mild steel plate.

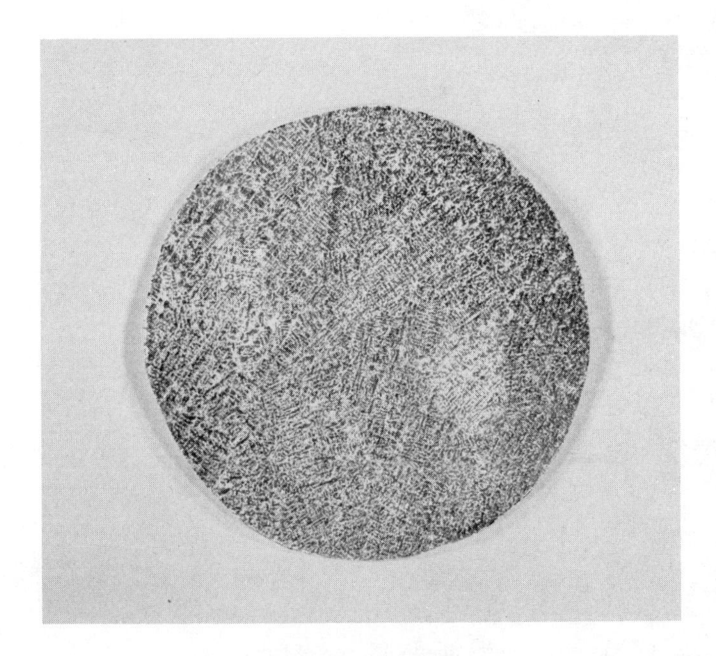

Fig. 4.18 Ingotism in a heavy steel forging.

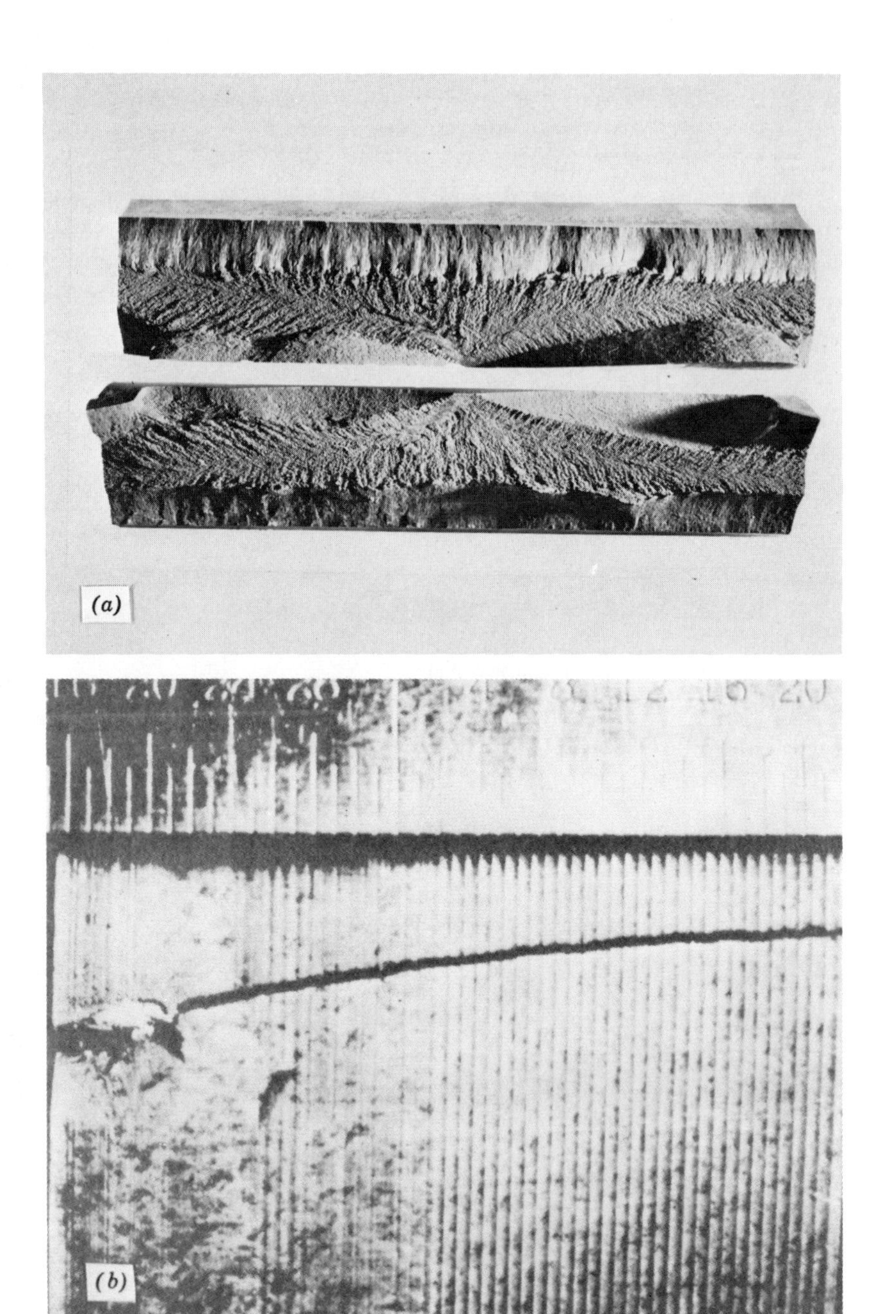

Fig. 4.19 Examples of failures due to fracture of inclusions.

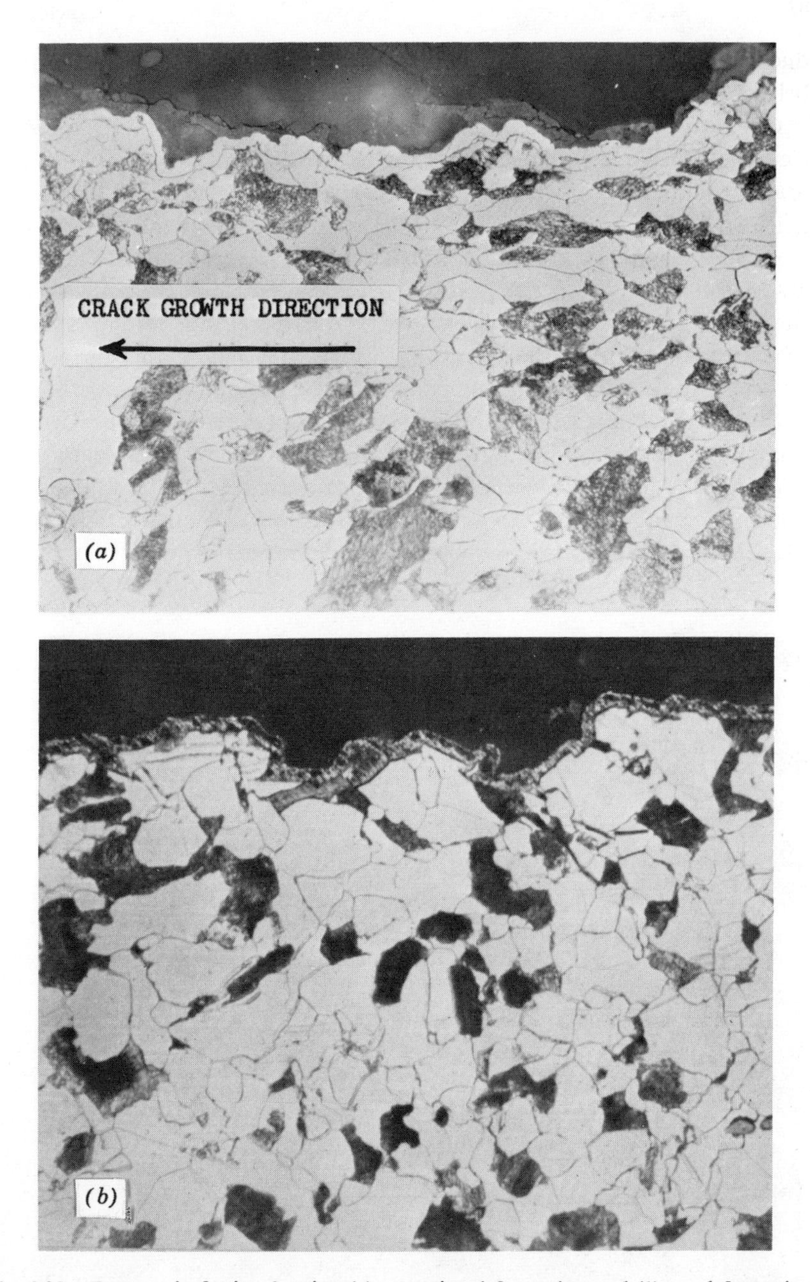

Fig. 4.20 Fracture in ferrite showing (*a*) extensive deformation and (*b*) no deformation.

due to testing at low temperature. Note the flat cleavage fracture in the brittle fracture and the considerable distortion in the ductile fracture.

One obvious feature that can be microscopically discerned is cleavage.[15,16] Since cleavage is associated with separation along crystallographic planes, its microscopic apperance is very flat (Fig. 4.21). Steps (Fig. 4.22) that are related to a change in orientation of the fracture plane at a grain boundary are often associated with the main facet. Cleavage is associated with low energy failure. Secondary cracking (Fig. 4.20*b*) below the fracture surface is sometimes observed in a cleavage failure. Flat fracture is seen both at the fracture surface and below the surface. The presence of cleavage cracks indicates a brittle failure.

In ductile failures which have associated plastic deformation, it is sometimes possible to see slip nucleation (Fig. 4.23)[17] and grain deformation (Fig. 4.24). Although these features do not generally explain failures, their appearance adds further strength to the possibility of ductile failure. They illustrate that in a ductile fracture, plastic deformation precedes fracture.

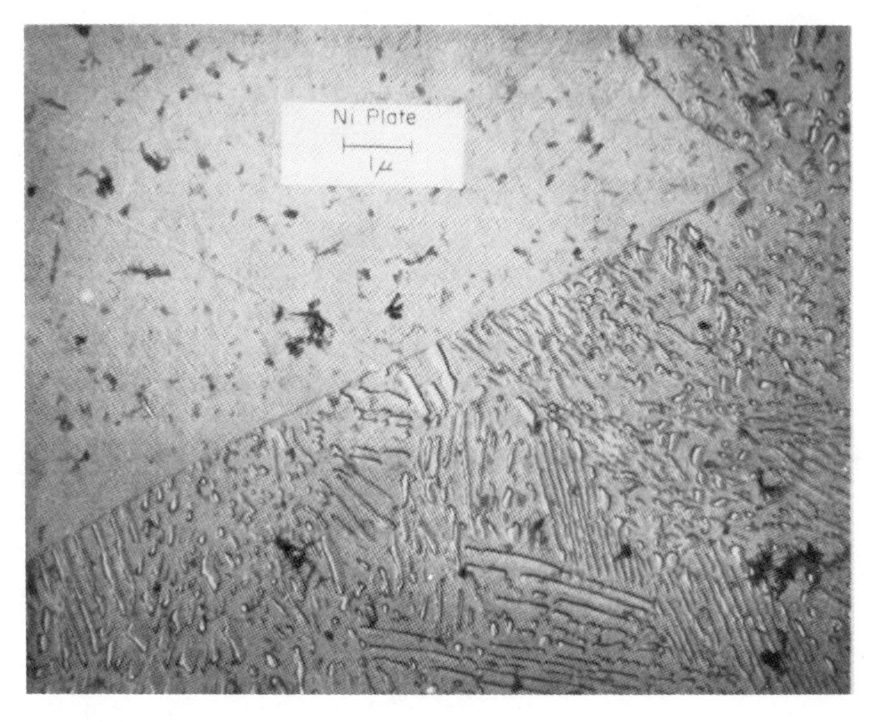

Fig. 4.21 Cleavage fracture in iron (Courtesy of A. Turkalo, Transactions, AIME, Vol. 218, 1960, p. 24.).

Fig. 4.22 Steps on a cleavage fracture.

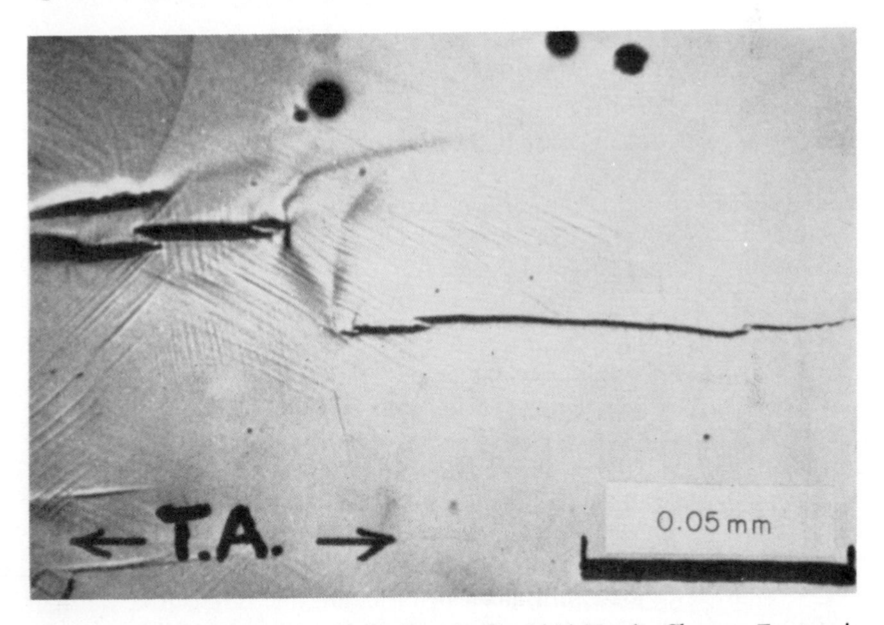

Fig. 4.23 Slip nucleation in a ductile fracture (Ryukichi Honda, Cleavage Fracture in Solid Crystals of Silicon Iron, J. Phys. Soc. Japan, Vol. 16, 1961, pp. 1309–1321.).

Fig. 4.24 Grain deformation associated with a ductile fracture.

The relation of the fracture path to the grain structure can often be used to interpret a failure. Fractures may be classified as transgranular (i.e., across the grains), or intergranular (i.e., between grains). Transgranular fractures may be ductile or brittle, depending on the amount of grain deformation (Fig. 4.20). However, integranular fractures are brittle since the separation of grains is usually due to the presence of a brittle interface. Many processing or service conditions can cause intergranular fractures:

1. *Temper embrittlement in steel.* Fracture occurs in relatively short segments of flat fracture in a tempered martensitic structure[18] (Fig. 4.25).
2. *Grain boundary corrosion.* This occurs because of preferential chemical attack of grain boundary precipitates or segregates[19] (Fig. 4.26). Stress corrosion, which requires both stress and chemical attack, is related to grain boundary corrosion, which is mainly chemical. Stress corrosion can be either transgranular[20] or intergranular,[21] as illustrated in Fig. 4.27.
3. *Elevated temperature creep.* The grain boundary cavities shown in Fig. 4.28 are associated with metal deformation due to creep at elevated temperatures.[22] At these temperatures, metals tend to deform by the movement of grains past one another, which leads to the formation of cavities.

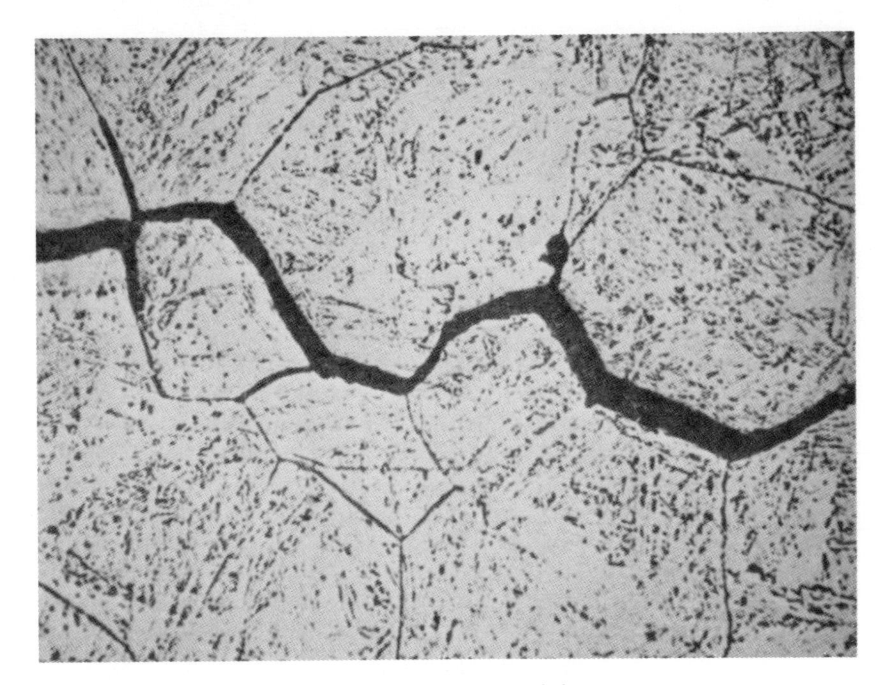

Fig. 4.25 Intergranular fracture due to temper embrittlement.

Exposure to a reactive atmosphere and elevated temperature can often be shown microscopically by alteration of the microstructure. Two opposite effects in steel, carburization and decarburization,[23] are shown in Fig. 4.29. Both conditions can develop during heat treatment or service. Both have been associated with failures, either by causing an undesirable brittle case (carburization) or by decreasing fatigue life (decarburization).

Loading conditions can also cause microstructural changes. Figure 4.30 shows transformation due to adiabatic shear, which is characteristic of failure caused by a high rate of loading, for example, an explosion. It occurs along maximum shear stress planes prior to fracture. The fracture associated with this type of failure shows a characteristic heavily deformed surface along the shear planes.

4.3 ELECTRON FRACTOGRAPHY

Electron fractography has become increasingly important in post-fracture analysis, mainly because of its high resolution capability when compared to microscopy. However, there are several inherent difficulties.

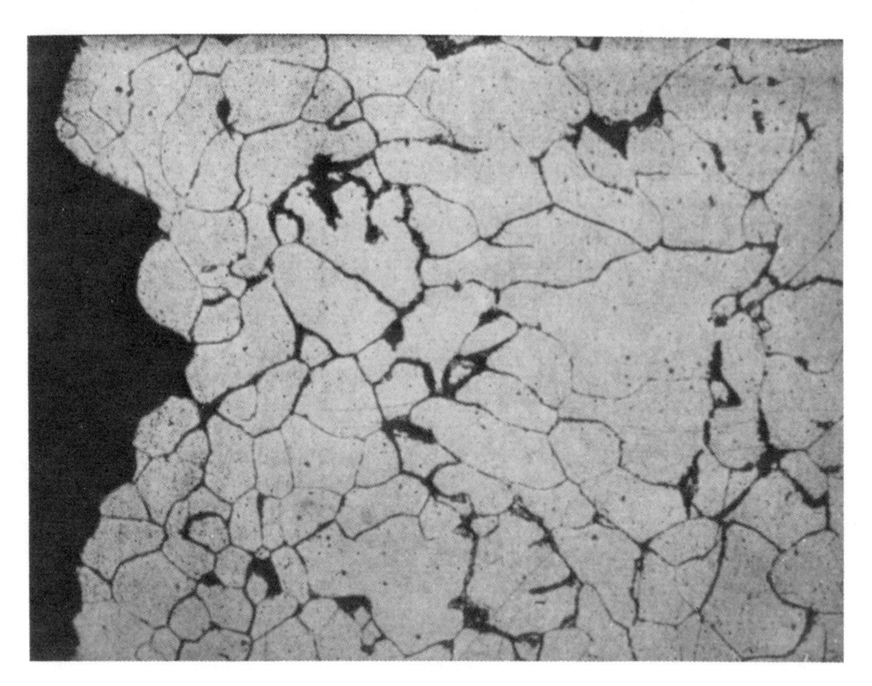

Fig. 4.26 Intergranular cracking due to grain boundary corrosion (From The Principles of Physical Metallurgy by G. Doan. Copyright 1953. Used with permission of McGraw Hill Book Company.).

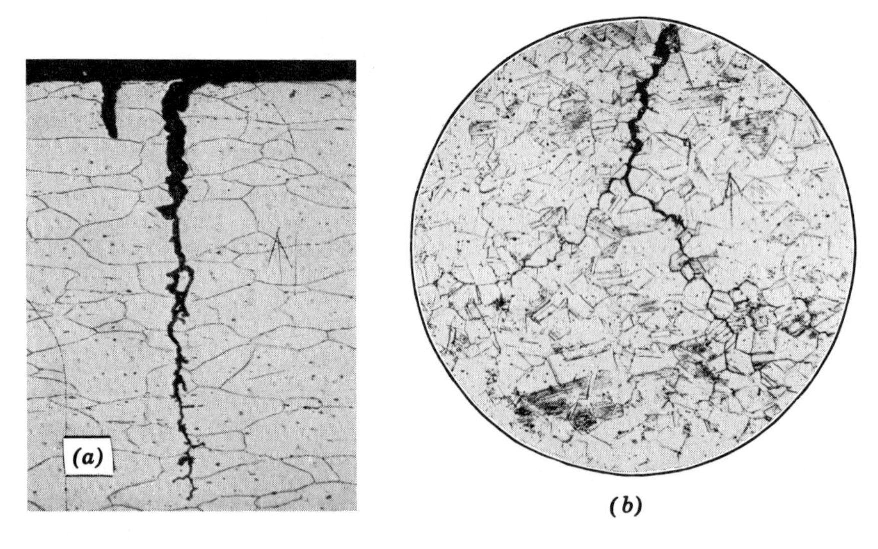

Fig. 4.27 Examples of stress corrosion cracking showing that it can occur (*a*) transgranularly or (*b*) intergranularly.

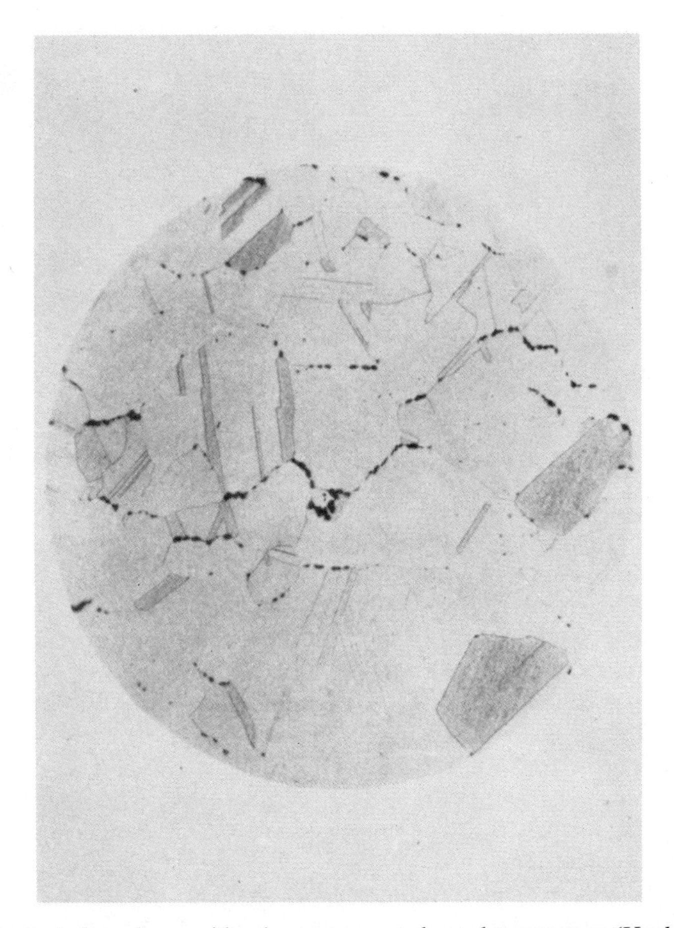

Fig. 4.28 Grain boundary cavities due to creep at elevated temperature (Used with the permission of Pergamon Press, Inc.).

First, access to an electron microscope is necessary. Fortunately, they are readily available throughout the country at universities or private consulting firms. Also, handling problems are markedly increased. Handling fracture surfaces or rubbing them against any abrasive material surface must be avoided. This tends to obliterate the fine details that are being examined.

Until recently it was not possible to examine the fracture surface directly. Rather, replicas made of the fracture surface, which duplicated the topography, were examined. More recent advances, including the Scanning Electron Microscope (SEM), have made it possible to examine the fracture

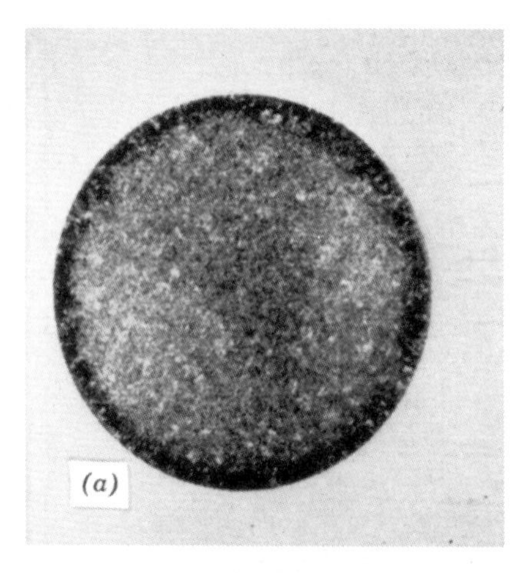

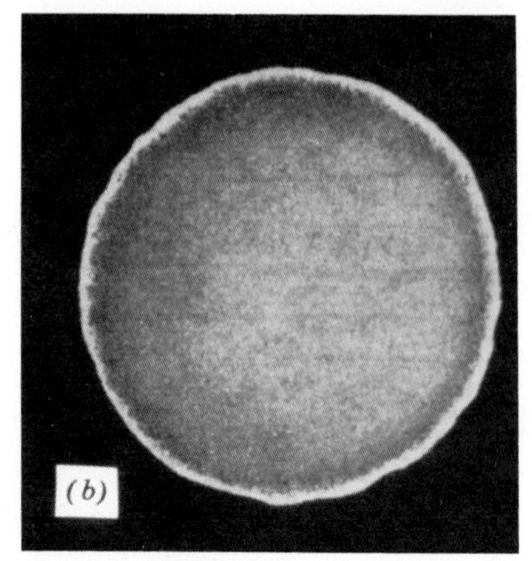

Fig. 4.29 (*a*) Carburization and (*b*) decarburization in steel (From the Principles of Metallographic Laboratory Practice by G. Kehl. Copyright 1949, McGraw Hill. Used with permission of McGraw Hill Book Company.).

surface itself. At the present time, higher magnifications can be employed with the replication method than with SEM.

The most common technique is two-step replication. An example of the steps involved are:

1. Impress the fracture features onto a thin sheet of plastic.
2. Strip the plastic from the surface.
3. Shadow the plastic in a vacuum with a heavy metal, such as chromium, to aid in intepretation.
4. Shadow the plastic in a vacuum with carbon.
5. Dissolve the plastic with acetone, leaving the carbon-heavy metal replica, which is viewed in the electron microscope.

Usually, the fracture surface itself is examined, although it is sometimes useful to examine nonfracture areas similar to the profile microexamination previously discussed. Examination of the fracture surface is actually an examination of its topography. There is no necessity to interpret microstructure, although microstructural features, such as grain boundaries, can often be identified by the manner in which the fracture occurs.

Transgranular fracture can usually be separated from intergranular fracture. Intergranular fractures are usually brittle, absorbing low energy during fracturing, and transgranular fractures can be ductile or brittle. The most common features associated with ductile fracture are dimples, which are formed by plastic deformation as part of the fracturing process. A microcrack, for example, a microvoid at an inclusion (Fig. 4.31), starts

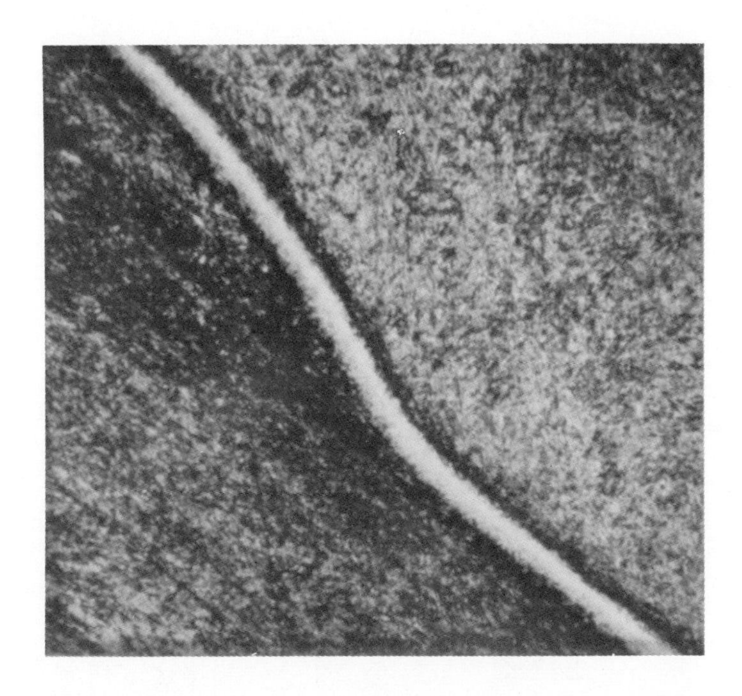

Fig. 4.30 Adiabatic shear due to application of high strain rate of loading.

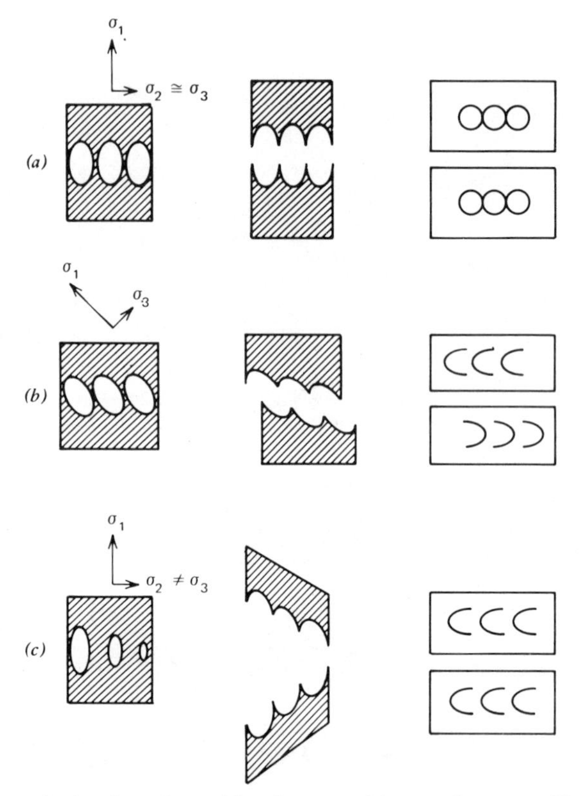

Fig. 4.31 Schematic showing microvoid coalescence: (*a*) normal rupture, (*b*) shear rupture, and (*c*) tearing (From C. Beachem and R. Pelloux. Used with permission of the American Society for Test Materials.).

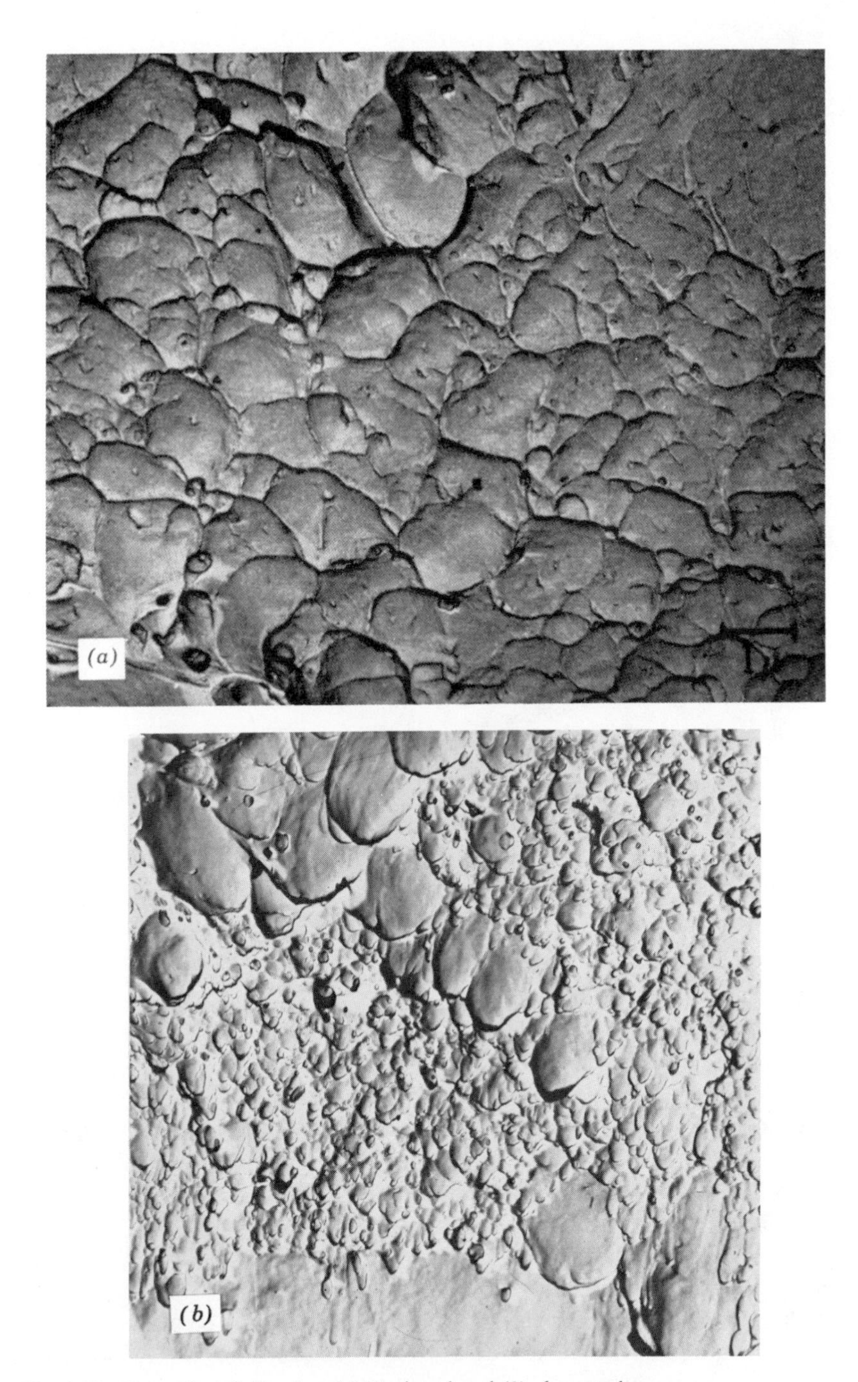

Fig. 4.32 Examples of dimples: (*a*) Equiaxed and (*b*) elongated.

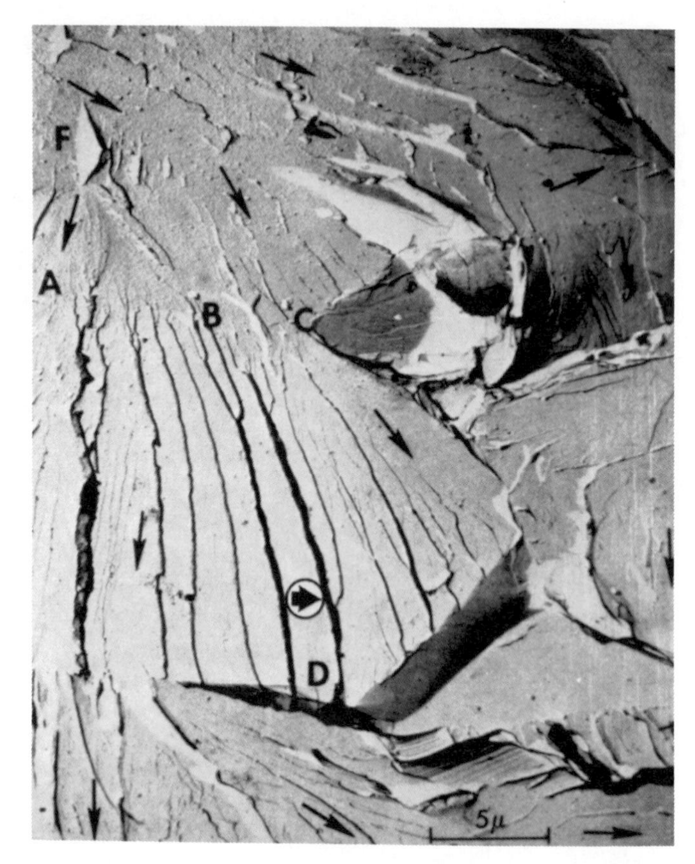

Fig. 4.33 Typical cleavage fracture in iron (Courtesy of R. Pelloux.).

the fracture which continues by plastic deformation and connection of the microvoids.[24] When the fracture is caused by tensile overload,[25] the dimples are equiaxed (Fig. 4.32*a*). When it is caused by shearing or tearing, elongated dimples are formed (Fig. 4.32*b*).

Transgranular fracture can also be brittle. Cleavage failure occurs by separation along crystallographic planes causing flat fractographic features. (Fig. 4.33).[26] There are several features that identify cleavage,[27–29] namely, herringbone pattern, tongues, and river pattern (Fig. 4.34). The latter are caused by a simultaneous progression of the cleavage crack on several planes. At grain boundaries the fracture plane changes because of crystallographic orientation differences. Cleavage, however, is not only associated with transgranular fracture, but also with brittle particles, as shown in

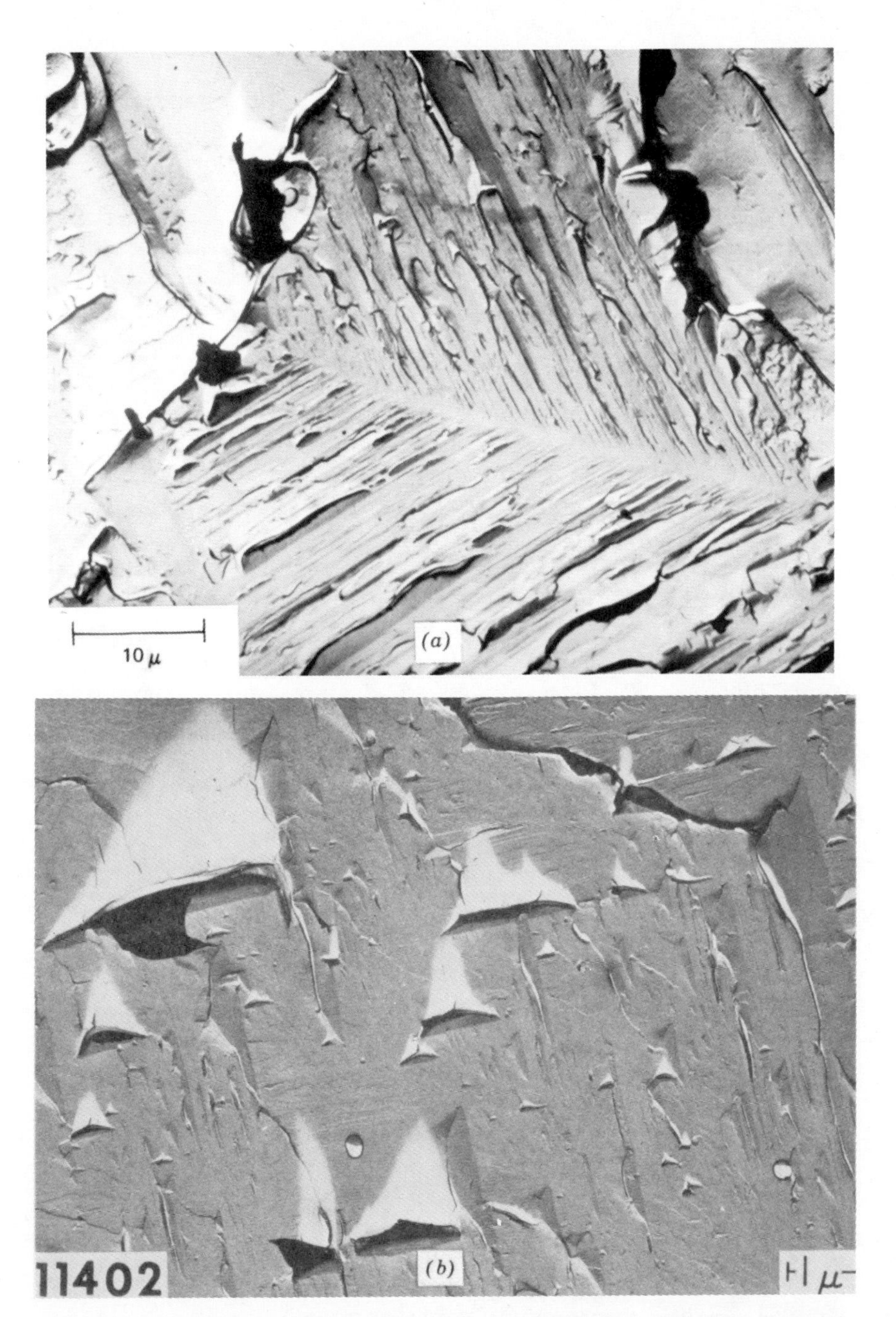

Fig. 4.34 Typical features identified as cleavage: (*a*) herringbone, (From H. Burghard and N. Stoloff. Used with permission of the American Society for Testing Materials.). (*b*) tongues, (Courtesy of C. Beachem.), and (*c*) river pattern.

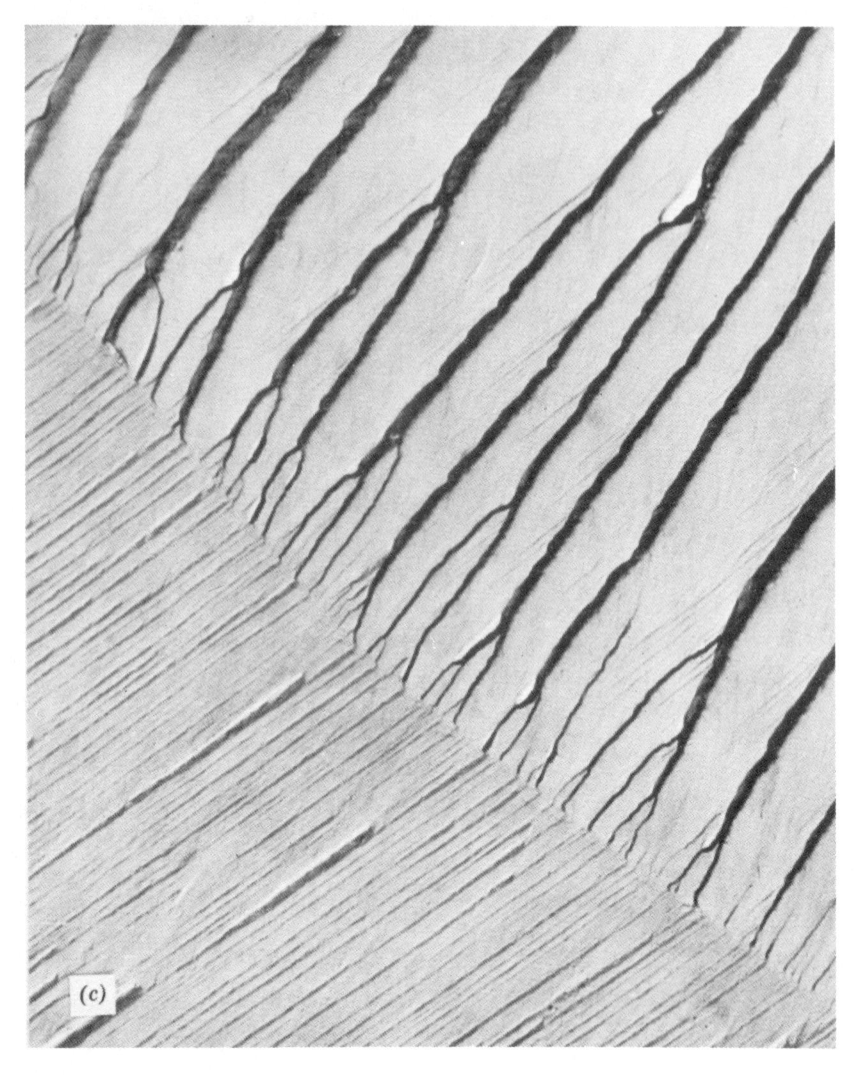

Fig. 4.34 (Continued)

Fig. 4.35, which displays the fracture of an intermetallic particle.[30] These local areas can fail ahead of the main crack, because of the stress fields associated with the crack, and weaken the structure. They are often seen in an otherwise ductile fracture.

Some metals fail in a brittle manner, but do not cleave, for example, martensite in steel. These fractures are identified as quasi-cleavage (Fig.

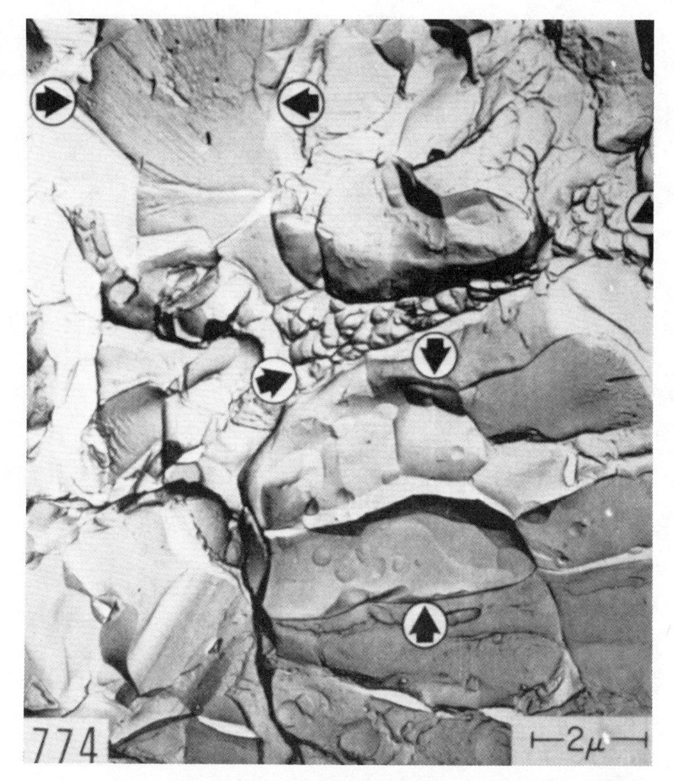

Fig. 4.35 Cleavage of an intermetallic particle, an inclusion (From C. Beachem and R. Pelloux. Used with permission of the American Society for Testing Materials.).

4.36). They are similar to cleavage because they are fairly flat but their features are usually smaller.[31]

Intergranular fractures can occur for a variety of causes,[32,33] but it is generally possible to define fractographically the outline of grains, grain boundaries, and triple points (Fig. 4.37). Temper embrittlement in steel (Fig. 4.38),[34] hydrogen embrittlement, and stress corrosion cause intergranular fracture (Figs. 4.39 and 4.40). The similarity in their fracture appearances, and other considerations, have led some investigators to propose that stress corrosion cracking, at least in steel, occurs by a hydrogen embrittlement mechanism. The differences in the fracture appearances are subtle; the features of the hydrogen-embrittled surfaces are better defined because the stress-corrosion surfaces suffer from the corrodent. In all the fractures illustrated, the features are almost identical.

A fracture mode that has been widely studied with fractography is

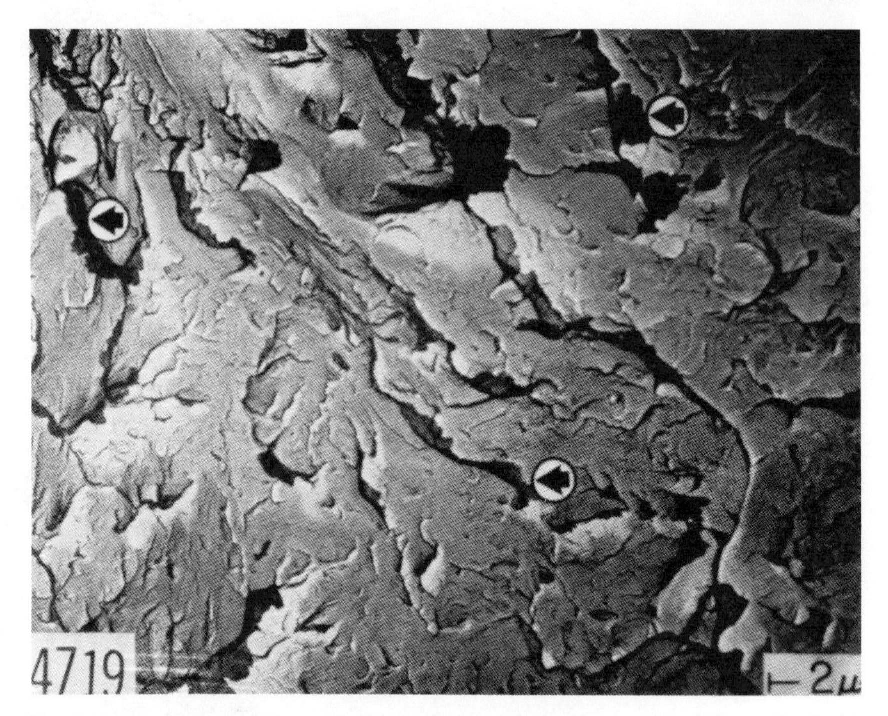

Fig. 4.36 Quasi-cleavage in martensitic steel (From C. Beachem and R. Pelloux. Used with permission of the American Society for Testing Materials.).

fatigue. The most obvious features of this mode are fatigue striations (Fig. 4.41), which represent the local progress of a crack; each striation represents one cycle.[35–37] Not all fatigue fractures show striations. They are common in aluminum but not in high strength steel.

Two general types of fatigue striations have been identified,[38] brittle and ductile (Fig. 4.42). Generally, striations form normal to the crack growth direction; however, occassionally they are seen in other directions. Figure 4.43 shows the striations moving opposite to the crack direction, because of the influence of an inclusion.[39] In this case, overload fracture occurred at the inclusion, ahead of the crack front. Then, fatigue fracture proceeded from the main crack front and the inclusion, until they connected.

Although striations are associated with fatigue, there are other fractures that are similar and should not be mistaken for striations, for example, Wallner lines (Fig. 4.44)[40] and fracture of lamellar structures such as pearlite (Fig. 4.45).

Fig. 4.37 Intergranular fracture showing triple point at grain juncture (Courtesy R. Pelloux.).

Electron fractography is a useful tool, but there are many artifacts that are misleading (Fig. 4.46)[41,42] In addition, single fractographs can be deceptive. The most brittle fracture can still show dimples (Fig. 5.9). Electron fractography is relatively expensive. If properly used, it will help solve problems; if improperly used, it can be confusing and misleading.

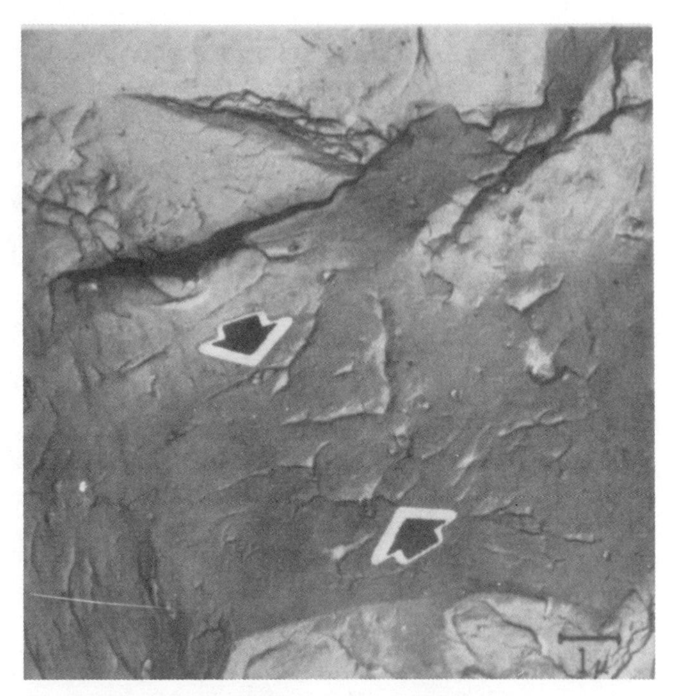

Fig. 4.38 Temper embrittlement fracture surface.

Fig. 4.39 Hydrogen embrittlement fracture surface.

Fig. 4.40 Fracture surface due to stress corrosion.

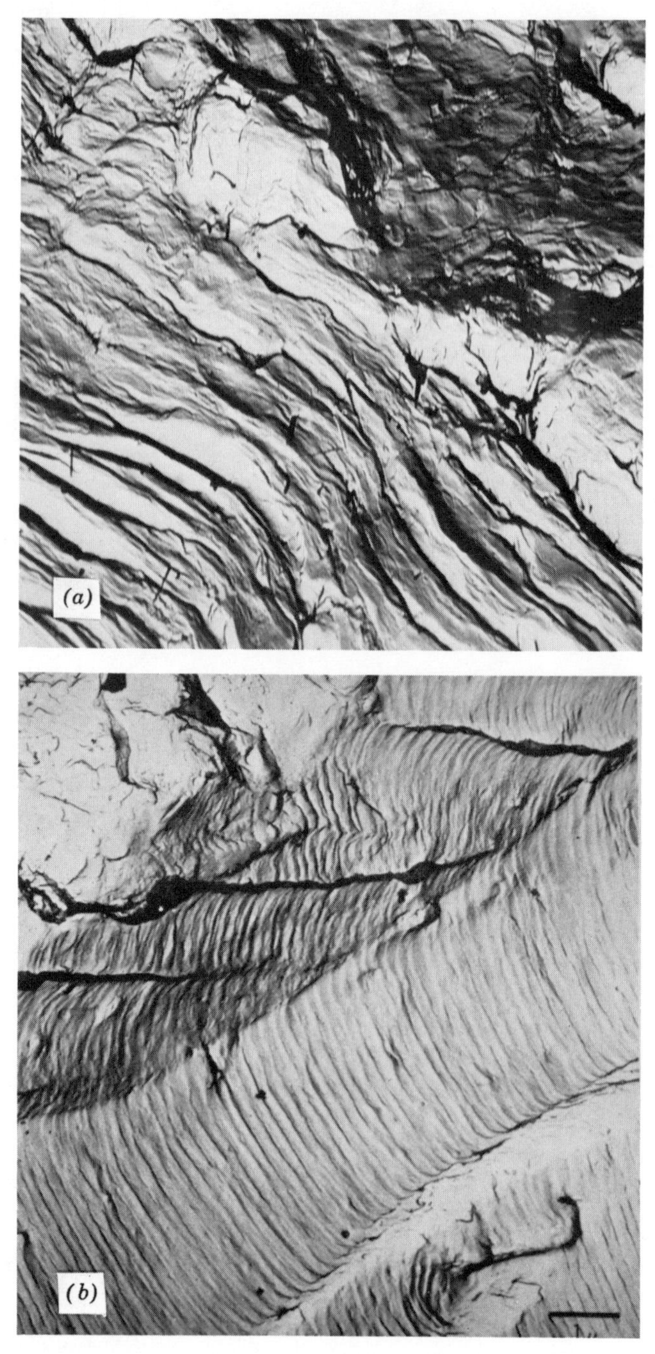

Fig. 4.41 Fatigue striations in (*a*) steel, (*b*) aluminum, and (*c*) titanium (Courtesy C. Beachem.).

Fig. 4.41 (Continued)

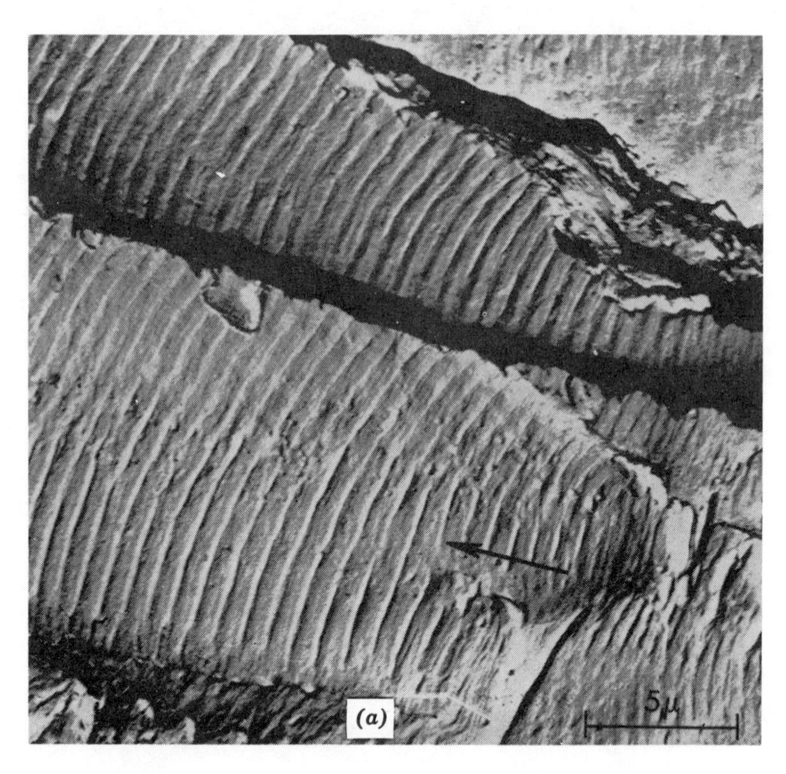

Fig. 4.42 Examples of (*a*) ductile and (*b*) brittle striations (From C. Beachem and R. Pelloux. Used with permission of the American Society for Testing Materials.).

Fig. 4.42 (Continued)

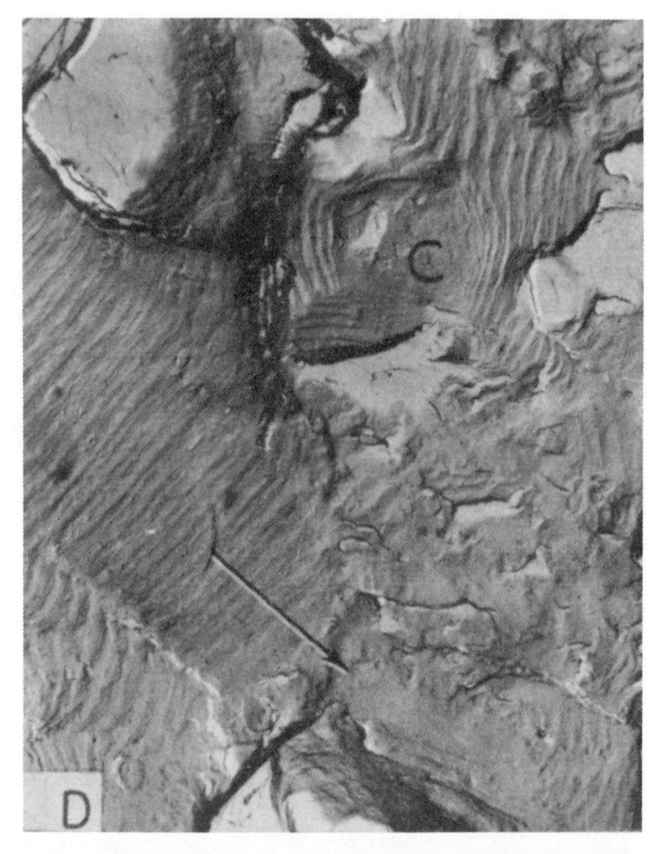

Fig. 4.43 Fatigue striations showing local crack growth in opposite direction to macroscopic crack growth (Reproduced by permission from Transactions, ASM, American Society for Metals, 1964.).

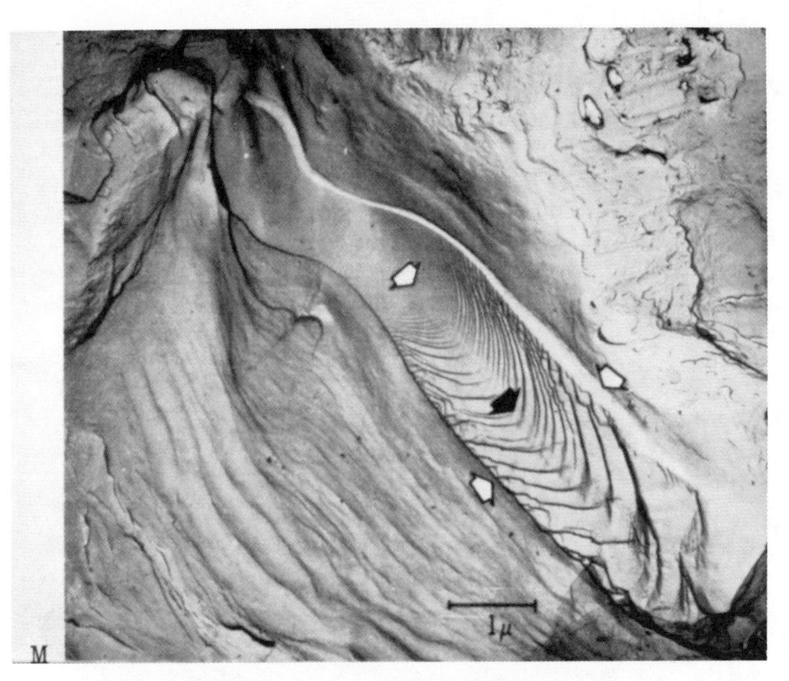

Fig. 4.44 Wallner lines that resemble striations.

Fig. 4.45 Fracture through a lamellar pearlite microstructure.

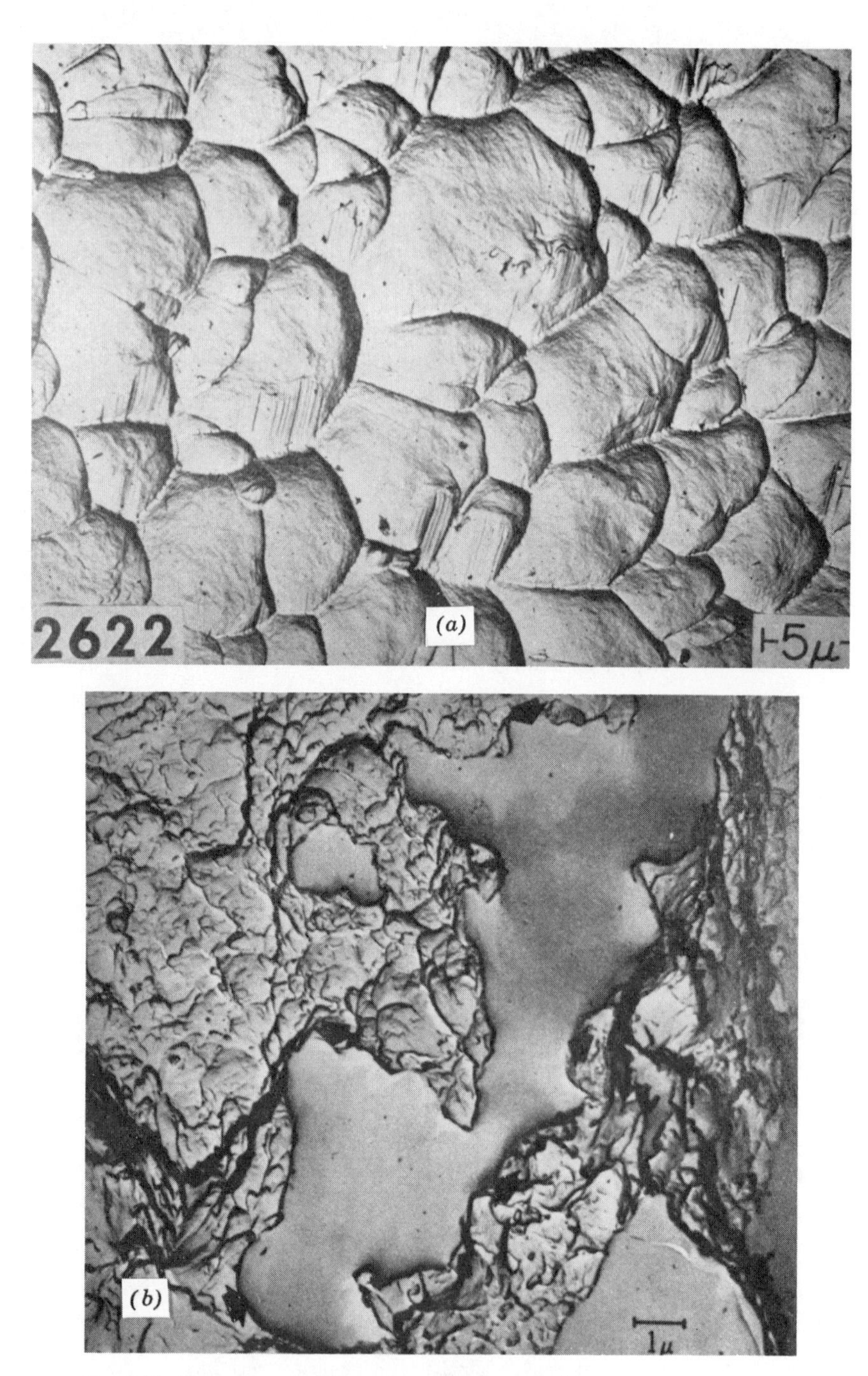

Fig. 4.46 Examples of artifacts due to faulty technique.

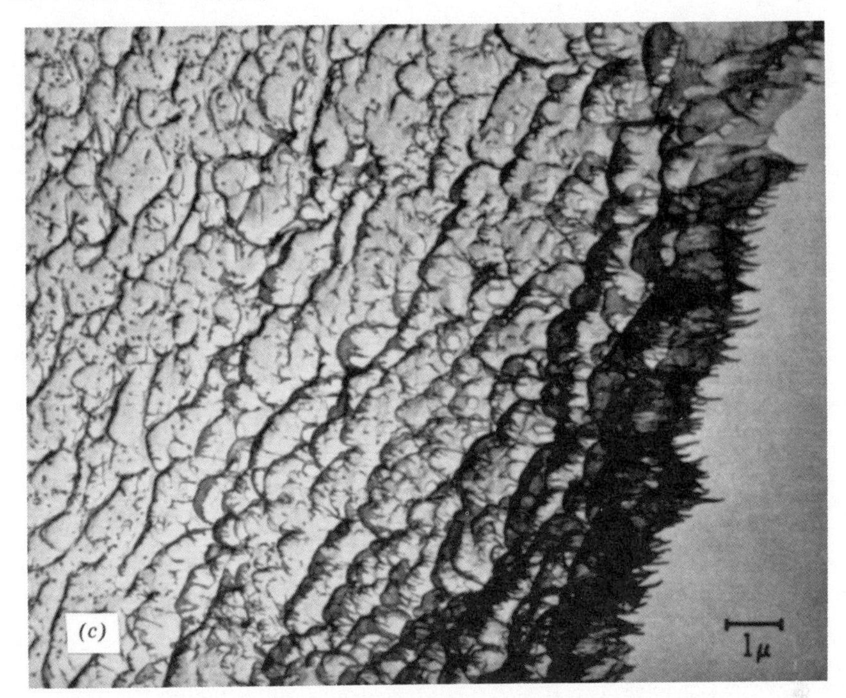

Fig. 4.46 (Continued)

REFERENCES

1. J. Bluhm, *Fracture Mechanics*, SAE, Paper 655c, 2 (1963).
2. D. Fegredo and R. Thurston, *Can. Met, Quart.*, Vol. 5, No. 3, 207 (July–September 1966).
3. H. Rogers, *Trans. AIME*, Vol. 218, 498 (1960).
4. F. Larson and F. Carr, *Trans. ASM*, Vol. 55, 599 (1962).
5. D. Wulpi, *How Components Fail*, ASM, 1966, p. 15.
6. D. Wulpi, *op. cit.*, p. 42.
7. D. Wulpi, *op. cit.*, p. 48.
8. F. Heiser and W. Mortimer, *Effect of Size and Orientation on Fatigue Crack Growth Rate in 4340 Steel*, Watervliet Arsenal Report WVT–7112, March 1971, pp. 34–35.
9. *The Tool Steel Trouble Shooter*, Bethlehem Steel Company, 1952, p. 33.
10. *The Tool Steel Trouble Shooter*, *op. cit.*, p. 91.
11. *Metals Handbook*, ASM, 1958, p. 252.
12. G. Doan, *The Principles of Physical Metallurgy*, McGraw-Hill, New York, 1953, p. 285.
13. G. Doan, *op. cit.*, p. 111.
14. F. Heiser, *Metallurgical Analysis of Material Failures*, Watervliet Arsenal Report, December 1967.

15. A. Turkalo, *Trans. AIME*, Vol. 218, 24 (1960).
16. J. Low, *Fracture*, Wiley, New York, 1959, p. 68.
17. R. Honda, *J. Phys. Soc. Jap.*, Vol. 16, 1309 (1961).
18. T. Davidson, A. Reiner, J. Throop, and C. Nolan, *Fatigue and Fracture Analysis of the 175mm M113 Gun Tube*, Watervliet Arsenal Report WVT–6822, November 1968, p. 64.
19. G. Doan, *op. cit.*, p. 224.
20. H. H. Uhlig, *Corrosion Handbook*, Wiley, New York, 1966, p. 1106.
21. H. H. Uhlig, *op. cit.*, p. 1108.
22. J. Low, *The Fracture of Metals*, MacMillan, New York, 1963, p. 60.
23. G. Kehl, *The Principles of Metallographic Laboratory Practice*, McGraw-Hill, New York, 1949, p. 211.
24. C. Beachem and R. Pelloux, *Fracture Toughness Testing and Its Application*, ASTM, STP 381, 1965, p. 227.
25. *Electron Fractography Handbook*, Air Force Materials Laboratory, Technical Report, ML–TDR–64–416, 1965, p. 2–3.
26. R. Pelloux, *Analysis of Fracture Surfaces by Electron Microscopy*, Boeing Scientific Research Laboratories Report DI–82–0169–RI, December 1963.
27. H. Burghard and N. Stoloff, *Cleavage Phenomena and Topographic Features*, ASTM, June 1967, p. 25.
28. C. Beachem, *The Interpretation of Electron Microscope Fractographs*, NRL Report 6360, January 1966, p. 9.
29. T. Johnston, R. Davies, and N. Stoloff, *Phil. Mag.*, Vol. 12, 305 (1965).
30. C. Beachem and R. Pelloux, *op. cit.*, p. 215.
31. C. Beachem and R. Pelloux, *op. cit.*, p. 218.
32. R. Pelloux, *op. cit.*, p. 234.
33. *Electron Fractography Handbook*, *op. cit.*, p. 6–21.
34. T. Davidson, A. Reiner, J. Throop, and C. Nolan, *op. cit.*, p. 69.
35. F. Heiser, *Anistotropy of Fatigue Crack Propagation in Hot Rolled Banded Steel Plate*, Ph.D. Thesis, Lehigh University, 1969.
36. *Electron Fractography Handbook*, *op. cit.*, p. 1.47.
37. C. Beachem, *Electron Microscope Fracture Examination to Characterize and Identify Modes of Fracture*, NRL Report 6293, September 1965, p. 57.
38. C. Beachem and R. Pelloux, *op. cit.*, pp. 236–237.
39. R. Pelloux, *Trans. ASM*, Vol. 57, 511 (1964).
40. *Electron Fractography Handbook*, *op. cit.*, p. 2.10.
41. T. Johnston, R. Davies, and N. Stoloff, *op. cit.*, p. 27.
42. *Electron Fractography Handbook*, *op. cit.*, p. 1.34.

5

Ductile and Brittle Fracture

5.1 INTRODUCTION

Ductile and brittle are qualitative terms that have a variety of meanings depending on the background of the observer and the scale of observation. The table below illustrates the meaning of each for three levels of observation.

Appearance	Brittle	Ductile
Gross	No plastic deformation	Gross plasticity, large deformation at fracture
Macroscopic	Flat	Shear (shear lips)
Fractographic	Cleavage	Dimples

There are varying degrees of brittleness; the trend from brittle to ductile behavior changes according to the following pattern:

brittle ⟶ ductile
flat fracture → shear lips
plane strain → plane stress
cleavage ⟶ dimples

5.2 DUCTILE FRACTURE

Ductile fractures are high-energy fractures. An inherent part of this process is plastic deformation. Unlike brittle fractures, they are characterized by stable crack propagation. If the load that causes a crack to propagate is removed, the crack stops.

5.2.1 Deformation

Ductile fractures are associated with plastic deformation, which requires a shear deformation.

Slip occurs on an atomic scale and can be viewed as the movement or sliding of one layer of atoms over another. Shear stress is required primarily to break the atomic bonds. Figure 5.1 is a schematic diagram showing these requirements.[1]

To compute the stress necessary to cause slip (τ_{max}), it is assumed that the stress varies sinusoidally with the distance between stable atomic positions. It is then possible to determine τ_{max} according to the following:

$$\tau = \tau_{max} \sin \frac{2\pi x}{b}$$

and

$$\tau = G\gamma$$

$$\tau = G \frac{x}{a}$$

where

τ = shear stress
τ_{max} = maximum shear stress
b = distance between atomic postions in the direction of slip
a = distance between atomic positions normal to the direction of slip

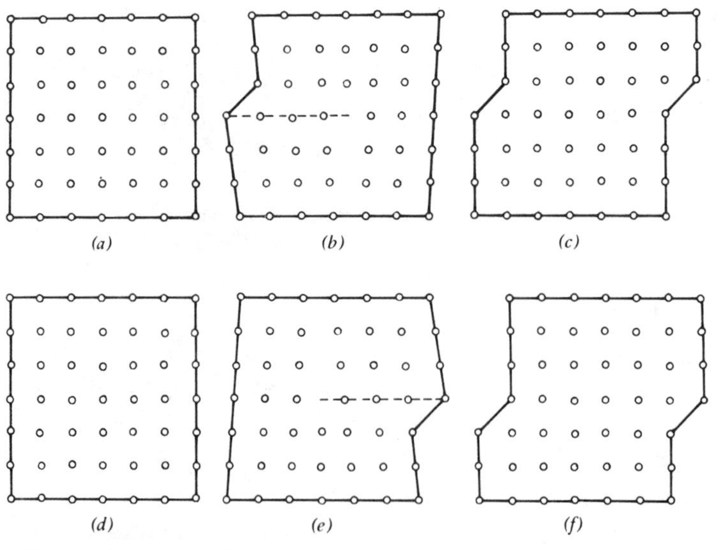

Fig. 5.1 Schematic showing slip process (Used with permission of The Royal Society).

x = distance in direction of slip
γ = shear strain
G = shear modulus

For small movements,

$$\sin \frac{2\pi x}{b} = \frac{2\pi x}{b}$$

therefore,

$$\tau = \tau_{\max} \frac{2\pi x}{b} = G \frac{x}{a}$$

then

$$\tau_{\max} = \frac{Gb}{2\pi a}$$

Letting

$$a = b$$

$$\tau_{\max} = \frac{1}{2\pi} G$$

Substituting observed values for G, values of 10^5 to 10^6 psi for $\tau_{\max}$ are obtained. However, values of only 10^2 to 10^3 psi are actually observed. The discrepancy is caused by the presence of dislocations that reduce the stress necessary for metal flow. Slip occurs by movement of only one atom at a time, rather than movement of a sheet or layer of atoms in a group.

Since slip is atomic in origin, it is affected by crystal structure and occurs on certain atomic planes and in certain crystal directions, depending on the crystal structure. Each combination of slip plane and direction is a slip system. There are three structures common to metals. Each has a number of slip systems common to itself: (a) face-centered cubic—12 systems, (b) body-centered cubic—48 systems, and (c) close-packed hexagonal—3 systems.

When a load is applied, it is possible to resolve the stress into normal and shear stresses. Figure 5.2 shows the direction of shear and normal stresses under tensile and torsional loading. Slip occurs where the resolved shear stress is highest. To compute the critical resolved shear stress (Fig. 5.3)

ϕ = angle between normal to slip plane and axis of load
λ = angle between slip direction and axis of load

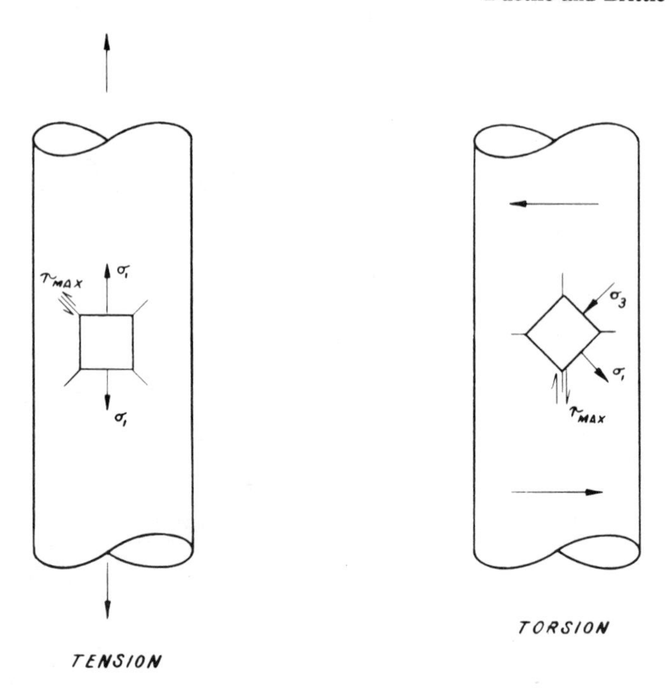

Fig. 5.2 Stress system under tensile and torsional loading.

$$A \text{ (slip plane)} = \frac{a}{\cos \phi}$$

$$P \text{ (slip direction)} = P (\cos \lambda)$$

where A = area normal to loading axis

and P = load

$$\tau = \frac{P \text{ (slip direction)}}{A \text{ (slip plane)}}$$

$$\tau = \frac{P}{A} (\cos \lambda)(\cos \phi)$$

Substituting various values for λ and ϕ shows that the critical resolved shear stress reaches a maximum at 45°. A similar calculation could be made for resolved normal stress on a plane. In this case, there is no dependence on slip direction. If the resolved normal stress were much higher than the resolved shear stress, brittle fracture would probably occur before slip.

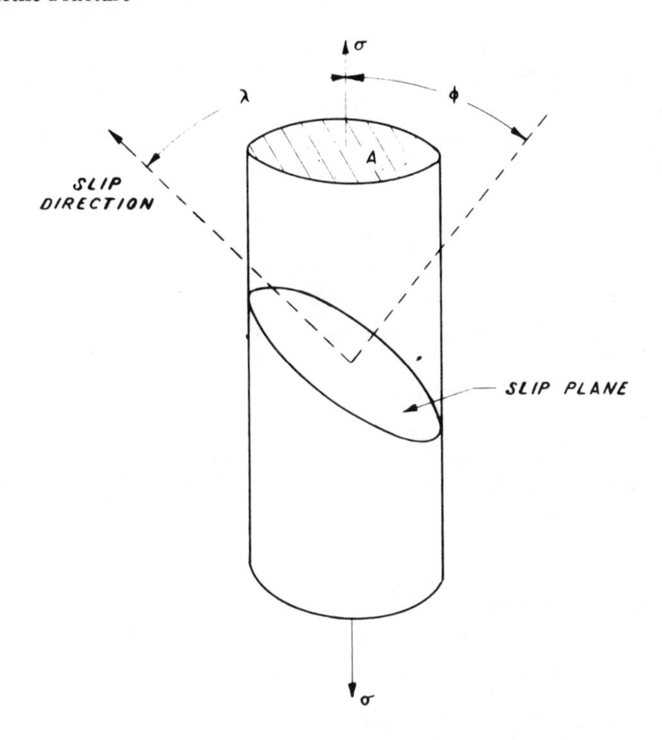

Fig. 5.3 Determination of critical resolved shear stress under tensile loading.

5.2.2 Fracture Appearance

Much of the work on ductile fracture has involved interpretation of tensile bar fractures. There are two principle types of fractures in tensile bars, that are ductile.

SHEAR FRACTURE

In this type of fracture, shear or plastic deformation is localized on certain planes and directions. Fracture occurs by one half of a bar sliding off the other (Fig. 2.9). These fractures are usually associated with high strain rates.

CUP-CONE

Cup-cone fractures are the more common type of ductile fracture. They occur primarily by the connection of voids at the center of the bar (cup) and then by shear at the outside surfaces (cone) (Fig. 2.11). This is referred to as void-coalescence and is manifested fractographically by dimples (Fig. 4.32). A necessary feature of fracture by void coalescence is plasticity. In fact, almost all fractures, ductile or brittle, are plastically induced, although the degree of plasticity is markedly different.

5.2.3 Crack Growth

Once a flaw reaches a critical size fracture occurs, whether ductile or brittle. It is possible to differentiate between high energy (ductile) and low energy (brittle) by the way a crack progresses towards is critical stage.

Assume that two pieces of metal exist with a starting crack size in each case of C_0 (Fig. 5.4) and gradually increase the stress. Fracture occurs when the stress reaches a critical point. In a brittle metal, when the critical stress is reached, unstable crack growth (fracture) occurs rapidly, with no increase in stress. In a ductile metal, slow growth (tearing) occurs with increasing stress until the crack is long enough to cause rupture. A measure of toughness is the amount of slow growth; the more slow growth, the tougher the material.[2]

5.2.4 Identifying Ductile Fracture

Ductile fractures occur at stresses above the material's yield strength. This may mean that one of several things has occurred.

1. The material was not strong enough.
2. The service conditions (loads) differed from those anticipated by the designer.
3. Abnormal loading conditions were applied.

The latter two reflect design problems. However, they are different in that (2) concerns normal loading conditions and a drastic mistake by the designer, whereas (3) concerns loading from an outside source, for example, an explosion in a pressurized vessel.

A ductile fracture occurs with large amounts of deformation. It is usually possible to differentiate between a ductile and a brittle fracture by a macroscopic examination. One obvious feature is deformation. However, there are other characteristics, one of which is the number of fragments. In a ductile fracture, much of the available energy goes into plastic deformation. In a brittle fracture, the energy is absorbed by the creation of surfaces, that is, cracking. Consequently, brittle fractures usually show more fragments than ductile fractures.

Ductile fractures also show characteristic features microscopically. They are almost invariably transgranular (Fig. 4.16). With ductile transgranular fractures, it is often possible to see deformed microscopic phases. Figure 4.20, which is a cross-section of tough ductile banded mild steel, displays

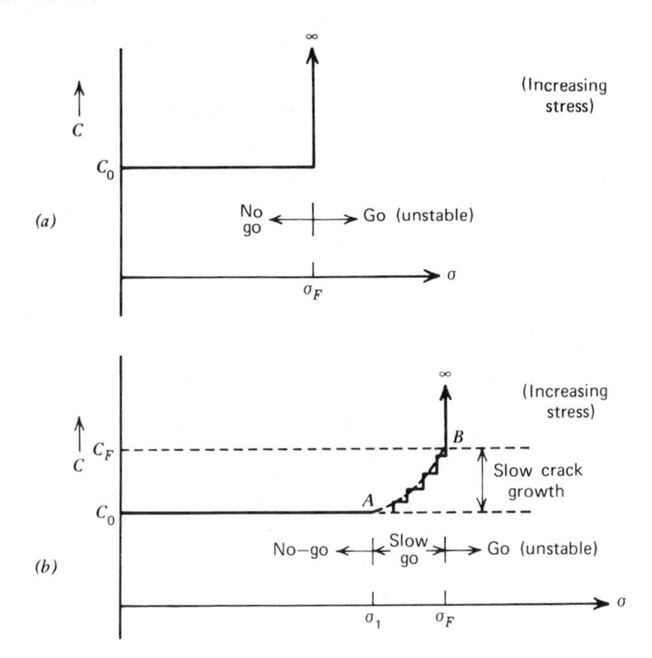

Fig. 5.4 Comparison of (*a*) brittle and (*b*) ductile overload on crack growth.

this phenomenon. The bands are alternating layers of ferrite and pearlite. The deformation in the ferrite is clearly shown.

Fractographically, ductile fracture is usually considered to be due to microvoid coalescence. Depending on the stress system applied, this can occur by opening, tearing, or shearing (Fig. 4.31). These result in two types of dimples, equiaxed and elongated (Fig. 4.32). The differentiation between tearing and shearing is subtle, and mating fracture surfaces must be replicated. In a failure analysis, it is usually sufficient to differentiate between dimples and other features.

Although microvoid coalescence is the main method of ductile rupture and dimples are usually seen on the fracture surface, there are cases when this may not occur. In a massive shear fracture, typified by the tensile bar shown earlier, the fracture surface may be so smeared that is is essentially fractographically featureless.

5.3 BRITTLE FRACTURE

Brittle fractures occur at stresses far below the yield strength, are usually associated with flaws, are often catastrophic, and usually occur without warning. The absence of gross plastic deformation distinguishes them from ductile fractures.

Many reported service failures have been brittle fractures. The Liberty ships that cracked in half during World War II, bridge failures, (e.g., one in Belgium in 1951 only three years after fabrication), and the molasses tank failure in Boston in 1919 were all attributed to a brittle condition. Often, failures result in the loss of lives and the initiation of lawsuits. The molasses tank failure caused the death of twelve people and took six years to settle. It was the first recorded incidence of a chevron pattern on the fracture surface. Recognition of this fracture pattern helped to identify the fracture origin.

Brittle failures can occur in brittle materials, or ductile materials, under certain loading conditions. High strain rates (e.g., impact), triaxial stresses, or low temperatures can cause a normally ductile material to behave in a brittle manner.

5.3.1 Theoretical (Atomic) Fracture Strength

Fracture is the breaking of bonds between atoms. In an ideal brittle fracture, all the energy is absorbed in the creation of new surfaces, that is, none is absorbed in plastic deformation.

If the force to separate atoms varies sinusoidally with distance, it can be shown that

$$\text{Energy to fracture} = \sigma_{\max} \frac{\lambda}{\pi}$$

where λ = sinusoidal wave length

and $\sigma_{\max}$ = maximum stress

Since fracture creates two new surfaces,

$$\sigma_{\max} \frac{\lambda}{\pi} = 2S$$

where S = surface tension

$$\sigma_{\max} = f(E, a_0, S)$$

where E = Young's modulus

a_0 = atomic spacing

Then, from Hooke's law

$$\sigma = E\epsilon$$

$$\sigma = E \frac{x}{a_0} \text{ (for small displacements)}$$

Differentiating with respect to x

$$\frac{d\sigma}{dx} = \sigma_{max} \cos\left(\frac{2\pi x}{\lambda}\right)\frac{2\pi}{x}$$

when

$$x \rightarrow 0, \cos\frac{2\pi x}{\lambda} \rightarrow 1$$

and

$$\frac{d\sigma}{d\lambda} = \frac{2\pi}{\lambda}\sigma_{max}$$

However,

$$\frac{d\sigma}{dx} = \frac{E}{a_0}$$

Therefore,

$$\frac{E}{a_0} = \frac{2\pi}{\lambda}\sigma_{max}$$

Since

$$\frac{2\pi}{\lambda} = \frac{\sigma_{max}}{S}$$

$$\frac{E}{a_0} = \frac{\sigma_{max}{}^2}{S}$$

$$\sigma_{max} = \frac{ES^{1/2}}{a_0}$$

or

$$\sigma_{max} = 0.1E$$

When Young's modulus is 30×10^6 psi, σ_{max}, the fracture strength for steel, should be 3×10^6 psi or 3,000,000 psi, a value that is far beyond that observed, except in rare cases in whiskers. The conclusion is similar to that for plastic deformation; there are defects in the material. The strength of a material can be increased by eliminating the defects, as in small whiskers, or by preventing their movement, as in piano wire.

One of the first tests showing the effect of defects on strength was conducted by Leonardo da Vinci. Testing glass fibers in tension, he noted that the strength decreased as the length of the fiber increased, and was higher for polished fibers. These observations were attributed to imperfections. As the length of the fiber increased, the probability that a deleterious imperfection was present increased; as surface finish improved, the probability decreased. This was the initial concept of material imperfections.

5.3.2 Fracture Strength in the Presence of Cracks

GRIFFITH (SURFACE ENERGY)

The first work in quantitatively measuring fracture strength was done by Griffith, and eventually led to the concepts of fracture mechanics. His theory considered energy to fracture a material with a penny-shaped

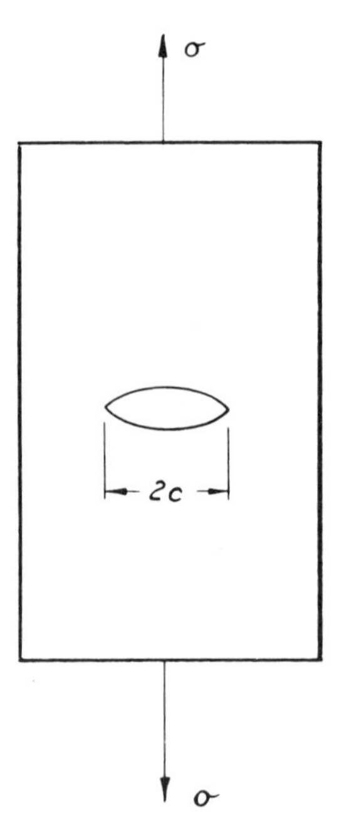

Fig. 5.5 Griffith's penny-shaped crack.

crack of length $2c$ (Fig. 5.5). He developed an energy balance with three parts: (a) strain energy without a void, (b) surface energy, since fracturing is the creation of new surfaces, and (c) energy released when a crack moves. Based on this,

total energy = strain energy + surface energy − released energy

$$U_T = U + 4ctT - \frac{\pi c^2 \sigma^2 t}{E}$$

As the crack moves strain energy is released, but surface energy is consumed. If there is a balance, the crack is stable; if there is an imbalance, the crack is unstable and will move.

The condition for instability is

$$\frac{\partial U_T}{\partial c} = 0$$

Differentiating U_T with respect to c

$$0 = 4tS - \frac{2\pi\sigma^2 ct}{E}$$

$$2S = \frac{\pi\sigma^2 c}{E}$$

and

$$\sigma = \left(\frac{2SE}{\pi c}\right)^{1/2} \qquad \text{(Griffith)}$$

OROWAN (PLASTIC DEFORMATION)

The Griffith relation considers only brittle fracture, with no allowance for slow growth or deformation. It is based solely on the use of energy to create new surfaces, that is, fracture. Modifying Griffith's work, Orowan introduced an energy of plastic deformation, P. When coupled with the above equation, this yielded

$$\sigma = \left(\frac{2E}{\pi c}(S + P)\right)^{1/2}$$

However, since energy of plastic deformation is usually much greater than surface energy in metals, the latter can be ignored to yield

$$\sigma = \left(\frac{2EP}{\pi c}\right)^{1/2} \qquad \text{(Orowan)}$$

IRWIN (G_c)

Both of the above relations are difficult to apply. Measuring difficult-to-measure quantities, such as surface tension, and fairly ill-defined concepts, such as plastic deformation energy, P, is required. The practical application of these concepts and fracture mechanics started with Irwin. He considered five forms of energy:

1. Work due to external force—input.
2. Strain energy release—input.
3. Kinetic energy—dissipation.
4. Surface energy—dissipation.
5. Plastic deformation—dissipation.

Irwin considered that kinetic energy at instability was small, and that plastic deformation was more important than surface energy. Since brittle fractures were of interest, deformations were small and, therefore, work due to external forces was small. This meant that only two forms of energy were important, namely, strain energy release, U, and plastic deformation, P.

Irwin then defined a new term, G, sometimes called strain energy release rate, where

$$G = \frac{1}{2}\left(\frac{\partial U}{\partial A}\right) = \text{rate of change of energy with crack growth}$$

In terms of the energy function previously mentioned

$$\frac{U}{A} \text{ is proportional to } \frac{P}{A}$$

Therefore,

$$G \, \alpha \, \frac{P}{A}$$

As with Orowan's analysis, the plastic deformation energy is not well-defined analytically, but can be determined experimentally. It was shown that

$$U = \frac{\pi c^2 \sigma^2}{E}$$

and, therefore,

$$G = \frac{\pi c \sigma^2}{E}$$

Instability, that is, fracture, occurs when $G = G_c$ and G_c = critical strain energy release rate. From the above equations it is evident that

$$G_c = \frac{\pi \sigma^2 c}{E}$$

or, in terms of stress

$$\sigma = \left(\frac{G_c E}{\pi c}\right)^{1/2} \qquad \text{(Irwin)}$$

The form of Irwin's equation is identical to that derived by Griffith and Orowan. They differ only in the significant energy term. Irwin's G is a defined quantity, not a basic material property.

All materials generate a plastic zone of material at the crack tip. Originally, G_c was measured in fracture toughness testing, using a precracked specimen. It was observed that G_c increased with notch radius. The notch root radius was actually a function of the precracking stress. As the stress increased, the notch root values increased, as did the apparent G_c.

STRESS APPROACH (FRACTURE MECHANICS)

An alternative to the energy approach has been based on stress calculations. Westergard and others showed that stresses in the vicinity of a crack tip had a common function, regardless of the type of loading condition. The general equations are as follows:

$$\sigma_x = \frac{K}{(2\pi r)^{1/2}} \cos\frac{\theta}{2}\left(1 - \sin\frac{\theta}{2}\sin\frac{3\theta}{2}\right)$$

$$\sigma_y = \frac{K}{(2\pi r)^{1/2}} \cos\frac{\theta}{2}\left(1 + \sin\frac{\theta}{2}\sin\frac{3\theta}{2}\right)$$

$$\sigma_z = \nu(\sigma_x + \sigma_y)$$

The common factor is the stress intensity factor, K. It reflects the effect of a crack on actual stresses. In this approach (fracture mechanics), fracture occurs when K reaches K critical.

Once K is determined from a standard specimen, it can be applied to a complex component if the K-calibration (i.e., the relation of stress, crack length, and geometry to K) is known.

$$K = f\,(\text{load, crack length, geometry})$$

For an infinite plate loaded in tension at infinity

$$K = \sigma\sqrt{\pi a}$$

This relation is similar to that developed in the energy approach where

$$\sigma\sqrt{\pi a} = (G_c E)^{1/2}$$

Therefore,

$$K_c = (G_c E)^{1/2}$$

or

$$K_c^2 = G_c E$$

5.3.3 Crack Growth

In the previous section, the reaction of brittle and other materials to loads, in the presence of a crack, was shown (Fig. 5.4). Brittle fractures occur when crack length suddenly increases, without an increase in stress. This is basically different from ductile fractures, in which crack extension requires increasing stress.

5.3.4 Effect of Plastic Zone Size

Heavy sections sometimes fail at lower stresses than thinner sections. The relative stress state and the associated plastic zone size can cause this phenomenon.

In a heavy section

$$\text{plane strain, } r_p = \frac{K^2}{2\pi YS}(1 - 2\upsilon)^2$$

where r_p = plastic zone size

and υ = Poisson's ratio

In a thin section

$$\text{plane stress, } r_p = \frac{K^2}{2\pi YS}$$

In the plane strain state the plastic zone size is smaller, indicating that more energy is used for fracturing than for plastic deformation.

5.3.5 Identifying Brittle Fracture

Brittle fractures can usually be distinguished from ductile fractures by an almost complete absence of plastic deformation. Fractures are often flat and shiny, with little or no evidence of shear lips. They occur at stresses below the yield strength.

It is often possible to trace a fracture path to its origin by interpreting chevron markings (Fig. 4.3) on the fracture surfaces. With a brittle fracture, the fracture origin is often a notch or a small crack.

A brittle fracture, like an outright ductile fracture, is obvious. However, since brittle fractures can occur in ductile materials, the material's properties should be evaluated to determine whether the material was inadequate, or the loading conditions were abnormal. Many undesirable conditions do exist in the use of materials. Some applications require high strain rates, low temperatures, or notches. The mechanical property to evaluate is that which most closely reflects the severity of the actual conditions. Thus notched bar or precracked tests are used as an indicator of brittle behavior.

Microscopically, brittle fractures may be transgranular or intergranular. Cleavage fracture is transgranular. Within a grain the fracture is flat, as shown in the fracture in ferrite (Fig. 5.6). Another common characteristic of brittle cleavage fracture is also shown. There are many flat fractures in ferrite grains not associated with the fracture surface.

Intergranular fractures are almost invariably brittle (low energy) (Fig. 4.21). They usually result from precipitation or segregation at grain boundaries. It is often difficult to resolve whether a fracture is intergranular or transgranular by examining the fracture surface. This is especially true if tempered martensitic steel is involved, since it is difficult to resolve grain boundaries. It is best to analyze secondary cracks because both sides of the crack are available for investigation. The fracture paths in the primary and secondary cracks are probably the same.

Three types of electron microscopic fracture surfaces are generally associated with brittle fractures: cleavage (Fig. 4.33), quasi-cleavage (Fig. 4.36), and intergranular (Fig. 4.37). Ductile dimpling is seen even on the most

Fig. 5.6 Cleavage fracture in ferrite.

brittle fracture. Figure 5.7 shows several fractographs taken from a steel Charpy bar broken at low temperature with 15% fibrosity, which is almost completely brittle. There were no shear lips; yet, ductile dimples were seen.[3]

A predominantly ductile fracture can have a brittle origin, due to the presence of a large inclusion, or stress corrosion. Figures 4.30 and 5.8 show a ductile fracture in a thick wall cylinder subjected to internal pressure. The fracture started at a piece of entrapped slag. By analyzing the fracture surfaces, it was possible to locate the general region of the origin. Then, using electron fractography, cleavage fracture was identified at the origin. Finally, using the electron microprobe, the material that cleaved was identified as a nonmetallic slag inclusion.

5.4 FACTORS AFFECTING BEHAVIOR

In a failure analysis, determination of whether a fracture occurred in a ductile or a brittle manner is usually only the beginning. It is also necessary to determine what caused the material to fracture. In ductile fractures, the main mechanical property of interest is the yield strength. If the material

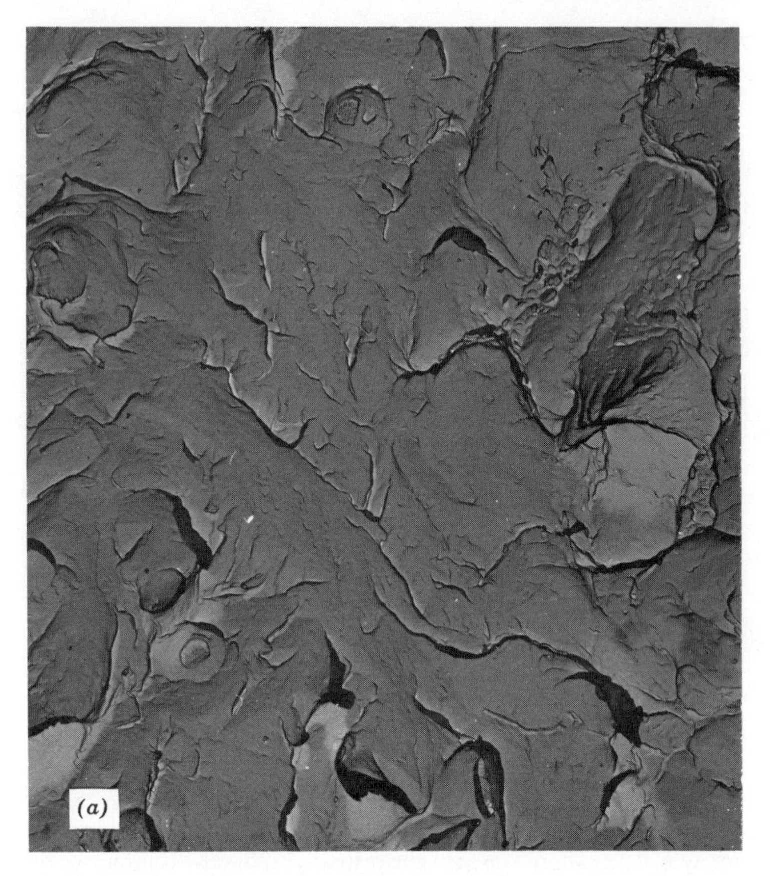

Fig. 5.7 Brittle and ductile features on a steel Charpy bar that fractured with 15% fibrosity and no shear lips.

meets the design requirements, either the designer improperly calculated the loads, or the loading condition was higher than expected. If the material does not meet the requirements, the material manufacturer is at fault. It would then be advisable to check the microstructure or the heat treatment.

Since brittle fractures are more insidious and detrimental, they have received more attention. Almost invariably, when a service failure is discussed it is a fatigue failure (Chapter 6) or a brittle failure due to an imperfection. As a result, special tests have been devised to measure the brittle tendencies of materials and conditions. Metals react differently to changes in loading rates, temperature, and notches. Since most studies are based

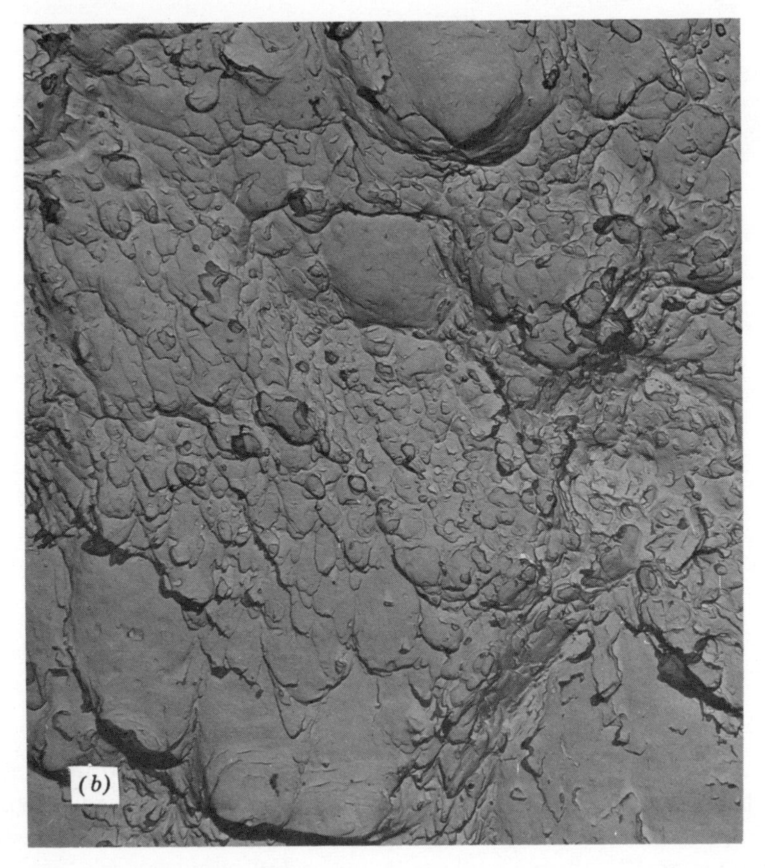

Fig. 5.7 (Continued)

on standard tests, the reaction of metals in these tests, and what this indicates, is discussed in the following section.

5.4.1 Fracture Toughness Behavior

The fracture toughness of a metal is usually measured as K_{IC} (plane strain fracture toughness), which is a material property. At thicknesses less than critical, K_C, which is a function of material and stress state is measured. It is often measured if the actual application involves thin sheet, such as in airplane structures.

Some data indicate that K_{IC} changes with both temperature and strain rate. As the test temperature decreases, K_{IC} decreases. In some steels, a transition with changing temperature has been observed. As the strain rate

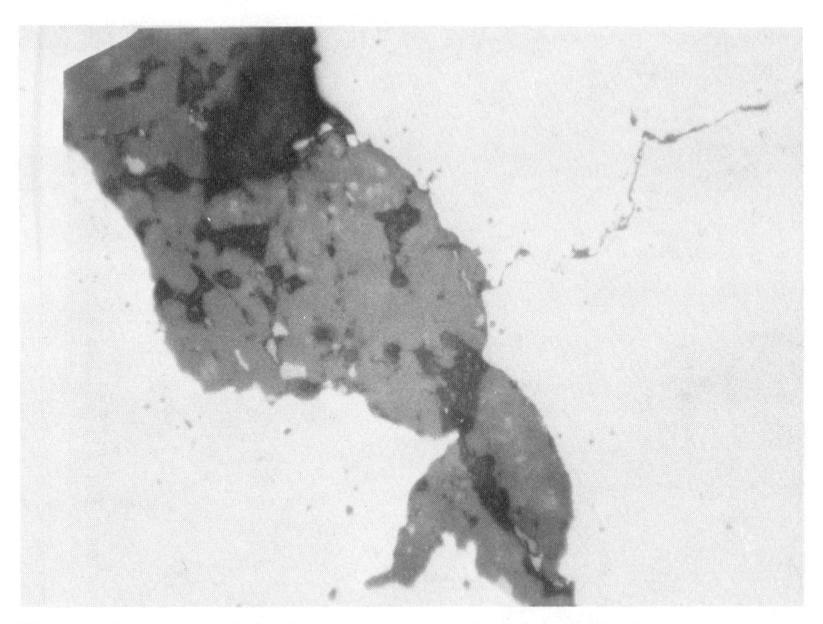

Fig. 5.8 Entrapped brittle slag particle that initiated a ductile fracture in a thick wall cylinder.

increases, K_{IC} often decreases, although it has also been reported that at very high strain rates K_{IC} actually increases because of localized adiabatic heating at the crack tip. In addition to the mechanical effects of testing conditions on K_{IC}, there are effects due to metallurgical factors.

ORIENTATION

Because of mechanical fibering, the mechanical properties of deformed metals are anisotropic. An example of this effect is shown below for maraging steel.[4] Specimens were taken as shown in Fig. 5.9, with the following results:

Orientation	G_c (measured) (in.-lb/in.^{2})	K_c (calculated) (ksi-$\sqrt{\text{in.}}$)
A	245	85
B	230	83
C	310	96
D	150	67

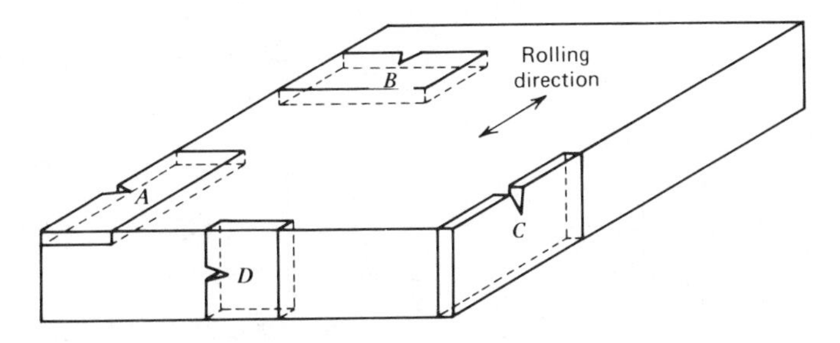

Fig. 5.9 Layout of specimens for study of effect of orientation G_c and K_c (From R. Wei, ASTM STP 381. Used with permission of the American Society of Testing Materials).

These observations reveal the importance of evaluating a material with respect to its orientation to the maximum service load. Depending on the orientation of a specimen, the same material could appear ductile or brittle.

HEAT TREATMENT AND MICROSTRUCTURE

Yield strength and K_{IC} are generally inversely related. Since yield strength decreases with increased tempering temperature, K_{IC} increases with increased tempering temperature, as in Fig. 5.10 for 4340 steel. Optimum conditions are achieved by a combination of *YS* and K_{IC}.[5]

Although *YS* and K_{IC} are usually inversely related, this is not universally true. In some metals, particularly steel, the same strength can be achieved with different microstructures. Sometimes the toughness associated with strength varies with the microstructure. With the same yield strength, the K_{IC} of annealed steel is lower than that for quench and tempered steel.

CHEMICAL COMPOSITION

Minor elements often have a major effect on fracture toughness. This is shown in Fig. 5.11, which shows the effect of varying the sulfur content in quench and tempered 4345 steel from 0.008 to 0.049%. As the sulfur increased, the fracture toughness for a given strength decreased significantly.[6]

5.4.2 Notched Bar and Tensile Behavior

The notched bar impact test (primarily Charpy V-notch) and the tensile test are used in many specifications. Although most of the data measuring brittle tendencies have been developed with impact tests, some embrittling conditions are more effective in the slower tensile test. Much of the information is phenomenological; however, it can be applied under the appropriate conditions.

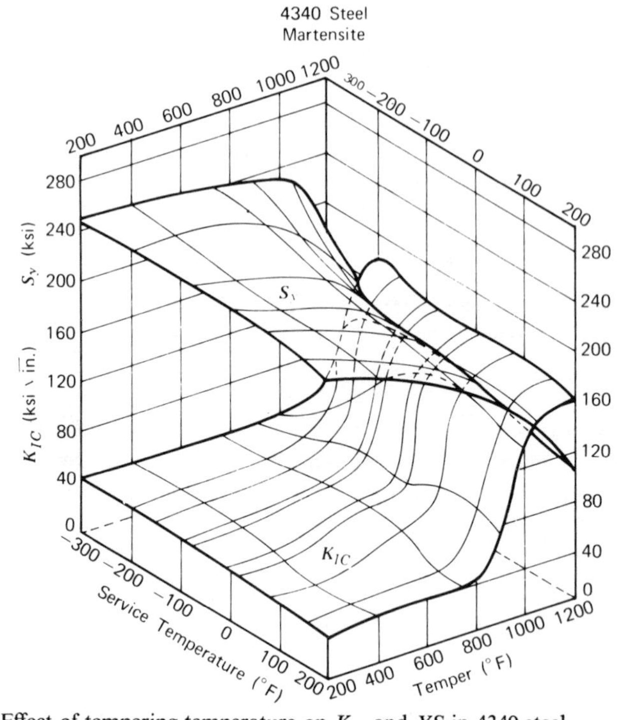

Fig. 5.10 Effect of tempering temperature on K_{IC} and YS in 4340 steel.

MANUFACTURING PRACTICE

The method used to manufacture a material often affects the impact data even at a constant yield strength. Figure 5.12 shows mild steel plate produced by these processes:[7] semikilled (as rolled), and normalized. Even though the yield strengths were comparable, the transition temperature of the normalized steel was lower.

Although manufacturing methods affect properties, it is often difficult to isolate what phase in the process is the cause. This problem is seen in the manufacture of large, thick wall forgings. Several manufacturers may use the same manufacturing process, will produce approximately the same microstructure, and yet, will generate impact values that may differ by 20 to 30 ft-lbs at a given temperature, or by 40 to 50°F in transition temperature.

COMPOSITION

Most of the Charpy impact testing has been applied to steel; the test is seldom applied to nonferrous materials. In steel, the alloying elements have

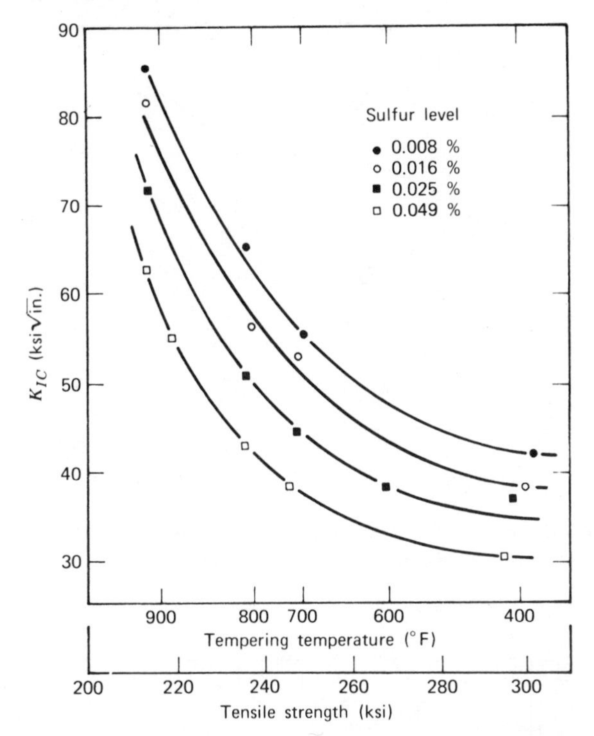

Fig. 5.11 Effect of sulfur content on fracture toughness in 4345 steel (From R. Wei, ASTM STP 381. Used with permission of the American Society for Testing Materials).

a large influence on transition temperatures. Increasing most elements increases the transition temperature; only Mn and Ni do not. One of the difficulties in comparing effects of alloying (or, in fact, any parameter) is that the effect observed may not be directly due to the parameter. If strength is affected, any changes in impact data may be caused by the change in strength. Figure 5.13 shows how each element affects fracture appearance transition temperature (FATT). The largest influence is due to P and C. Since P is seldom seen in the percentages given, carbon is the most influential element for controlling FATT.[8]

GRAIN SIZE

Grain size influences impact energy, FATT, and tensile fracture stress.[9] Generally, as the grain size increases (i.e., grains get larger), Charpy transition temperature increases, and fracture stress decreases (Fig. 5.14).[14] Changes in grain size can be effected by deoxidation practice and finishing

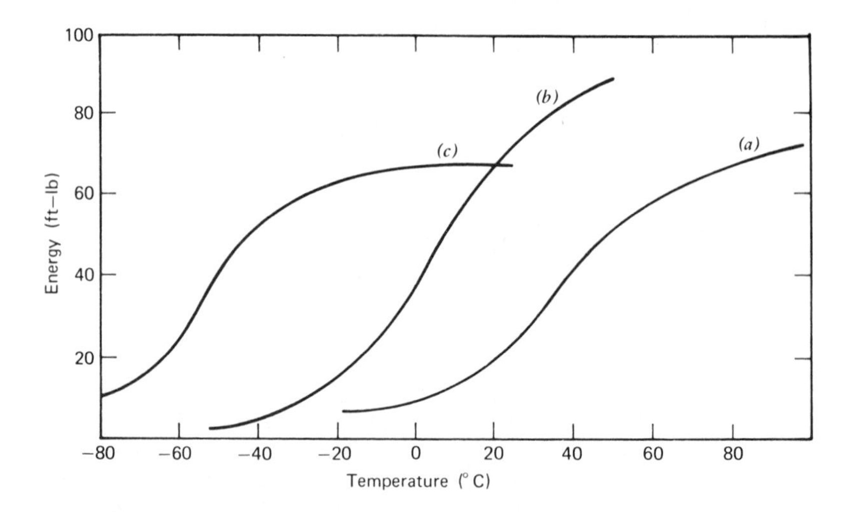

Fig. 5.12 Effect of manufacturing practice on impact data for mild steel. (*a*) Semikilled, as rolled. (*b*) 0.15% C, 1% Mn, semikilled, as rolled. (*c*) 0.15% C, 1% Mn, normalized.

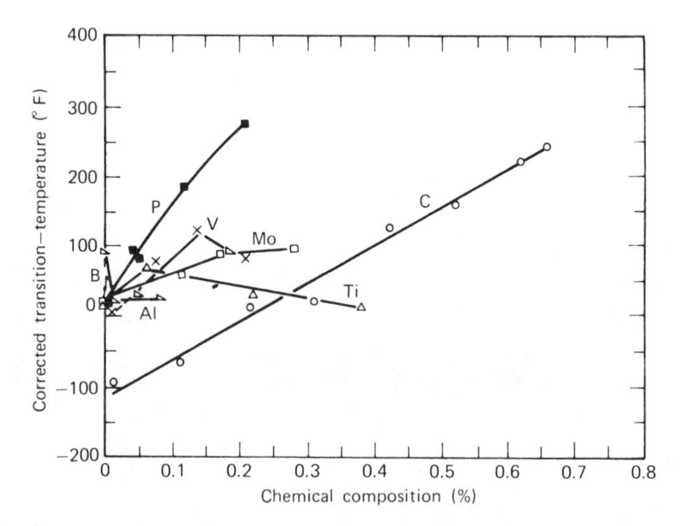

Fig. 5.13 Effect of various alloying elements on FATT in quench and tempered steel (Reproduced with permission, from Transactions, ASM, American Society for Metals, 1951).

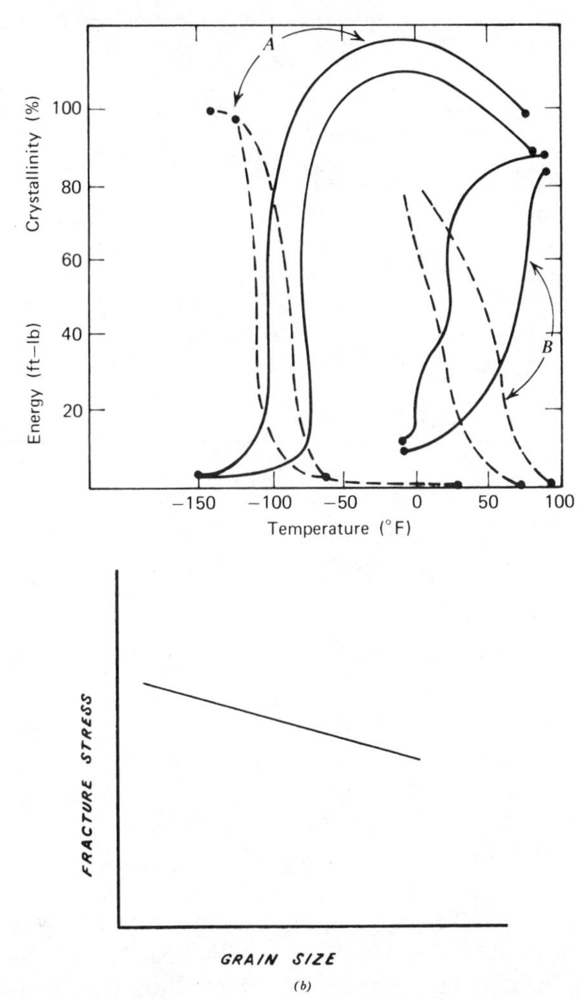

Fig. 5.14 Relation of grain size to (*a*) C_v transition temperature (A = fine grain, B = coarse grain) and (*b*) fracture stress (schematic) (Courtesy of C. Tipper, The Brittle Fracture Story, Cambridge University Press, 1962).

temperature. Certain elements, for example, V, retard grain growth. As the finishing temperature during rolling or forming increases, the grain size also increases. High temperature homogenization, or high temperature austenitization, may cause excessive grain growth. The influence of grain size has led to an extensive study of the development of ultrafine grain size. It has been possible to markedly improve ductility and toughness.

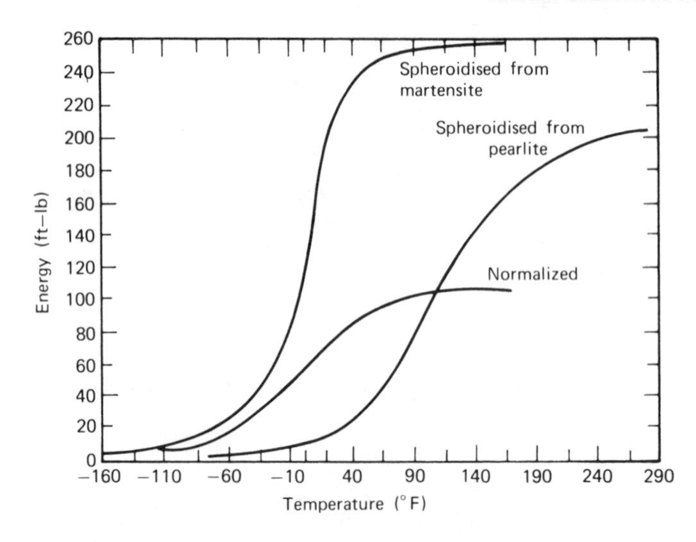

Fig. 5.15 Effect of shape of carbide particle on FATT (From J. Rinebolt. Used with permission of the American Society for Testing Materials).

MICROSTRUCTURE

Microstructure often has a noticeable effect, as does the shape of carbide precipitates in steel. For example, similar strengths and microstructural elements can be obtained in a .30% C steel by normalizing, spheroidizing from pearlite, and spheroidizing from martensite. In the normalized condition the carbides are more angular, which results in lower energy to fracture and higher FATT (Fig. 5.15).[11]

Because heavy steel sections are difficult to quench to martensite, bainite often results. Tempered upper bainite, tempered lower bainite, and tempered martensite were compared by isothermal heat treating techniques (Table 5.1). The data for a common hardness show that tempered marten-

Table 5.1 Comparison of the Toughness of Martensite and Bainite at Equal Hardness Levels

	Hardness (R_c)	$C_v(RT)$ (ft-lb)
Tempered martensite	45	35
Tempered lower bainite	44	20
Tempered upper bainite	45	15

site was the toughest.[12] They also show an effect on tempering, reflected in a variation in yield strength for the same hardness. This variation emphasizes that hardness values cannot be used to reliably estimate yield strength unless all the pertinent details, including microstructure, are available, and a conversion chart has been developed.

ORIENTATION

The orientation of the test bar in a formed product also influences both the impact energy and FATT, and the tensile *RA*. The effect on *RA* was reported as early as 1932 for aluminum forgings. An example from a banded mill steel plate is shown in Table 5.2. By changing orientation, the upper shelf energy varied from 6 to 120 ft-lbs.[13] When directionality is considered, three orientations are generally considered:

1. *Longitudinal.* Where the orientation of the test bar is parallel to the forming direction.
2. *Transverse.* Where the orientation of the test bar is perpendicular to the forming direction.
3. *Short transverse (plate).* Where the orientation of the test bar is perpendicular to the rolling direction and rolling plane.

Table 5.2 Change in Upper Shelf Energy with Change in Orientation of Test Specimen for Mild Steel

	Orientation[a]	Energy (ft-lbs)
Banded pearlite-ferrite microstructure	WT	105
	RW	32
	TR	19
Homogenized pearlite-ferrite microstructure	WT	129
	RW	40
	TR	22
Banded martensite-ferrite microstructure	WT	111
	RW	23
	TR	13
Homogenized martensite-ferrite microstructure	WT	115
	RW	27
	TR	11

[a] WT—Crack arrester orientation.
RW—Crack divider orientation.
TR—Short transverse orientation.

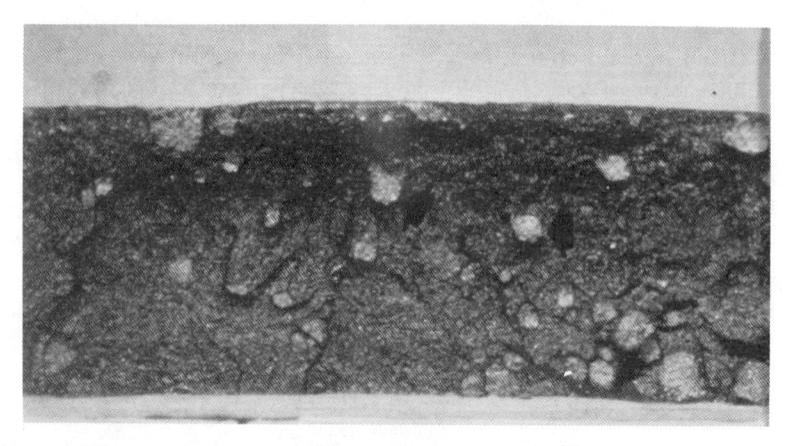

Fig. 5.16 Fish-eyes indicative of hydrogen embrittlement (Reproduced with permission from Metal Progress, American Society for Metals, 1969).

These definitions are not sufficient. The orientation relation between the applied load and the crack plane and direction must be considered. In plates, impact values are higher if the cracks propagate through the thickness than if they propagate in the length or width direction, regardless of whether the bars are longitudinal or transverse.

When considering the orientation of test bars in a failure analysis, thought should be given to the orientation of the failure. The test bar should be aligned so that the stresses or strains on it are in the same orientation as those on the fracture.

EMBRITTLEMENT

Hydrogen Embrittlement. Excessive amounts of hydrogen can cause a decrease in tensile ductility, without affecting impact data. For hydrogen to be effective, a relatively slow loading rate is required. Hydrogen embrittlement can be shown by conducting both tensile and Charpy tests on the suspect material and on the same material subjected to a low temperature heat treatment. If the original material were hydrogen embrittled, there should be no effect on C_v, but an increase in RA.

Flaking in steel forgings, that is, hairline cracks, is due to a combination of internal stresses and concentration of hydrogen.[14] When these fractures are opened, they often show a characteristic marking called "fish-eyes" (Fig. 5.16). Flaking is not too prevalent since the introduction of vacuum degassing techniques that remove the hydrogen from the original melt prior to pouring into the ingot.

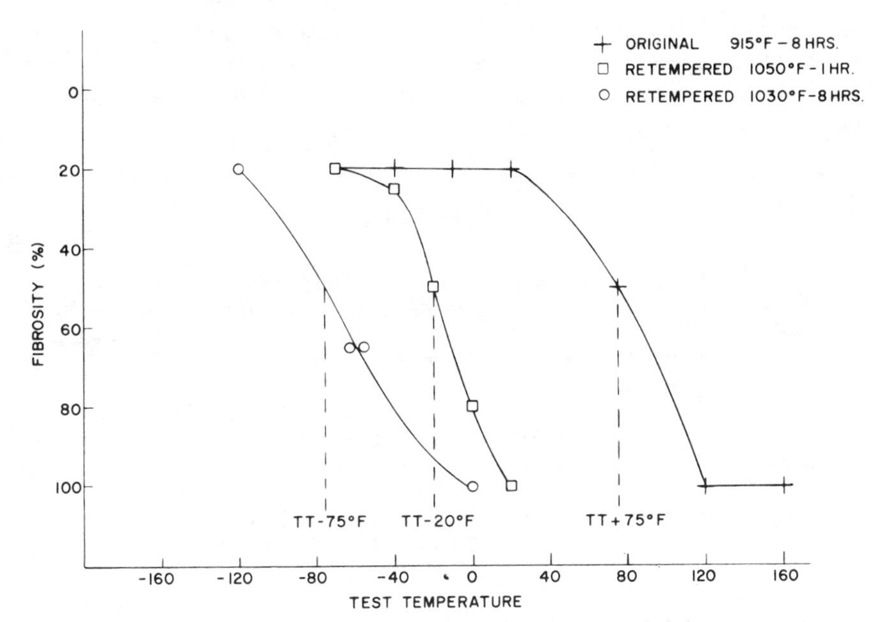

Fig. 5.17 Change in transition temperature caused by temper embrittlement.

Temper Embrittlement. Tempering some steels in the general range of 850 to 1000°F results in temper embrittlement, which is manifested by an increase in impact transition temperature. The exact cause of temper embrittlement is not known, but has been associated with tramp elements, for example, Sn. It is a time-temperature controlled phenomenon. A certain degree of embrittlement may result from slow cooling through the embrittling range from a higher tempering temperature. Quenching from tempering is often employed to avoid embrittlement.

One technique used to show a temper-embrittled material is to develop transition curves from the base material, and from the same material heat treated under controlled conditions to avoid embrittling. If the material were temper embrittled, a decrease in FATT should result (Fig. 5.17).

Temper embrittlement cracks are intergranular. Since other types of embrittlement also cause intergranular fracture, the best experimental evidence for temper embrittlement is a comparison of FATT. This is particularly true if the tempering practice is not known. If it is known, probability of embrittlement can often be estimated without any testing.

Blue Brittleness. When worked steel is heat treated in the vicinity of 400°F, the C_v versus tempering temperature passes through a minimum, even though the yield strength continues to decrease smoothly. This phenomenon is related to strain aging. It is not too prevalent, since the low tempering temperatures necessary are seldom employed. If the required yield strength necessitates tempering at low temperatures, a change in material or design should be considered.

REFERENCES

1. G. Taylor, *Proc. Roy. Soc. (Lond.)*, Vol. 145A, 1934, p. 369.
2. A. Tetelman and A. McEvily, *Fracture of Structural Materials*, Wiley, New York, 1967, p. 59.
3. F. Heiser, *Impact Transition Temperature in 175mm M113E1 Gun Tube Material*, Watervliet Arsenal Report WVT–7109, February 1971, p. 26.
4. R. Wei, *Fracture Toughness Testing and Its Application*, ASTM STP 381, 1965 p. 287.
5. J. Throop and G. Miller, *Optimum Fatigue Crack Resistance*, Watervliet Arsenal Report WVT–7006, January 1970, p. 8.
6. R. Wei, *op. cit.*, p. 286.
7. Admiralty Advisory Committee on Structural—A1 Steel, Report No. P. 2, 1960.
8. J. Rinebolt and W. Harris, *Trans. ASM*, Vol. 43, p. 1191 (1951).
9. J. Hodge, R. Manning, and H. Reichbold, *Trans. AIME*, Vol. 185, 1949, p. 233.
10. C. Tipper, *The Brittle Fracture Story*, Cambridge University Press, 1962, p. 132.
11. J. Rinebolt, *Symp. ASTM*, 1953, p. 203.
12. R. DeFries, C. Nolan, and T. Brassard, *Some Observations on the Relationship Between Microstructure and Mechanical Properties in Large Cylindrical Gun Tube Forgings*, Watervliet Arsenal Report WVT–7018, March 1970.
13. F. Heiser, *Anisotropy of Fatigue Crack Propagation in Hot Rolled Banded Steel Plate*, Ph.D. Thesis, Lehigh University, 1969.
14. A. Phillips and V. Kerlins, *Met. Prog.*, May (1969) 81.

6

Fatigue

6.1 INTRODUCTION

Fatigue is failure under repeated loads. There are three phases in a fatigue fracture: crack initiation, crack propagation, and fracture. These phases are not completely separable. The process may be described as the formation of a crack, because of repeated local plasticity, its progression until a critical size is reached, whereupon it fails. Fatigue accounts for about 90% of all service failures.

A part can be subjected to various kinds of loading conditions, including fluctuating stress, fluctuating strain, fluctuating temperature (thermal fatigue), or any of these in a corrosive environment or at elevated temperatures. Most service failures occur as a result of tensile stresses.

Three general types of stress cycling[1] are shown in Fig. 6.1: complete reversal (Fig. 6.1*a*)—where the stress fluctuates around a mean of zero with a constant amplitude; repeated (Fig. 6.1*b*)—where the stress fluctuates around a mean not equal to zero but with a constant amplitude; and complicated (Fig. 6.1*c*)—where both the alternating and mean loads change, either randomly or with a definite pattern.

Fatigue failure starts prior to the initiation of a crack. With repeated loading, localized regions of slip (plastic deformation) develop. Woods showed that a series of intrusions and extrusions developed during the stress cycling (Fig. 6.2).[2,3] He explained that when slip occurs, it is seen at the free surfaces as a step caused by the displacement of metal along a slip plane. When the stress is reversed, the slip that occurs could be the exact negative of the original slip, completely overriding any deformation effects. However, it does not, so that a residual deformation remains. This deformation is accentuated by repeated loading, until a discernible crack finally appears.

The initial cracks form along slip planes. The crack growth is crystallographically oriented along the slip plane for a short distance. This is some-

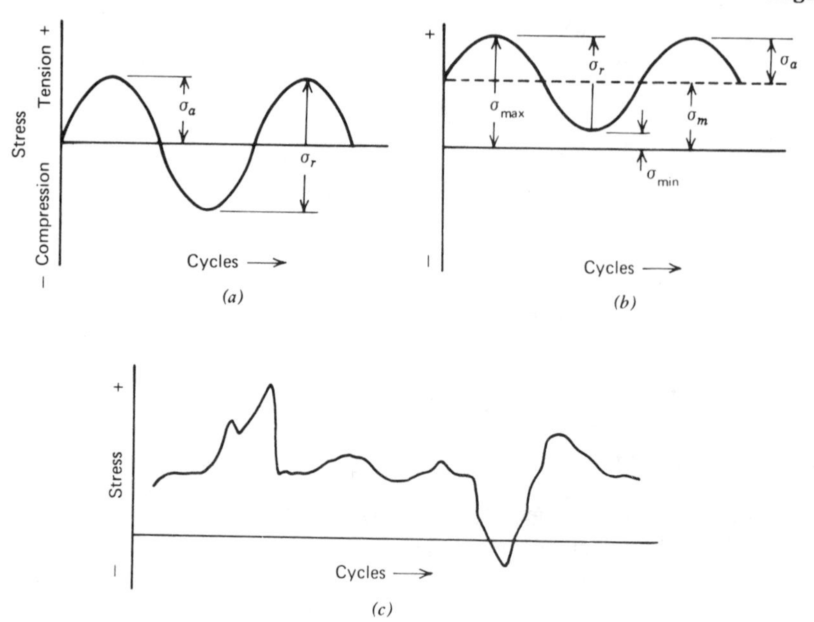

Fig. 6.1 Schematic showing general types of fatigue stress cycling (From Mechanical Metallurgy by G. Dieter. Copyright 1961, McGraw Hill. Used with permission of McGraw Hill Book Company).

times referred to as Stage I crack growth.[4] Eventually, the crack propagation direction becomes macroscopically normal to the maximum tensile stress. This is referred to as Stage II crack propagation, and it comprises most of the crack propagation life.

The relative cycles for crack initiation and propagation depend on the stress applied. As the stress increases, the crack initiation phase decreases.[5] At very low stresses (high cycle fatigue), therefore, most of the fatigue life is utilized to initiate a crack. At very high stresses (low cycle fatigue), cracks form very early. Fatigue may be divided into two categories, high cycle and low cycle. The separation is not clear-cut. Generally, the low cycle region is that which results from stresses that are often high enough to develop significant plastic strains.

There are visual differences between high cycle (low stress) and low cycle (high stress) fatigue. In the latter, deformation resembles that seen with unidirectional loading. Strain hardening can occur and the slip bands are coarse.[6] In high cycle fatigue, the slip bands are usually very fine.

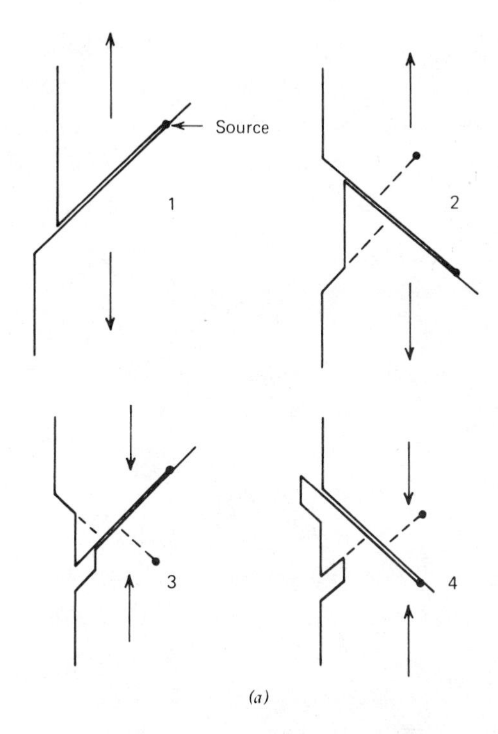

Fig. 6.2 Extrusions and intrusions in heavy slip regions prior to crack initiation. (*a*) Schematic. (Used with permission of the Royal Society). (*b*) Fatigue surface. (From D. Hull, J. Inst. of Metals. Used with permission of The Institute of Metals).

6.2 S-N DATA

Fatigue data is usually presented in the form of *S-N* curves, where applied stress (*S*) is plotted against cycles to failure (*N*). In *S-N* curves, the total cycles to failure, that is, cycles to initiate plus cycles to propoagate, are included.

As the stress decreases, cycles to failure increase (Fig. 6.3). In steel, there is a stress limit below which fatigue failure does not occur, the fatigue or endurance limit. In aluminum and other nonferrous alloys, there is no fatigue limit; a finite life exists at any stress level. Since infinite life tests are impossible, the endurance limit is that stress which will not cause failure in 10^7 cycles.

When characterizing the fatigue properties of a material, two considerations are possible, either the life at a specific stress or the endurance limit. The service conditions dictate the practice. To design a component that will

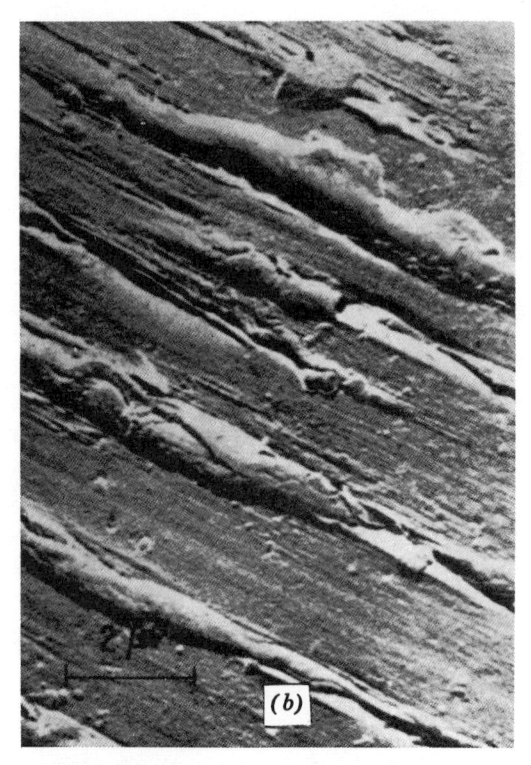

Fig. 6.2 (Continued)

last indefinitely, the endurance limit must be considered. If a component is not expected to last indefinitely, for example, if there are keyways in highly-stressed locales or stress conditions that can cause early cracking, finite life considerations apply.

6.3 MECHANICAL FACTORS AFFECTING FATIGUE LIFE

6.3.1 Stress Concentration

Stress raisers reduce fatigue life. They can be mechanical (e.g., fillets or keyways) or metallurgical (e.g., porosity or inclusions). Fatigue failures often start at stress raisers, usually at or near the surface. However, if loading conditions are appropriate, stress raisers under the surface can initiate fracture.

The effects of notches are evaluated by comparing notched versus unnotched *S-N* data. To determine an *S-N* curve for a notched specimen,

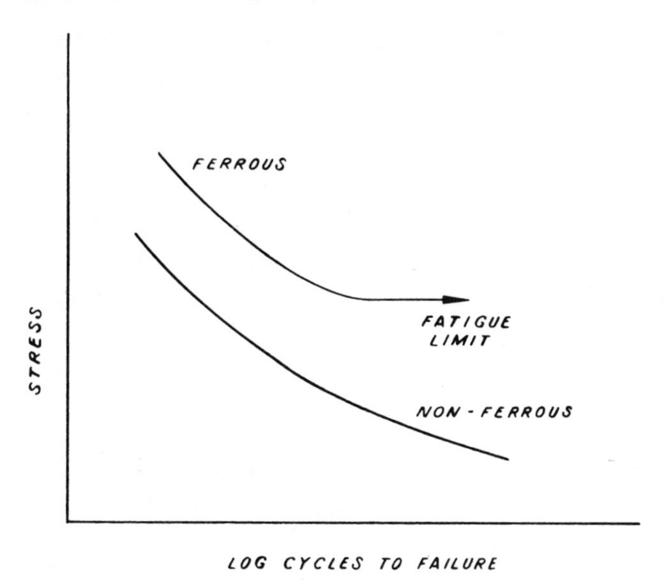

Fig. 6.3 Typical *S-N* curves for ferrous and nonferrous materials.

the net stress is plotted. By comparing data, it is then possible to calculate a fatigue-notch factor (K_f) where

$$K_f = \frac{\text{fatigue limit, unnotched}}{\text{fatigue limit, notched}}$$

From the K_f, the notch sensitivity, q, can be calculated.

$$q = \frac{K_f - 1}{K_t - 1}$$

where

$$K_t = \text{stress concentration factor}$$

As tensile strength, notch radius, and section size *increase*, and as grain size *decreases*, q increases.[7] Figure 6.4 shows the effect of notch radius on notch sensitivity for several materials. The effect of tensile strength is contrary to that which predominates in smooth specimens, namely, that as strength increases, fatigue life increases. It does agree with fracture mechanics concepts, since increasing the strength decreases the critical crack size, and failure occurs at a shorter crack length.

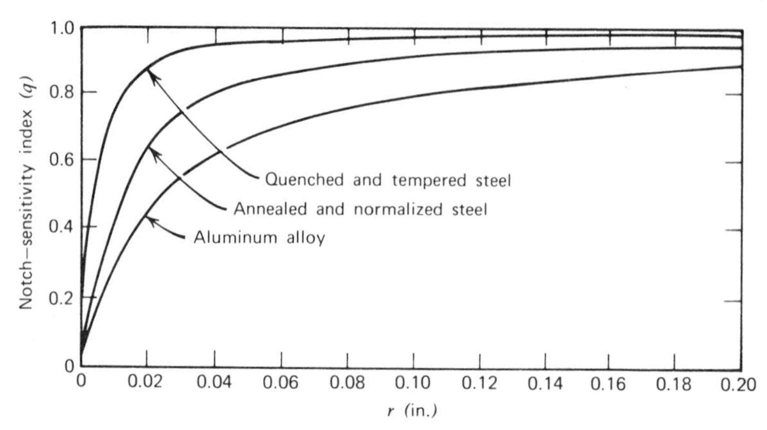

Fig. 6.4 Effect of notch radius on notch sensitivity, q (From Metal Fatigue by P. Peterson. Copyright 1959, McGraw Hill. Used with permission of McGraw Hill Book Company).

6.3.2 Size

As specimen size increases, fatigue life sometimes decreases. There are several reasons for this phenomenon. Fatigue failure usually starts at the surface. The additional surface area of larger specimens starts at the surface. The additional surface area of larger specimens increases the probability that a flow exists, which will initiate failure and reduce the time for crack initiation. Increasing the specimen size also decreases stress gradients so that more material may be highly stressed.

The data observed are not completely consistent; some investigators did not list any size effect. There may be an effect of the type of test used. In plain carbon steel, smooth bars show no effect whereas notched bars are affected.

6.3.3 Surface Effects

In most tests and service applications, the maximum stress occurs at the surface. Fatigue life, therefore, is sensitive to surface conditions. Although surface finish is important, there are other factors to consider, such as surface properties and surface residual stress.

SURFACE FINISH (SURFACE ROUGHNESS)

Machining marks are small notches. As the surface finish becomes coarser, the depth of the notches increases. Therefore, as the surface becomes rougher, the fatigue life decreases.

The results of studies on the effect of surface finish are qualitative;

however, they do indicate trends. The following example, for SAE 3130 under completely reversed stress at 95,000 psi, illustrates this effect:[8]

Finishing operation	Surface finish (μ in.)	Fatigue life (cycles)
Lathe	105	24,000
Partly hand polished	6	91,000
Hand polished	5	137,000
Ground	7	217,000
Ground and polished	2	234,000

These data show that both the surface finish and the finishing operation are important. The exact reason for this result is not known. In some operations, the height of the notches may be changed although the notch radius is not; in others, both the notch height and root radius are changed. Another possibility is that some techniques introduce residual stresses which affect fatigue life.

SURFACE PROPERTIES

Surface or processing conditions cause changes in material properties that can significantly affect fatigue life. These effects may be divided into those which decrease life and those which increase life.

Electroplating often decreases fatigue life. When a plate is deposited, tensile stresses are developed. These can cause cracking (Fig. 6.5) which is, in effect, instantaneous fatigue crack initation. In addition, in a plating bath hydrogen is liberated at the cathode, that is, the material on which the plating is deposited. When combined with the tensile stresses, this can lead to hydrogen embrittlement.

Decarburization of steel readily occurs in heat treating without atmosphere, or at moderately high temperature service conditions. It also occurs during solidification in some casting practices, notably investment casting. Decarburization reduces fatigue life.

Conversely, carburizing increases fatigue life, as do nitriding and flame and induction surface hardening. This occurs either by a strengthening of the surface material or by the generation of residual compressive stresses. A combination of these two effects probably occurs.

SURFACE RESIDUAL STRESSES

As with all service applications, the gross stress condition is the total effect of all the individual stresses. Compressive residual surface stresses

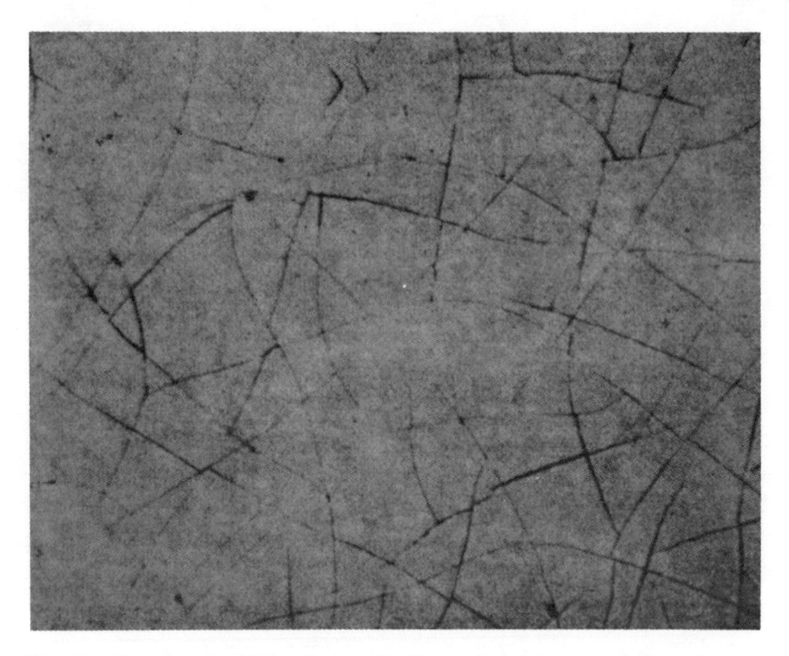

Fig. 6.5 Cracking in electroplated deposit (Reproduced by permission, from Metals Handbook, Vol. 2, American Society for Metals, 1964).

decrease the effect of applied tensile stresses, resulting in increased fatigue life. This is probably the most effective way to increase fatigue life.

Residual compressive stresses can be developed when plastic deformation is not uniform. In practice, this is accomplished by selectively plastically deforming the outside surface of a component. When deformation stresses are removed, the elastic area under the plastic region causes compressive stresses at the surface.

One common practice for developing residual stresses is shot peening, which can be applied to a complex component, or to a specific area of a part. Residual stresses are developed in some high pressure cylinders by the application of autofrettage, where the inside diameter is plastically expanded to a predetermined size. After the pressure is released, the elastic stresses develop significant compressive stresses at the bore. This technique is used in high pressure cylinders to increase fatigue life.

There is a danger in utilizing these techniques, particularly shot peening. Surface compressive stresses are balanced by residual internal tensile stresses, as shown in the simplified diagram in Fig. 6.6. It is possible thereby to develop subsurface crack initiation.

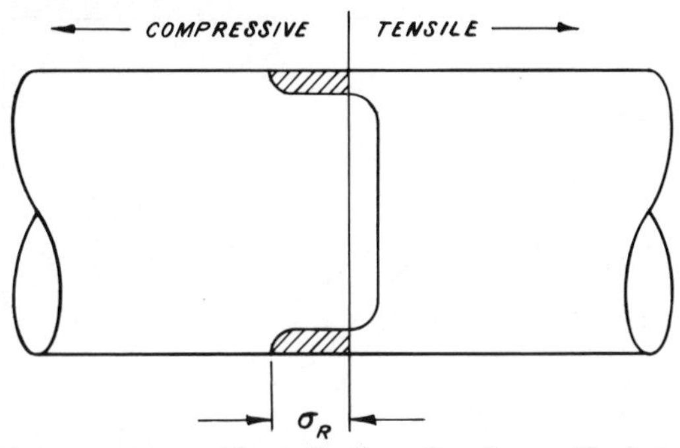

Fig. 6.6 Residual stress pattern with application of surface residual stresses.

If surface stresses are tensile, early cracking can ensue. The most common occurrence is the development of surface cracks due to grinding. If the grinding cut is too deep, tensile stresses can be developed. In extreme cases, cracking may develop immediately (Fig. 4.14). Even if cracks do not develop, the stress state may cause short fatigue life.

6.3.4 Mean Stress

The stress amplitude exerts the major influence, however, the mean stress also affects fatigue life.[10] The term often used to express mean stress is R, where $R = \sigma_{\min}/\sigma_{\max}$. Therefore

$\sigma_{\min}$	$\sigma_{\max}$	R
− (compressive)	+ (tensile)	−
0	+	0
+	+	+

For example, if the stress is completely reversed, that is, $\sigma_{\min} = -\sigma_{\max}$, then $R = -1$. As R becomes positive the fatigue limit decreases, as illustrated in Fig. 6.7.

The same fatigue life can be attained with different combinations of mean stress (σ_m) and stress amplitude (σ_a).[11,12] For the same fatigue life, as σ_m increases, σ_a decreases. This observation has led to the concept of constant life diagrams, such as the Goodman diagram. These diagrams show the combination of σ_m and σ_a that results in a given life or sustains an infinite number of cycles without failure.

The basic tenets of the various constant life diagrams vary. In the Goodman diagram (Fig. 6.8), σ_a becomes smaller as σ_m increases until $\sigma_m = UTS$. Some analyses use the yield strength for the zero life point.

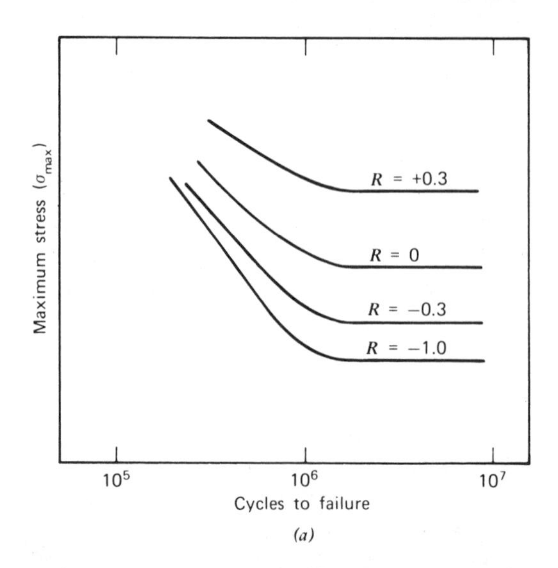

Fig. 6.7 Effect of mean stress on fatigue life ($R = \sigma_{min}/\sigma_{max}$) (From Mechanical Metallurgy, by G. Dieter, Copyright 1961, McGraw Hill. Used with permission of McGraw Hill Book Company).

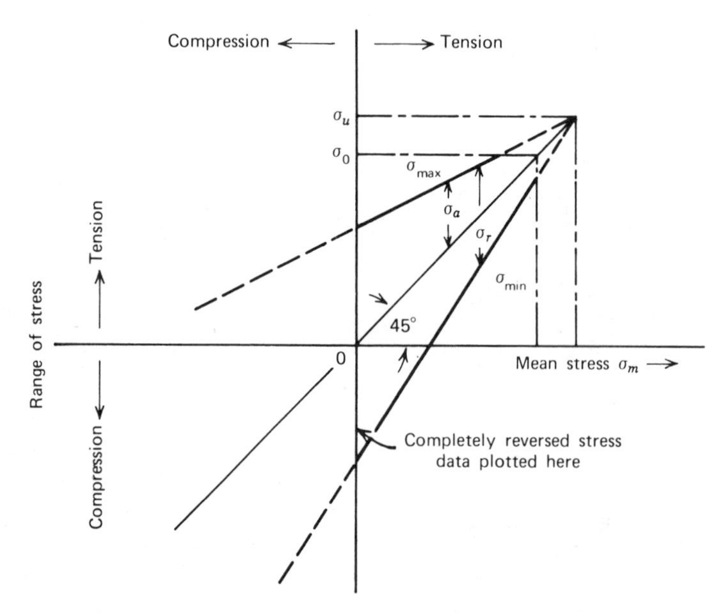

Fig. 6.8 Typical Goodman diagram (From Fatigue Design Handbook. Copyright 1968, SAE. Used with permission of the Society for Automotive Engineers).

The data for these diagrams are derived from S-N curves. They are used to estimate fatigue life or to determine the stresses for a desired life. Most diagrams show combinations for a certain life. If enough data are available, a family of curves for a range of fatigue lives can be generated.

6.4 METALLURGICAL FACTORS AFFECTING FATIGUE LIFE

Fatigue is sensitive to structural design. Major improvements in fatigue life are attainable through improvements in design, whereas increases in fatigue life through metallurgical changes are generally limited.

6.4.1 Grain Size

Fatigue is affected by microstructure, that is, it is structure sensitive. Although it is influenced by grain size, the results are variable. For nonferrous materials and annealed steel, the fatigue strength increases as grain size decreases. However, in unnotched heat-treated steel, there is no effect. Apparently, the effects of the microconstituents present override those produced by grain size.

6.4.2 Microstructure

Much of the data published on the effects of microstructure deal with steel. The same strength can be achieved in steel with a variety of microstructures. At a constant strength, the fatigue life with a coarse pearlite is lower than that with a spheroidal microstructure because of the shape of the carbide particles.[13] The rounded carbides in the spheroidized specimen have lower stress concentrating effects, which result in a longer life. This observation can be generalized to other types of microstructures. The fatigue life with a coarse, angular microstructure is lower than with a fine, rounded microstructure.

In steel, the longest fatigue life is achieved with a tempered martensite microstructure. Slack quenching, which results in a mixed microstructure of martensite and bainite or ferrite, decreases fatigue life. As the tempering temperature decreases and strength increases, the material becomes more sensitive to notches and, therefore, more sensitive to surface conditions. More care must be taken in the design and preparation of the part.

6.4.3 Orientation

Working of metals causes mechanical fibering, that is, the alignment of grains, chemically segregated areas, and inclusions in the rolling direction. Fatigue life is anisotropic; it is lower when transverse to the rolling direction. The effect varies with the material and test conditions, but becomes more pronounced with increased load,[14,15] increased yield strength,[16,17] decreased ductility,[18] decreased cleanliness,[19] and the change from smooth to notched bar.[15,16] A general relation describing the anisotropy of fatigue life shows that transverse life is approximately 0.6 to 0.7 of the longitudinal life. Although such relations are desirable and sometimes useful, it must be remembered that they are general, and specific cases can vary significantly.

Inclusions are one of the main causes of the fatigue life anisotropy. The orientation difference is less with vacuum-melted steel than with electric furnance air-melted steel, primarily because of the decreased inclusion count. The orientation of inclusions also controls crack growth rate.

6.5 CRACK PROPAGATION

Because most structures have inherent flaws, or develop cracks relatively early in life, considerable interest has developed in the crack propagation stage of fatigue. In this stage, the crack grows from an easily discernible to a critical size. From the data, a prediction of fatigue life can be developed. As with fracture mechanics concepts, these studies present a potential method for design information in complex structures from simple tests. Compared to the development of *S-N* curves, more data are developed from one crack growth test than from one standard fatigue bar test.

The rate at which a crack grows, da/dN, is a function of material properties, crack length, and applied stress. Paris[20] showed that these data could be interpreted in fracture mechanics terms as

$$\frac{da}{dN} = C(\Delta K)^m$$

where

K = range of stress intensity factor
C = material constant
m = material constant, originally suggested as 4

Since crack growth is a function of the stress intensity factor, the specimen used to determine da/dN must be such that a K-calibration is known. Since these tests evaluate only crack propagation, a crack starter (e.g., a notch) is introduced. The extension of the crack is then monitored as a function of cycles. In many tests, crack length is measured visually with a telescope. Other types of measurements have also been successfully employed, including electrical resistivity change and ultrasonics.

The crack growth rate is the change in crack length, with respect to N (cycles); if a is plotted against N, it is the slope of the line (Fig. 6.9) at any point. To relate growth rate to ΔK, da/dN is evaluated at selected crack lengths. Then, ΔK is calculated for that length, assuming that a is constant and only the stress varies. This is shown below for a simplified K-calibration:

$$K = \sigma\sqrt{a}$$
$$\Delta K = \Delta\sigma\sqrt{a}$$
$$\Delta K = \frac{\Delta P}{A}\sqrt{a}$$

where

$$\Delta\sigma = \text{stress range} = \frac{\Delta P}{A}$$
$$\Delta P = \text{load range}$$
$$A = \text{area}$$
$$a = \text{instantaneous crack length}$$

Two data points are developed for each value of a, $(da/dN)_a$ and $(\Delta K)_a$. Thus log da/dN can be plotted against log ΔK, with m as the slope of the least squares best fit line through the data.

Although the relation shown requires a linear fit of the log-log plot,

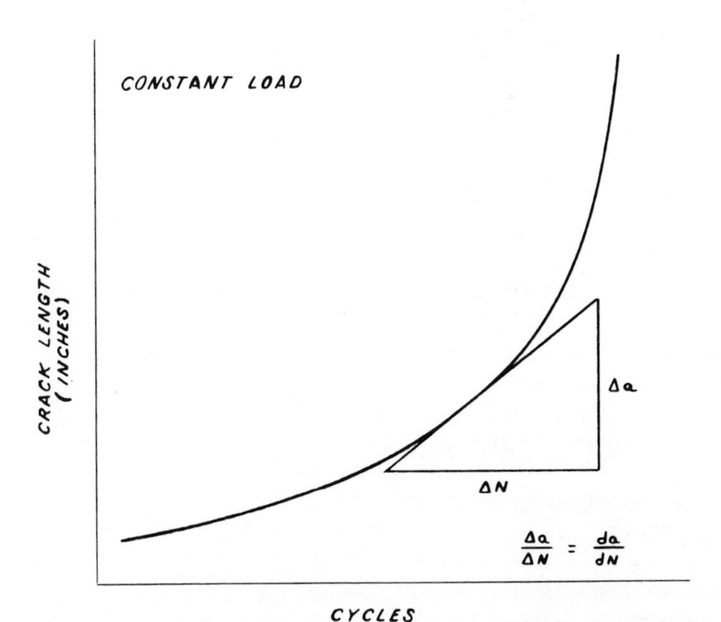

Fig. 6.9 Typical crack length versus cycles curve showing determination of da/dN.

data have been presented suggesting that a sigmoidal relation is more appropriate.

In those data showing nonlinear relations, the change in slope usually occurs at the higher values of K, where the remaining life is short. For the majority of the useful fatigue life, it is valid to assume the relation between $\log da/dN$ and $\log \Delta K$ is a straight line.

6.6 FACTORS AFFECTING CRACK GROWTH RATE

In the fracture mechanics interpretation, crack growth rate can be affected through an effect on C or m. Changes in C alter the position of the data, whereas changes in m reflect variations in slope.

6.6.1 Factors Affecting C

Relations have been developed for calculating C, which consider dislocation theory, structural size parameters, and mechanical properties. Those involving dislocations or structural size parameters are difficult to apply since they involve features that are not readily measurable or definable. Of more practical interest are those involving mechanical properties that are quantitatively measurable. Some of the properties that influence C are shown below:

Property	Effect
$E\uparrow$	$C\downarrow$
$YS\uparrow$	$C\downarrow$
$\sigma_f\uparrow$	$C\downarrow$
$\epsilon_f\uparrow$	$C\downarrow$
$K_{IC}\uparrow$	$C\downarrow$

The optimum mechanical properties are a combination of high strength and high toughness or ductility. Generally, as strength increases, toughness and ductility decrease, therefore some compromise point that maximizes the product of strength and ductility or toughness will be reached. This observation has been used to establish an optimum tempering temperature of 1050 to 1100°F for 4340 steel.[21]

6.6.2 Factors Affecting Slope (m)

Some of the factors that affect m are:

1. *Orientation.* Orientation affects fatigue, not strictly by the relation of the test bar to the forming direction but rather by the relation of the crack plane and direction to the forming direction.[22] This is shown in Fig. 6.10

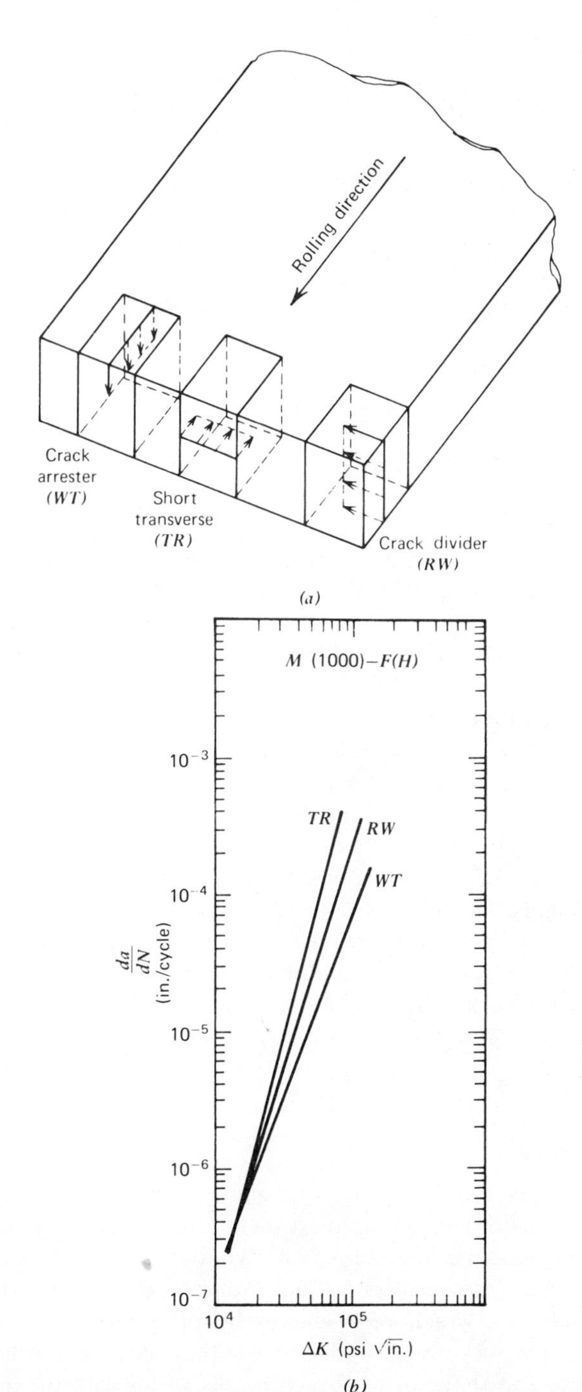

Fig. 6.10 Effect of orientation of test specimens on crack growth rate. (*a*) Orientation of test bars. (*b*) Fatigue crack growth rate.

for mild steel; Fig. 6.10*a* shows the relation of several specimens to the rolling direction. The fastest growth rate was seen in the short transverse orientation and the slowest was seen in the crack arrester orientation, even though the latter was a transverse specimen. This is due to an interaction with the aligned inclusions. In the *TR* orientation (Fig. 6.10*b*), inclusion-matrix separation extended the crack, whereas in the *WT*, it diverted the crack and slowed it down. The yield strengths were the same in all orientations.

2. *Toughness*. As the toughness (K_{IC} or C_v)[22,23] increases, m decreases. This was suggested by the nonlinear relation shown earlier. It was also shown in studies of mild steel and several high strength steels. The slope varies from 2.5 to 6.7, with no apparent relation to *YS*.[23]

3. *Yield strength*. A separate study of high-strength steels showed that m decreased as *YS* increased.[24] Since *YS* and K_{IC} are usually inversely proportional for a given material, these data suggest that m increases as K_{IC} increases. This is also contrary to the above information.

However, studies of the effect of tempering temperature in 4340 steel have also been made. These showed that as tempering temperature increased, *YS* and m decreased, which tend to support the observation about the effect of fracture toughness.[25]

4. *Size*. Conflicting data have also been presented on the effects of size. In one study, m decreased with increasing size,[26] and in another, m increased with increasing size.[27] Both are apparently plausible and have suggested a relationship with stress state. In some fatigue tests when the stress state becomes plane stress, m increases. Since increasing the specimen size tends to change the stress state to plane strain, m should decrease. However, fracture toughness decreases as specimen size increases, becoming a minimum where plane strain conditions apply, that is, K_{IC}. Therefore, an analogy with crack growth rate may exist, such that large plane strain specimens should show higher crack growth rates.

6.7 APPEARANCES OF FATIGUE

6.7.1 Macroscopic

Fatigue failures usually show little permanent deformation. As with static failures, ductility and toughness dictate the amount. As the material toughness increases, the tolerable crack size increases, as does the amount of shear lip. Macroscopically, fatigue failures often show beach marks (Figs. 4.6 and 4.9), which represent cyclic progression of the fracture. Characteristically, they are smooth because they have been rubbed.

The location and shape of the beach marks varies with the loading conditions. The fractures usually start at stress concentration regions such as fillets and keyways. Depending on the loading conditions, it is possible to have more than one initiation point,[28] as shown by the ratchet markings in Figs. 6.11 and 6.12. The location and typical fracture appearance for various conditions are shown schematically[29] in Fig. 6.13.

Fatigue fractures propagate normal to the principal tensile axis. This observation can be used to determine probable origins, or if a fatigue fracture has been observed, to identify maximum stress areas.

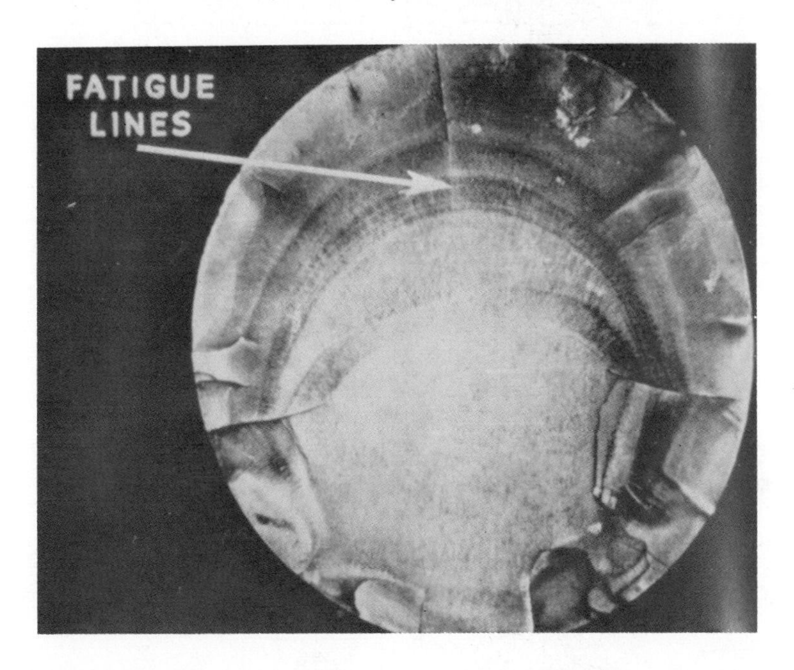

Fig. 6.11 Multiorigin torsional fatigue fracture. Note the ratchet marking (Courtesy Bethlehem Steel Corporation).

Fig. 6.12 Multiorigin fatigue failure due to internal pressure in thick wall cylinder. Note that the several origins eventually link to form one main crack.

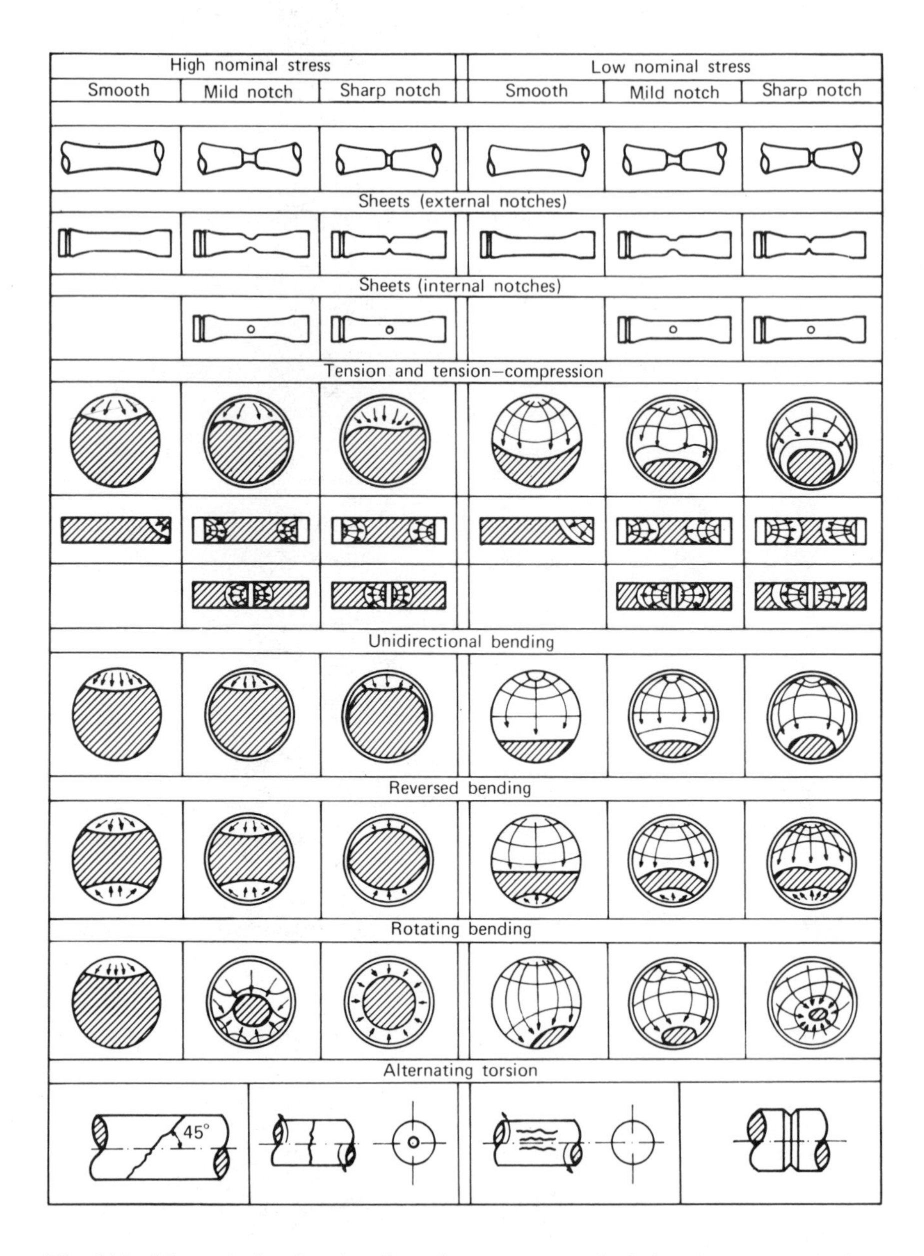

Fig. 6.13 Schematic showing the effect of stress state on the fatigue fracture origin, location, and appearance (From Fatigue Design Handbook, Copyright 1968, SAE. Used with permission of the Society for Automotive Engineers).

Any metal can fail in fatigue. Brittle materials can be distinguished from ductile materials in several ways, including the extent of plastic deformation. Since the tolerable crack length is less, the extent of the beach marking will be less. It is not possible to differentiate between brittle and ductile materials based on the texture in the beach-marked region. However, the texture of the fast fracture regions differs. In brittle material, the fracture is smoother, whereas in ductile material, the failure is fibrous. Both show markings characteristic of static overload. Chevron or river markings are often present. In many cases, it is possible to locate the origin by tracing the markings.

Fatigue failures can often be positively identified macroscopically. However, beach marks are not always seen. It is therefore sometimes necessary to utilize microscopic or fractographic analysis.

6.7.2 Fractographic

In fatigue failures light microscopy is of marginal use, since there is nothing characteristic of the fracture path to differentiate the failure from

Fig. 6.14 Dimples on the plane stress portion of a fatigue failure (From R. Hertzberg, ASTM STP 415. Used with permission of the American Society for Trsting Materials).

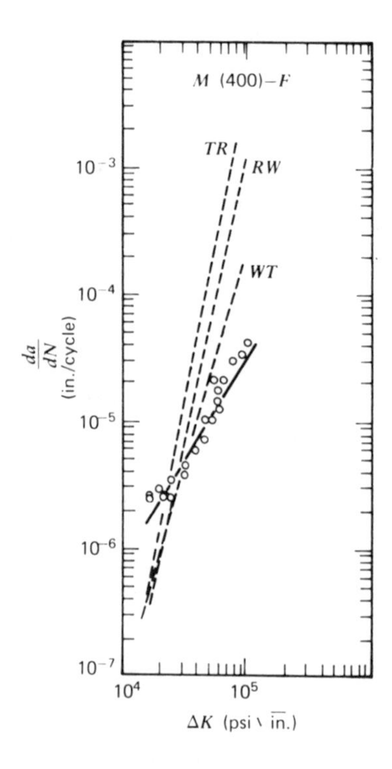

Fig. 6.15 Relation of macroscopic crack growth rate and striation spacing to stress intensity factor in mild steel plate.

static overloads. Fatigue failures may occur either transgranularly or intergranularly. Additional difficulties are also involved, since one half of the fracture surface is usually available. Striations are a fractographic feature that have been associated with fatigue (Fig. 4.41). These features represent local progression of the crack. Their exact cause has not been identified; it is not certain whether they are crystallographically or noncrystallographically related. It is possible that they may be either, depending on the system, since they are seen in noncrystallographic plastics as well as metals.

They are not seen clearly in all metals. Their appearance also depends on the stress state.[30] They occur more often in plane strain situations than in plane stress, where the predominant feature is dimples (Fig. 6.14).

Striations represent one cycle of loading. As with macroscopic growth rate, striation spacing can be related to ΔK. However, the relationship between da/dN and ΔK, and striation spacing and ΔK, are not the same (Fig. 6.15). This indicates that the crack front advances by a combination of striation formation and other fracture mechanisms. As ΔK increases, striation formation becomes less significant in the overall crack growth.

There is a risk involved in identifying striations.[31] Several features and

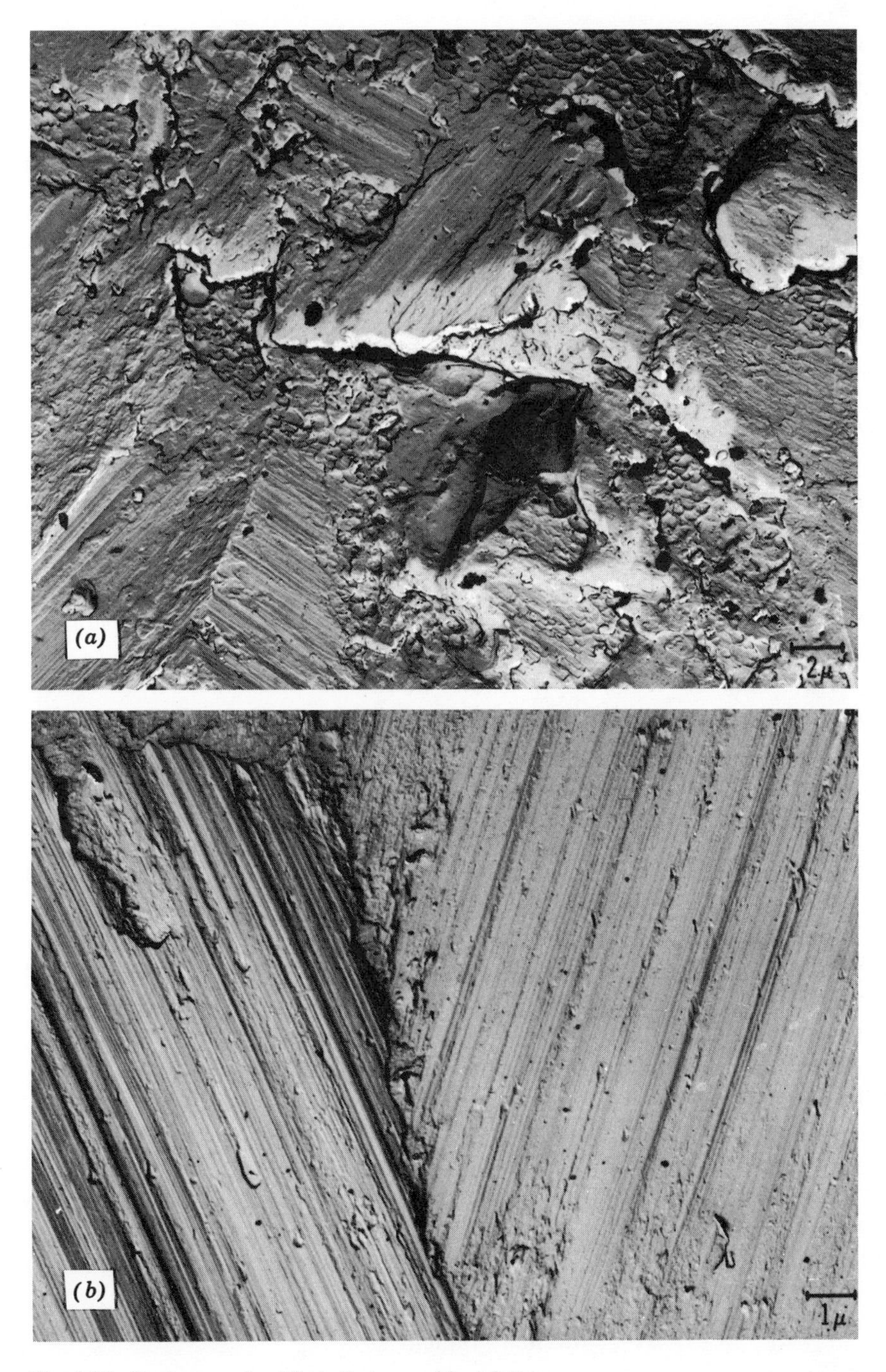

Fig. 6.16 Features and artifacts that resemble striations.

artifacts also resemble striations (Fig. 6.16). Therefore, care must be exercised in this situation whenever electron fractography is used.

6.8 SPECIAL SUBJECTS

There are several types of fatigue conditions that are discussed separately because of their significance. These are thermal and environment effects and contact fatigue.

6.8.1 Thermal Effects

Temperature effects can be considered from two different viewpoints, fatigue at elevated (or low) temperature or fatigue due to thermal cycling (thermal fatigue).

The effect of temperature on stress cycling fatigue is more or less consistent with its effect on strength. As temperature decreases, strength increases. In fatigue, as temperature decreases, the fatigue strength of unnotched specimens increases; but the materials, particularly steel, become more notch sensitive. As temperature decreases, critical crack size decreases and the material becomes more brittle. Thus low temperature fatigue failures also occur with smaller cracks.

Increasing the temperature creates additional problems. Physical metallurgical changes often occur, for example, secondary hardening in some steels or solutionizing of aluminum alloys. Creep, or the growth of material under static loading with relatively low loads, may also occur. In fatigue, this is more likely to occur as the cyclic frequency decreases. As temperature increases, the fatigue limit disappears and fatigue failures occur in a finite number of cycles for all stresses.

The preceding situations apply to stress cycling at a constant temperature. However, there is a temperature fatigue problem associated with components that are not subjected to significant mechanical stresses, such as turbine buckets and heat treating fixtures. When the temperature of a part is varied significantly, thermal stresses that cause fatigue can be developed, even though there are no additional mechanical stresses. The stresses due to thermal cycling are

$$\sigma = \alpha E \Delta T$$

where

σ = thermal stress
α = thermal coefficient of expansion
E = Young's modulus
ΔT = thermal gradient

As the thermal gradient is increased, the thermal stresses are increased. It is not necessary to subject the entire component to a large thermal gradient. If one face is held at a fixed temperature, and another face is heated and cooled, the thermal gradient within the component can cause cracking. This phenomenon, called heat checking, is seen in heavy wall gun tubes (Fig. 6.17) and can ultimately lead to fatigue failure.

As with all thermal effects, if phase changes occur the effects of temperature are accentuated. Thus steels are particularly vulnerable to thermal crazing since they undergo phase changes at relatively low temperatures. As the metal at the surface is heated, austenite can form. With a sufficient heat sink, a large thermal gradient can be set up, and the austenite quenched to martensite. The constant expansion and contraction due to phase changes creates sufficient stresses to cause early cracking.

Thermal fatigue failures are often identified by a network of cracks at approximately right angles. Mechanical fatigue at elevated temperatures does not generate the crack network. It will generally show jagged cracks, and perhaps some noticeable plastic deformation due to creep. Light microscopy can be helpful in both cases, since microstructural changes may be caused by the elevated temperature.

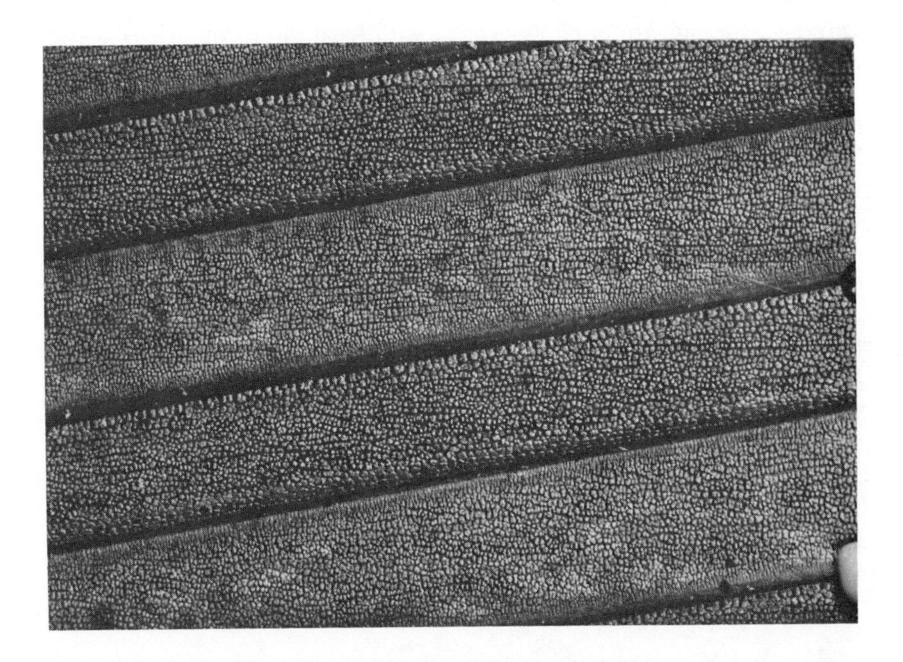

Fig. 6.17 Cracking in heavy wall gun tubes due to thermal stresses.

6.8.2 Contact Fatigue

Failure due to contact fatigue occurs when surfaces are continually in contact. The stress situation is more complex than in standard bending or axial fatigue. The contact areas tend to be small, so that the stresses are quite large. Three types of damage occur: surface pitting, subsurface pitting, and spalling.

SURFACE PITTING

Surface pitting may result from rolling and/or sliding due to shear stress caused by the contact stresses. If the condition results from pure rolling, the maximum shear occurs below the surface.[32] If there is sliding in addition to the rolling. the maximum shear stress moves towards the surface. Since shear stresses cause the pitting, cracks initiate at 45° to the surface (Fig. 6.18).[33]

The pit that develops is arrow shaped (Fig. 6.19) and points in the direction of rotation, that is, opposite to the rolling direction.[33]

SUBSURFACE PITTING

If the maximum shear stress occurs below the surface, subsurface pits may develop. This is the primary mode of fatigue failure in frictionless bearings. The pits are often associated with local points on intensification, such as inclusions. Clean metals often alleviate the problem.

Cracks often run both normal and parallel to the surface (Fig. 6.20).[34] However, they are often difficult to identify because subsequent operations can destroy the original shape.

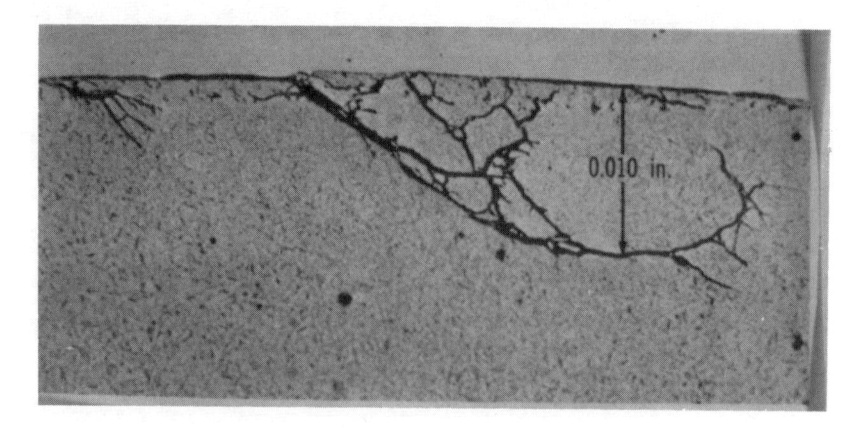

Fig. 6.18 Surface pitting showing formation of cracks at 45° to the surface (Reproduced by permission, from How Components Fail, American Society for Metals, 1966).

Fig. 6.19 Arrow shaped pit typical of contact fatigue (Reproduced by permission, from How Components Fail, American Society for Metals, 1966).

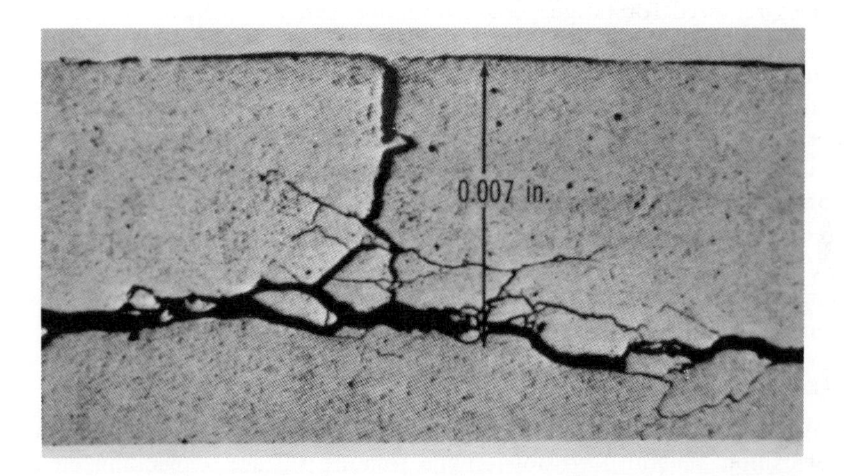

Fig. 6.20 Typical subsurface cracking in contact fatigue (Reproduced by permission, from How Components Fail, American Society for Metals, 1966).

SPALLING

Spalling is similar to subsurface pitting, except that the spalled areas are larger. In carburized parts, this is sometimes referred to as "subcase fatigue.[11,35] Cracking occurs near case-core boundaries if the shear strength at that region is lower than the shear stress. This type of failure can often be avoided by increasing case depth.

REDUCING CONTACT STRESS FATIGUE

Since the primary cause of contact fatigue is the concentrated load due to small contact area, distributing the load over a larger area reduces the tendency to failure. Resistance to surface damage can be decreased by increasing hardness. However, this decreases the ability of the metal to flow or wear, thereby distributing loads, and makes the surface finish more important.[36]

Since inclusions often affect this kind of fatigue, cleaner melts increase life. Some coating methods, such as phosphate coating and copper plating, can reduce pitting fatigue. If improperly done, however, hydrogen embrittlement can result. Finally, an effective method for reducing contact fatigue, especially surface pitting, is to lubricate adequately.

6.8.3 Corrosion Fatigue

Corrosion fatigue may be considered as a special case of general fatigue, with certain modifying effects resulting from the environment. The time or cycles required for fatigue crack initiation can be markedly reduced by corrosion processes that create pits or other surface damage. In addition, the fatigue crack propagation rate can be increased by a corrosive environment. Such behavior has been noted in many studies, even when the corrodent was relatively innocuous, such as moist air.

In corrosion fatigue, the total test time is an important factor. Since corrosion is a time-dependent phenomenon, fatigue tests employing lower frequencies exhibit greater degradation than do tests at higher frequencies, where the overall exposure time at stress is greater. Increased temperature also tends to increase crack propagation rate and consequently, to reduce the fatigue life.

RECOGNITION OF CORROSION FATIGUE

A component that has suffered corrosion fatigue damage may or may not exhibit general corrosion on the fracture surface, depending on the immersion duration. Similarly, an ordinary fatigue crack can corrode after the fact, so that immediate identification as corrosion fatigue, based simply on the presence of corrosion products, is not always possible. However,

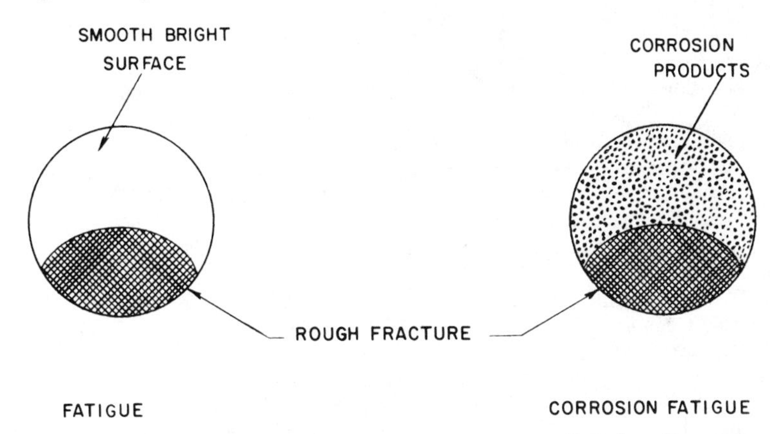

Fig. 6.21 Schematic illustration of the effect of corrosion on the fatigue fracture surface.

it is known that a corrosion fatigue specimen exhibits a different macroscopic fracture appearance than one tested in air (Fig. 6.21). The bars broken in air were bright and lustrous whereas those tested in a corrosive environment exhibited a dull surface with evidence of corrosion product.[37]

Another indication of corrosion fatigue is the presence of a number of cracks, rather than one. These cracks are usually perpendicular to the principal tensile stresses and originate at the surface where the stresses are at a maximum. In general, these fatigue cracks are transgranular in nature; however, some intergranular cracking is found, for example, in lead and aluminum systems that have been embrittled.

On a microscopic level, striations are often found by examination with electron fractographic methods. These striations are often less pronounced than those found in conventional fatigue, which is probably the result of environmental action on the already fractured surface, but may be related to a change in mechanism as well.

REFERENCES

1. G. Dieter, *Mechanical Metallurgy*, McGraw-Hill, New York, 1961, p. 298.
2. A. Cottrell and D. Hull, *Proc. Roy. Soc.*, Vol. A242, 211 (1957).
3. D. Hull, *J. Inst. Met.*, Vol. 86, 425 (1957).
4. P. J. E. Forsyth, *Proceedings of the Crack Propagation Symposium*, Vol. 1, Cranfield, England, 1962, p. 76.
5. *Fatigue Design Handbook*, SAE, 1960, p. 31.
6. W. A. Wood, *Phil. Mag.*, Vol. 3, July, 692–699 (1958).
7. P. Peterson, *Metal Fatigue*, McGraw-Hill, New York, 1959, p. 301.
8. P. G. Fluck, *Proc. ASTM*, Vol. 51, 584–592 (1951).

9. *Metals Handbook*, Vol. 2, 8th ed., ASM, 1964, p. 471.
10. G. Dieter, *op. cit.*, p. 323.
11. J. Goodman, *Mechanics Applied to Engineering*, 9th ed., Longmans, Green & Co., 1930.
12. *Fatigue Design Handbook*, SAE, 1968, p. 51.
13. G. E. Dieter, R. F. Mehl, and G. T. Horne, *Trans. ASM*, Vol. 47, 423–439 (1955).
14. H. A. Lipsitt, G. E. Dieter, G. T. Horne, and R. F. Mehl, NACA TN 3380, 1955.
15. J. Pomey and J. Ancelle, *Fatigue of Metals*, Cazaud, 1953.
16. B. B. Muvdi, G. Sachs, and E. P. Klier, *Proc. ASTM*, Vol. 57, 655 (1957).
17. R. Boyd, *Proc. Inst. Mech. Eng.*, Vol. 179, Pt. 1, No. 23, 733 (1964–1965).
18. J. T. Ransom and R. F. Mehl, *Proc. ASTM*, Vol. 52, 779, (1952).
19. H. N. Cummings, F. B. Stulen, and W. C. Schulte, *International Conference on Fatigue of Metals*, 1956, p. 439.
20. P. C. Paris, *Fatigue*, *An Interdisciplinary Approach*, Syracuse University Press, 1964, pp. 107–132.
21. J. F. Throop and G. A. Miller, Watervliet Arsenal Report WVT–7006, January 1970.
22. F. A. Heiser, *J. Basic Eng.*, June 1971, pp. 211–217.
23. G. A. Miller, *Trans. Quart.*, *ASM*, Vol. 61, No. 3, 442 (1968).
24. A. Brothers and S. Yukawa, *J. Basic Eng.*, Vol. 89, No. 1, 19 (1967).
25. A. A. Anctil and E. B. Kula, AMMRC Report TR–6915, June, 1969.
26. W. G. Clark, Jr., *Met. Prog.*, May 1970, pp. 81–86.
27. F. A. Heiser and W. Mortimer, *Met. Trans.*, Vol. 3, August 1972, 2119–2123.
28. *The Tool Steel Trouble Shooter*, Bethlehem Steel Co., 1952, p. 106.
29. *Fatigue Design Handbook*, *op. cit.*, p. 19.
30. R. Hertzberg, *Fatigue Crack Propagation*, ASTM, STP 415, 1967, p. 219.
31. *Electron Fractography Handbook*, Air Force Materials Laboratory, ML–TDR–64–416, 1965, p. 3.31.
32. W. E. Littman and R. L. Widner, ASME Paper No. 63—WA/CF–2.
33. D. Wulpi, *How Components Fail*, ASM, 1966, p. 31.
34. D. Wulpi, *op. cit.*, p. 32.
35. R. Pederson and S. L. Rice, SAE Preprint 22B, 1960.
36. J. T. Burwell, *Interpretation of Tests and Correlation with Service*, ASM, 1951.
37. M. Fontana and N. Greene, *Corrosion Engineering*, McGraw-Hill, New York, 1967, p. 107.

7

Corrosion

7.1 INTRODUCTION

Corrosion, in its broadest sense, has been defined by Fontana and Greene[1] as the degradation of a material by reaction with its environment, regardless of whether that material is a metal, polymer, elastomer, or ceramic. However, since we are primarily involved with metallurgical failures, the more restrictive definition of metallic dissolution applies. All corrosion processes have electrochemical reactions as their basis. Some are purely electrochemically whereas others result from the action of chemical plus mechanical factors, such as erosion, corrosion, and stress corrosion.

This chapter provides a condensed description of the various types of corrosion, the visible manifestations, the causes, and the recommendations for remedial action.

7.2 GENERAL CORROSION

General attack is the most significant form of corrosion, in terms of the economic consequences. It is characterized by a more or less uniform attack over the entire exposed surface, with only minimal variations in the depth of damage. A typical example of such attack is illustrated in Fig. 7.1. Whole segments of industry, such as those involved in coatings and cathodic protection, owe their survival to the threat presented by general corrosion. Such protection undoubtedly costs hundreds of millions of dollars annually.

From the standpoint of failure analysis, general corrosion is not considered a major problem, but is included here for the sake of completeness. In most engineered systems, the environment is usually sufficiently described so that gross errors in material selection are not common. Protective coatings and cathodic protection provide additional assurance in those cases where natural resistance might be marginal. When general corrosion does

Fig. 7.1 General attack of structural steel in sea water environment. Courtesy D. P. Slawsky.

occur, the rate of attack is usually predictable, and catastrophic failure does not often result.

7.3 GALVANIC CORROSION

One of the more common and serious types of corrosion occurs when two or more dissimilar metals are electrically coupled and placed in an electrolyte. The ensuing action is known as galvanic corrosion and results from the existence of a potential difference between the metals, which causes a flow of current between them. The more active metal undergoes accelerated corrosion whereas corrosion in the less active member of the couple is retarded or eliminated. Table 7.1, developed by Bauer and Vogel,[2] illustrates the change in corrosion rate that results from coupling iron to a series of metals.

7.3.1 Electromotive Series

In the design of a structure involving dissimilar metals, it is essential to know which metal in the couple will suffer accelerated corrosion. The basis for establishing the reactivity of various metals in the couple is the relative potential difference. Under reversible, noncorroding conditions, these potentials give rise to the standard electromotive series. The potential of each metal is determined in relation to the hydrogen electrode (H_2/H^+), which is arbitrarily assigned a value of zero.

Table 7.1 Comparison of the Corrosion Rates When Iron is Coupled to a Second Metal (1% NaCl Solution)[a]

Second Metal	Weight Loss (Iron) (mg)	Weight Loss (Second Metal) (mg)
Magnesium	0.0	3104.3
Zinc	0.4	688.0
Cadmium	0.4	307.9
Aluminum	9.8	105.9
Antimony	153.1	13.8
Tungsten	176.0	5.2
Lead	183.2	3.6
Tin	171.1	2.5
Nickel	181.1	0.2
Copper	183.1	0.0

[a] From Ref. 2.

These potentials are tabulated in Table 7.2. Metals whose potentials are more negative than hydrogen are more reactive and those more positive than hydrogen are less active.

7.3.2 Galvanic Series

When dealing with corrosion problems, it is important to distinguish between the electromotive series and the galvanic series. The former is valid only for pure metals in specific concentrations of their own salts; and metals are rarely applied under these idealized conditions. In studying galvanic corrosion, a somewhat similar series, based on the actual experience gained with a number of metals or alloys in a number of environments, is used. Table 7.3 illustrates the relative position for a number of alloys in a sea water environment. Ideally, such a series should be generated for each environment and temperature of interest. In practice, however, positions are approximated from the available data.

In each couple, the metal nearest to the active end of the series is the anode and undergoes accelerated corrosion, whereas the more noble member receives some measure of protection. Reexamining Table 7.1 shows that a metal either can be protected or suffer greatly increased attack, depending on the other member of the couple.

Table 7.2 Standard Electromotive Series

	Electrode Reaction	Standard Electrode Potential, E° (Volts), 25°C
Active ↑	$K = K^+ + e^-$	−2.922
	$Ca = Ca^{++} + 2e^-$	−2.87
	$Na = Na^+ + e^-$	−2.712
	$Mg = Mg^{++} + 2e^-$	−2.34
	$Be = Be^{++} + 2e^-$	−1.70
	$Al = Al^{+++} + 3e^-$	−1.67
	$Mn = Mn^{++} + 2e^-$	−1.05
	$Zn = Zn^{++} + 2e^-$	−0.762
	$Cr = Cr^{+++} + 3e^-$	−0.71
	$Ga = Ga^{+++} + 3e^-$	−0.52
	$Fe = Fe^{++} + 2e^-$	−0.440
	$Cd = Cd^{++} + 2e^-$	−0.402
	$In = In^{+++} + 3e^-$	−0.340
	$Tl = Tl^+ + e^-$	−0.336
	$Co = Co^{++} + 2e^-$	−0.277
	$Ni = Ni^{++} + 2e^-$	−0.250
	$Sn = Sn^{++} + 2e^-$	−0.136
	$Pb = Pb^{++} + 2e^-$	−0.126
	$H2 = 2H^+ + 2e^-$	**0.000**
	$Cu = Cu^{++} + 2e^-$	0.345
	$Cu = Cu^+ + e^-$	0.522
	$2Hg = Hg2^{++} + 2e^-$	0.799
	$Ag = Ag^+ + e^-$	0.800
	$Pd = Pd^{++} + 2e^-$	0.83
	$Hg = Hg^{++} + 2e^-$	0.854
	$Pt = Pt^{++} + 2e^-$	ca 1.2
↓	$Au = Au^{+++} + 3e^-$	1.42
Noble	$Au = Au^+ + e^-$	1.68

7.3.3 Contributing Factors

RELATIVE AREAS

Whitman and Russell[3] showed that the rate of galvanic corrosion was directly proportional to the ratio of the cathodic metal to the anodic metal. For any given current, the current density is a function of the area. Therefore, when a large cathode is coupled with a small anode, the current

Table 7.3 Galvanic Series of Metals and Alloys in Sea Water

	Metal
Active ↑	Magnesium
	Zinc
	Alclad 3S
	Aluminum 3S
	Aluminum 61S
	Aluminum 63S
	Aluminum 52
	Low carbon steel
	Alloy carbon steel
	Cast iron
	Type 410 (active)
	Type 430 (active)
	Type 340 (active)
	Type 316 (active)
	Ni-Resist (corrosion-resisting, nickel cast iron)
	Muntz metal
	Yellow brass
	Admiralty brass
	Aluminum brass
	Red brass
	Copper
	Aluminum bronze
	Composition G bronze
	90/10 Copper-nickel
	70/30 Copper-nickel—low iron
	70/30 Copper-nickel—high iron
	Nickel
	Inconel, nickel-chromium alloy 600
	Silver
	Type 410 (passive)
	Type 430 (passive)
	Type 304 (passive)
	Type 316 (passive)
	Monel, nickel-copper alloy 400
	Hastelloy, alloy C
	Titanium
	Graphite
	Gold
↓ Noble	Platinum

densities increase at the anode, that is, the corrosion rates, increase. Failure to observe this rule of reduced cathode/anode ratios can cause serious consequences. Greene[4] cites the example of a tank that, for economic reasons, was manufactured with carbon steel sidewalls welded to 18–8 stainless steel bottoms. To minimize product contamination, the steel sidewalls were coated with a phenolic resin. Since coatings invariably have pin-hole defects, this resulted in a situation combining a large cathode area (18–8 tank bottom) and a small anode (uncoated regions in the tank walls). The end result was perforation of the sidewalls above the weld.

POLARITY CHANGES

In certain systems, the polarity of metal couples can reverse. With iron-zinc couples, in certain water supplies containing nitrates and bicarbonates, the polarity reversal occurs at temperatures above 140°F. This phenomenon is not completely understood, but is believed to be associated with changes in the conductivity of the corrosion product. The zinc-iron system is the most common one in which reversal occurs, however, Schikorr[5] has recently described polarity changes occurring in a cast iron-aluminum alloy (9% Si) couple in automotive cooling systems. Other systems in which reversals have been reported are aluminum-iron and tin-iron.

ELECTROLYTE CONDUCTIVITY

In high-conductivity electrolytes, the corrosion rate of the anodic member is high, relatively uniform, and may be slightly displaced from the couple junction. With high resistance environments, the total weight loss suffered by the anode is generally lower and is confined to the immediate area of the

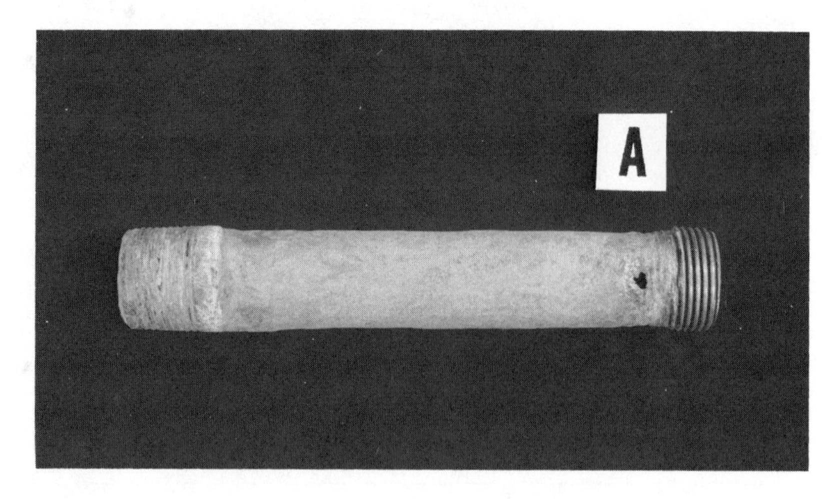

Fig. 7.2 Steel pipe coupled to a brass section showing the results of galvanic corrosion.

couple. This is illustrated in Fig. 7.2. The section at the end of the pipe, labeled *A*, was coupled to a length of brass pipe in a fresh water well. Note that thinning is most prominent in the section close to the joint, and that perforation has occurred in this region.

7.3.4 Design Practices

To avoid a galvanic corrosion problem, it is important to select, in the design stage, materials close together in the galvanic series. It is also necessary to keep contaminant metals from creating an inadvertant couple by deposition from elsewhere in the system.

If the use of dissimilar metals cannot be avoided, galvanic corrosion can be minimized, by keeping the anode large (in relation to the cathode), or avoided by electrically insulating the members of the couple with dielectric spacers, provided this isolation is complete (as shown in Fig. 7.3). Coatings, such as chromate paints or epoxy resins, may also be used if the area effects previously described are not created, and if these coatings are not subject to mechanical damage.

When galvanic corrosion cannot be avoided, its effects should be anticipated, and the anodic components should be either more massive or easily replaceable.

7.4 PITTING

Pitting is a form of localized corrosion in which the attack is confined to numerous small cavities on the metal surface. The cavities created may vary in size and form, however, it is commonly held that a true pit has a

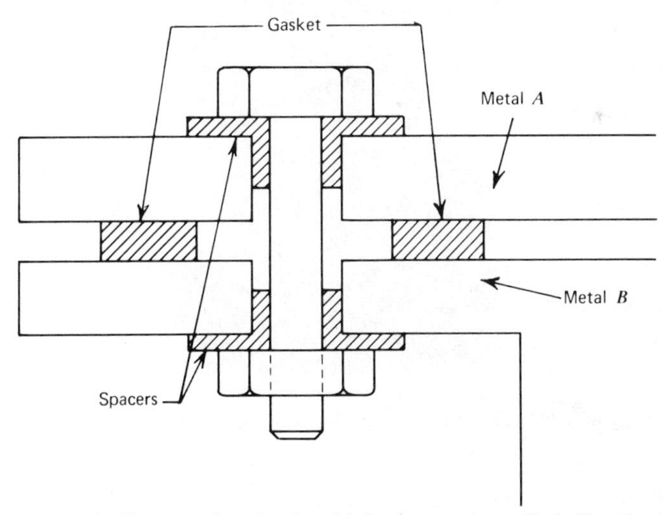

Fig. 7.3 Schematic diagram of an insulated joint between two dissimilar metals.

Fig. 7.4 Pitting in a 304 stainless steel bar after two years in an industrial-marine atmosphere.

length/depth ratio equal to or greater than 1. The plate in Fig. 7.4 shows the results of pitting attack on a stainless steel bar in a marine atmosphere.

Since only slight metal losses may result in perforation, failure can be quite rapid and can occur without warning, making pitting an extremely dangerous form of attack. Pits can also contribute to failure in another way. In highly-stressed components, these pits can act as stress raisers, which results in nucleation of fatigue cracks.

7.4.1 Environmental Factors

Pitting can occur with a number of metals, but the most common structural materials susceptible to this type of degradation are stainless steels and aluminum alloys. Pitting occurs most frequently in solutions of near neutral pH, containing chloride, bromide of hypochlorite ions. Fontana and Greene[6] have pointed out that pitting is intermediate between the condition of general corrosion and complete immunity. Consequently, those factors favoring general corrosion, low pH and increased temperatures, do not generally favor pitting attack. Greene and Fontana[7] have tabulated the effect of a number of these environmental variables on the rate of pitting attack. These are reproduced in Table 7.4.

Table 7.4 Effect of Environmental Variables on the Pitting Behavior of 18-8 Stainless with Specific Cathodic Reactions[a]

	Oxygen Reduction	Metal Ion Reduction
Examples	NaCl, NaBr, $CaCl_2$, $MgCl_2$, $AlCl_3$	$FeCl_3$, $FeBr_3$, $CuCl_2$, $HgCl_2$
Relative rate of pitting attack	Slow	Moderate to rapid
Increasing temperature	Corrosion rate increases, then decreases. Maximum point.	Corrosion rate increases. No maximum point.
Increasing chloride concentration	Corrosion rate increases, then decreases. Maximum point.	Corrosion rate increases. No maximum point.
Increasing O_2 content	Increases attack	Negligible effect
Effect of pH	Low pH—general attack; neutral pH—pitting; high pH—no attack.	Solutions stable only at low pH. Neutral and high pH—no attack.
Corrosion products	Usually insoluble	Usually soluble

[a] From N. D. Greene and M. G. Fontana, *Corrosion*, Vol. 15, 25t (1959).

7.4.2 Mechanism

Pitting behavior may be divided into initiation and propagation stages. The factors acting in the selection and initiation of pit sites are not clear and are somewhat controversial, but Bond's experiments[8] on high purity aluminum in NaCl solutions suggest that segregation is an important factor. The extremely pure aluminum (99.9999%) used in these experiments was relatively unaffected by pitting damage. In another study,[9] the same workers showed that the distribution of elements, rather than the nominal composition, was the important consideration. Pits occurred on cell boundaries and other highly segregated regions.

Once pitting has begun, however, its stability may be affected by the location of nearby pits, gravity, and velocity. The presence of pits suppresses the formation of nearby pits. This is generally attributed to the fact that a cathodic region surrounds an active pit (i.e., cathodic protection of the surrounding region).

The propagative mechanism is strongly dependent on the concentration of corrosion products within the pit, as shown in the following section. The existence of high velocities tends to purge the beginning pit and to retard pit growth. Similarly pit growth is greatest in the direction of gravity, since the concentrated solution within the pit may be easily retained.

Many theories have been proposed to explain the propagation behavior of pits but the one that has gained the widest acceptance is that pitting is an autocatalytic process. Metal dissolution occurs within the pit, and is accompanied by O_2 reduction near the mouth of the pit or on the metal surface (Fig. 7.5*a*). This results in an excess of positive charge ($M+$) at the pit base and migration of the chloride ion occurs to offset the imbalance (Fig. 7.5*b*). The net effect is an increase in metal chloride concentration in the pit. These salts are subject to hydrolysis, thereby increasing the hydrogen ion concentration (lowering the pH) within the pit and promoting increased dissolution rates. The reaction for this hydrolysis is

$$M^{+}Cl + H_2O \rightarrow MOH + H^{+}Cl^{-}$$

7.4.3 Determination of Pitting Attack

Visual examination of the surface, even if possible, does not reveal the extent of pitting damage since the pit orifice may be small, bordering on the microscopic, whereas penetration may be severe. Pitting can be best evaluated by making metallographic sections of representative components to determine maximum pit depth, or by ultrasonic inspection. Since the former method is destructive, it is best adapted to laboratory evaluations.

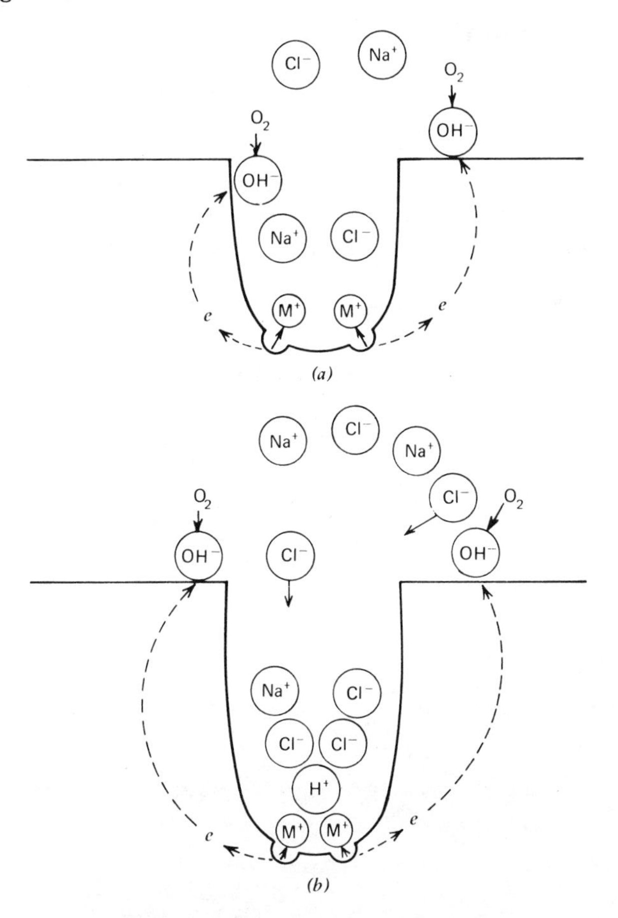

Fig. 7.5 (*a*) Initiation and (*b*) growth of a corrosion pit.

As Fontana and Greene[10] point out, the probability of finding deeper pits increases with increasing surface area, and laboratory results on small specimens must be used advisedly.

7.4.4 Prevention

The best protection against pitting attack is to select a material with adequate pitting resistance. Therefore, ample data regarding the behavior of candidate materials in pitting environments must be available. These data may be generated by laboratory evaluations, but should be followed by an examination of prototype hardware whenever possible.

7.5. CREVICE CORROSION

Crevice corrosion is a type of corrosive attack which is associated with the confined spaces or crevices formed by certain mechanical configurations, such as tapped joints, gasket interfaces, tubular sleeves, and so forth.

Crevices can exist in any assembly, but there appears to be a geometrical requirement. The crevice must be close fitting, with dimensions of only a few thousandths or less, for corrosion to occur. Although the limits of the gap have not been defined, crevice corrosion does not occur in larger spaces. Figure 7.6 shows a washer that has a slight bevel on one side. Note the presence of the corroded area near the edge of the bevel. France[11] has shown that the dimensions of crevices are also extremely important in components that are being anodically protected. Where the dimensions of the crevice are small, the resistance is large, and the lower regions of the crevice are unprotected.

Both approximating surfaces need not be metal for crevice corrosion to occur. It has been reported in crevices formed by a number of nonmetallic materials, such as polymers, glasses, and rubber, that are in contact with metal surfaces. This fact is of particular importance in applying and select-

Fig. 7.6 A 304 stainless steel washer with crevice corrosion near the beveled edge.

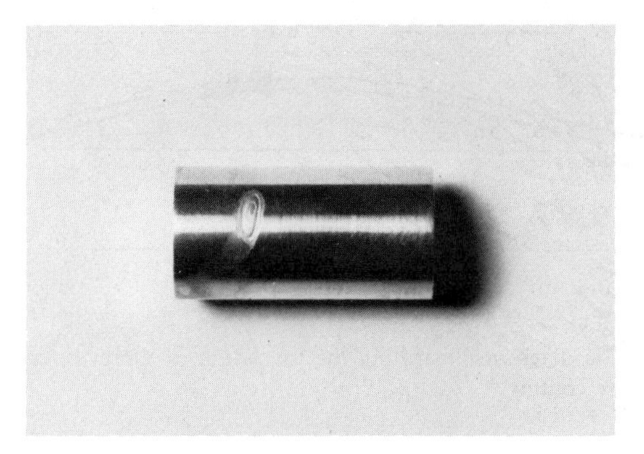

Fig. 7.7 Crevice corrosion resulting from contact between Hastelloy B and a glass rod.

ing gasketing materials. The corroded area on the Hastelloy B cylinder, shown in Fig. 7.7, occurred in the region where it made contact with a glass basket, in less than 600 hours, at 30°C, in 0.9% NaCl solution. This type of corrosion can also be associated with the shielded areas caused by the settling out of particulate solids on a surface or under marine growth. Bush[12] describes a variation of this condition, called "poultice" corrosion, common in the automotive field. Road salts and debris that collect on ledges and pockets are kept moist by weather and washing, and severe body corrosion ensues.

7.5.1 Mechanism

Initial investigations of the mechanism of crevice corrosion indicated that it was caused by a concentration cell that formed within the crevice, which was either a metal-ion type or one resulting from differences in oxygen concentration. Evans[13] was the first to demonstrate that the corrosion rate was increased by changes in the oxygen concentration using iron electrodes in a cell containing a partition. This occurs because of changes in potential, which are related to the oxygen concentration. Although these cells may be associated with crevice corrosion, they are not its primary cause. Shaefer et al.[14] have recently shown that the crevice corrosion mechanism may be separated into stages. The initial action in the site is the result of oxygen concentration differences, but the continued activity is dependent on an autocatylytic process, much like that used to explain pitting. Consider a section of steel in a sea water environment. Corrosion occurs because of an oxygen deficiency within the crevice. This increases the positive ion concentration in the crevice, and results in a momentary imbalance. To restore the balance of charge, the migration of Cl ions

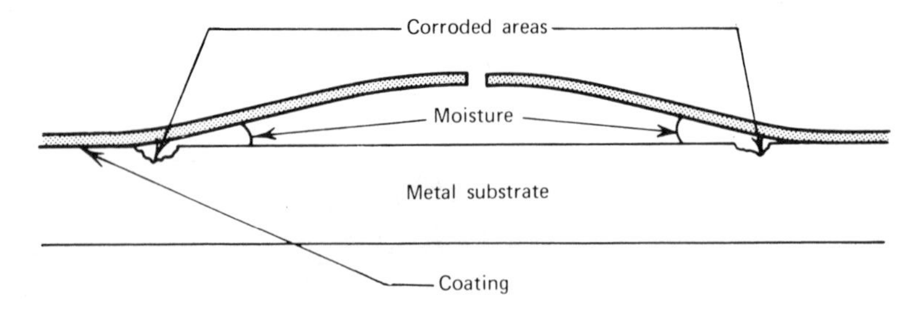

Fig. 7.8 Schematic diagram illustrating the formation of a crevice condition by a non-adherent protective coating.

occurs. Therefore, there is a high concentration of FeCl within the crevice. As a result of hydrolysis (see Section 7.4.2), an excess of hydrogen ions is produced, which stimulates further corrosion.

7.5.2 Preventative Measures

The most effective method for minimizing crevice corrosion is the elimination of the crevice itself. The means of accomplishing this varies, depending on the situation. For example, the following steps might be taken.

1. Design changes incorporating welded butt joints in lieu of bolted lap joints.
2. Using nonpermeable gaskets or seals. We believe that the use of materials with poor wettability improves the effectiveness of these gaskets.
3. Periodic removal of sediment and debris from tank bottoms.
4. Insuring adequate adhesion of coatings. A crevice is often created by the presence of a blister or nonadhering portion of a coating, as illustrated in Fig. 7.8.

7.6 SELECTIVE DISSOLUTION

Selective dissolution is the term applied when a phase is selectively attacked in an alloy, or when one element is preferentially dissolved from a solid solution. This type of corrosion occurs in several common alloy systems, such as the dezincification of brass alloys, preferential attack of ferrite in austenitic[15] and martensitic stainless steels,[16] or of ferrite in grey cast irons (graphitization). These are the most common examples; however, selective dissolution has also been reported in several other systems (under highly specific conditions), such as the loss of nickel from Monel,[17] of aluminum from aluminum bronze,[18] and of copper from 70–30 cupro-nickel alloy,[19]

Modern technology has created another group of materials in which selective dissolution can produce serious consequences, metallic composite structures. These are usually constructed so that the fibers and matrices possess widely differing chemical and mechanical characteristics. They may therefore be severely degraded and show sharp losses in strength if the reinforcing fibers are selectively attacked.

7.6.1 Form of the Attack

The form of the attack may vary; however, the severest degradation usually results when the attacked species or phase is present in a continuous network. There may be little change in the overall configuration or geometry of the component, but the mechanical properties are adversely affected in a very serious way. Microstructurally, if the resistant phase or species is also continuous, it is left behind with the corrosion products, as a porous network. Figure 7.9 (Schreir) shows the results of selective attack on grey cast iron. The graphite skeleton is left behind with some phosphide eutectic whereas the ferrite matrix has been dissolved.

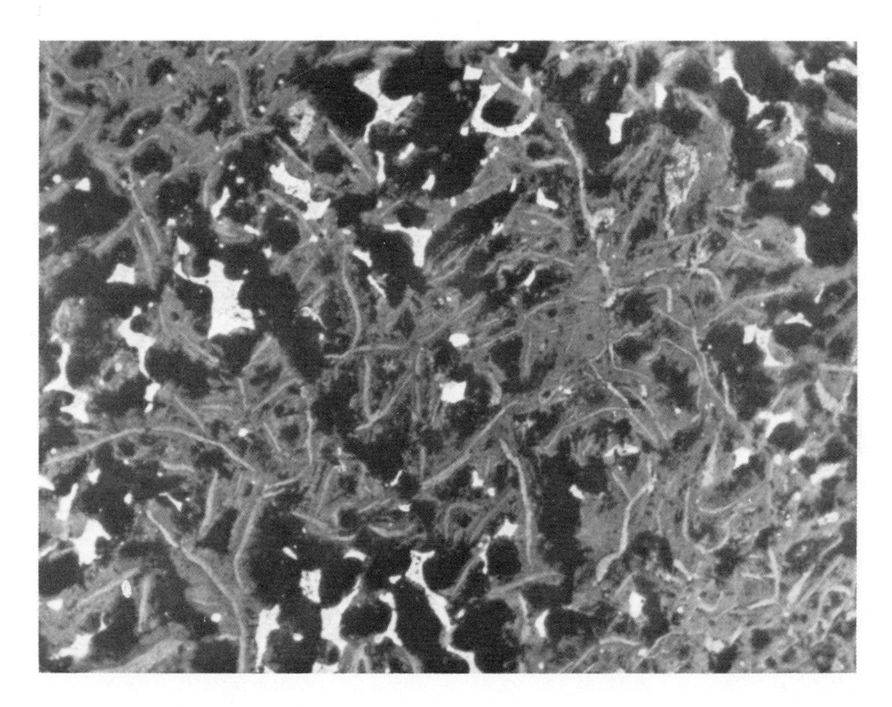

Fig. 7.9 Residual skeleton after selective attack of gray cast iron, containing graphite flakes and phosphide eutectic.

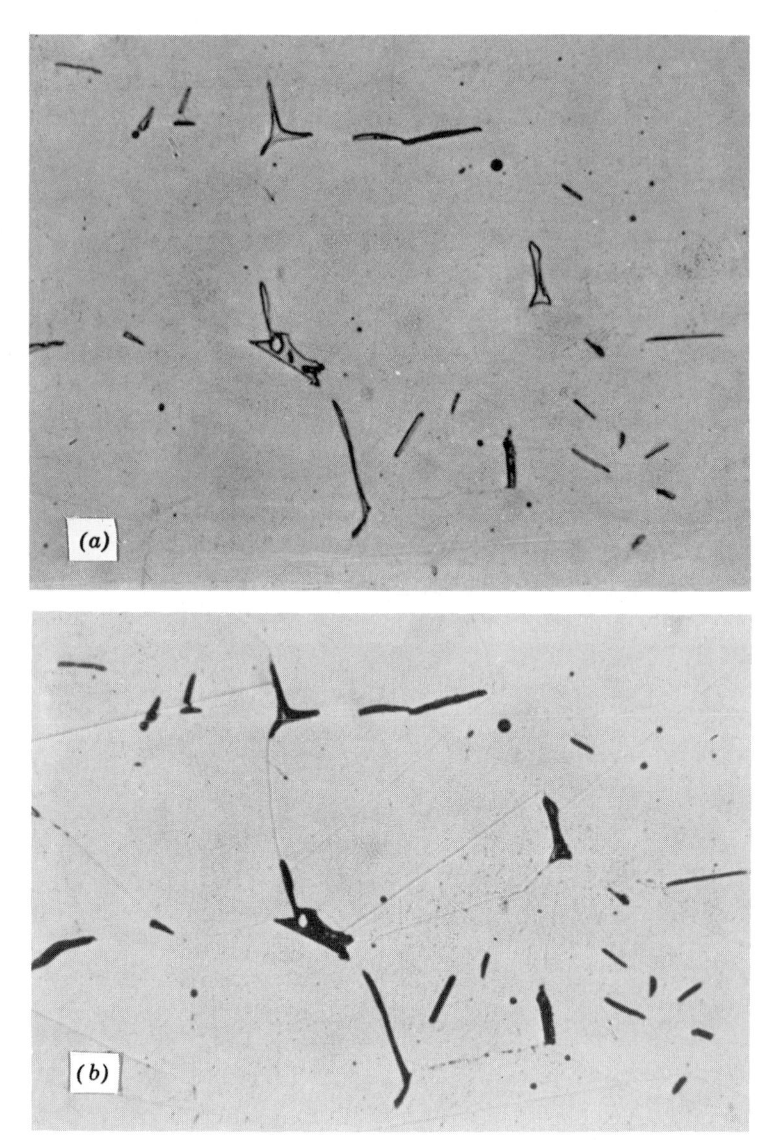

Fig 7.10 Selective attack of ferrite in an austenite matrix. (*a*) The structure of a defective rod end, lightly etched, to show the micro appearance of ferrite phase (500×). (*b*) The same area, after heavier etching, showing the ferrite is present along grain boundaries and along planes within the grains (500×). Courtesy D. Moore.

Similar results would be obtained upon microstructural examination of dezincified structures, except that the remaining phase would be largely a copper-copper oxide network.

Figure 7.10 shows ferrite, in an austenite matrix, in various stages of attack.

7.6.2 Mechanism

The precise mechanism for selective corrosion varies with both the alloy system and the environment; however, there are undoubtedly some unifying characteristics. Attack is usually more severe when the corroding phase is present as a continuous network rather than as isolated inclusions. The factors advanced by Green,[20] as significant in the corrosion of composite materials, are also applicable to selective corrosion in general. These are: (a) the specificity of the environment towards the constituents comprising the composite, as illustrated in Fig. 7.11, and (b) the nature of the galvanic cell set up between these constituents. Graphitization is partially the result of the latter cause. Since graphite is cathodic to ferrite in most environments, the corrosion of the ferrite is accelerated.

The mechanism of dezincification is well known, but it is still somewhat controversial. Two basic mechanisms have been advanced. One states that zinc-rich areas dissolve preferentially, and the other proposes that complete dissolution of the brass occurs with copper redepositing. Evans[21] indicates that both processes can occur under the proper circumstances. The rate of dezincification rises with increasing zinc content, varying from relatively uniform attack in the two phase high-zinc brasses, to plug-type attack on the low zinc content and brasses, as shown in Fig. 7.12. Other factors that increase the probability of dezincification are high temperature, a slightly acidic environment, and slow solution velocities.

7.7 INTERGRANULAR CORROSION

Excluding semiconductors, metal components are usually polycrystalline. In certain cases, the grain boundaries that exist in such aggregates are more suceptible to corrosive attack than the grain interior. The preferential dissolution suffered by these areas may be related to several factors, depending on the particular circumstances.

The primary cause of intergranular attack (IGA) is the presence of an inhomogeneous condition at the grain boundary. This may be the result of a segregation mechanism, as described in Chapter 8, or of intergranular precipitation. These conditions may also be modified by enhanced diffusion effects operating within the grain boundary, or by the selective adsorption of certain solutes, such as hydrogen.

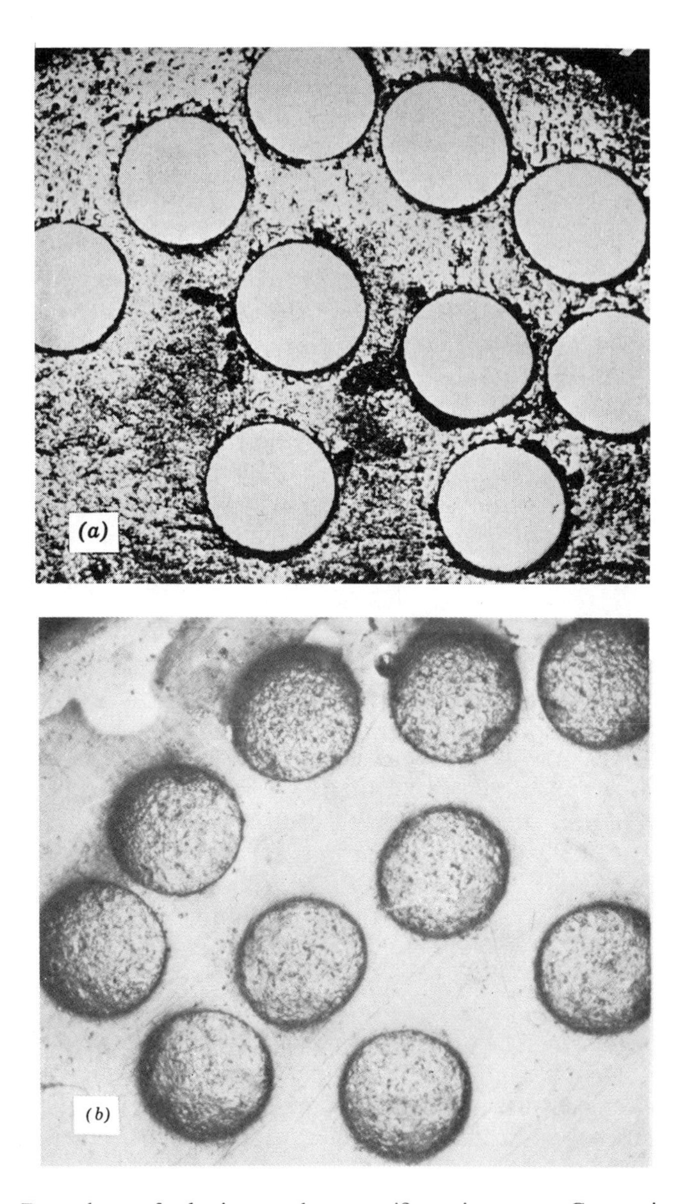

Fig. 7.11 Dependence of selective attack on specific environments, Cu matrix—Ta filament composite. (*a*) Selective attack of copper matrix in a Cu/Ta filament composite after brief exposure to 70% nitric acid (70X). (*b*) Selective corrosion of Tantalum in a Cu/Ta filament composite after brief exposure to 45% hydrofluoric acid (70×).

Fig. 7.12 Plug-type dezincification of brass pipe.

The overall effect of such preferential dissolution is that great damage to the structure can occur with only slight corrosive damage to the grain bodies themselves. Because dissolution is confined to such small regions, the actual weight losses are small, penetration rates are high, and destruction can be quite rapid. Figure 7.13 shows a typical example of intergranular attack, viewed from the surface, and in a cross-section.

7.7.1 Sensitization of Austenitic Stainless Steels

Intergranular attack on stainless steels is generally attributed to the depletion of dissolved chromium within the grain boundary. This is directly related to the precipitation of chromium carbides during sensitizing heat treatment. When unmodified 18–8 stainless steels are heated in the temperature range of 900 to 1500°F, or cooled slowly through it, the chromium-rich (Cr, Fe, C_6) carbide, which is relatively insoluble in this range, precipitates. The degree of sensitization is directly related to the temperature and duration of exposure. At the low end of the range, about 1000°F, times measured in hours are needed to effect a sensitized condition. The precipitation results in the depletion of chromium in the grain boundary region, as shown in Fig. 7.14. Active evidence that chromium depletion does occur has been presented by Alm and Kiessling,[22] using electron probe microanalysis.

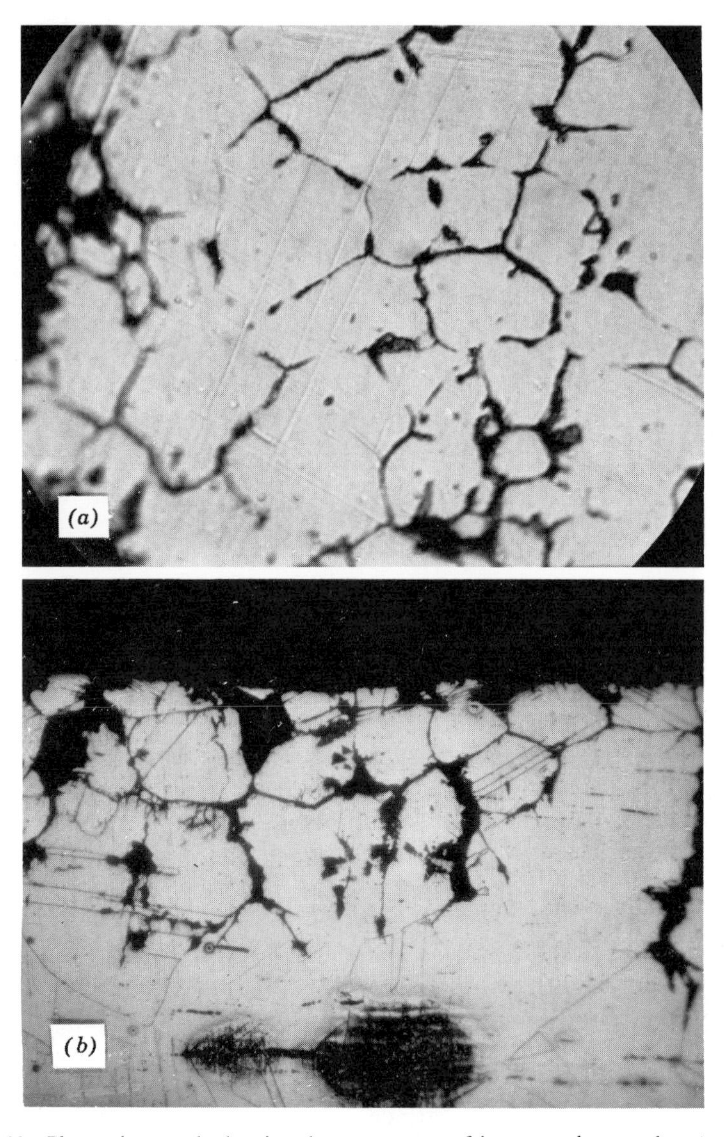

Fig. 7.13 Photomicrograph showing the appearance of intergranular attack. (*a*) surface (100×). (*b*) Cross-section (100×).

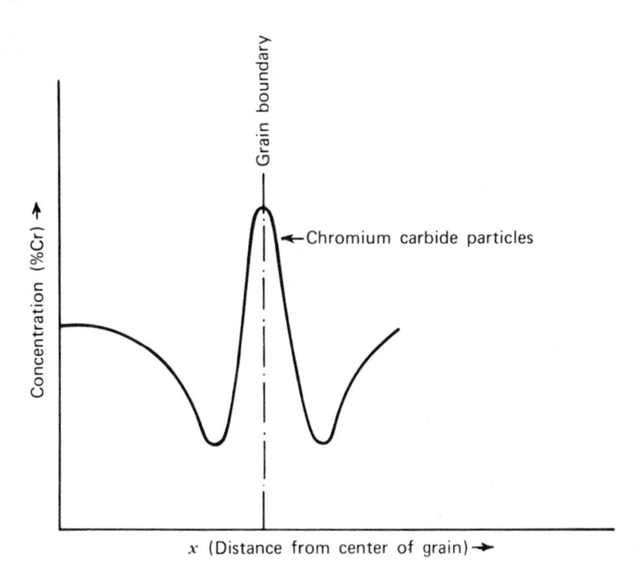

Fig. 7.14 Schematic diagram illustrating the effective depletion of chromium adjacent to the grain boundaries.

The effect of this depletion is that the grain boundaries are now compositionally different from the bulk of the grain. Higher corrosion rates result from the interaction of several factors: the poor corrosion resistance of the low chromium alloy formed, possible galvanic effects between the grain boundary and interior, and an unfavorable cathode/anode ratio.

Sensitization may result from any of the following.

1. Annealing or stress-relieving heat treatments.
2. Exposure at some time during the course of service, prior to exposure.
3. Fabrication procedures, generally welding or flame cutting, in which the adjacent zones may pass through the temperature range.

7.7.2 Environments Producing Intergranular Attack

A recent study[23] conducted by the Welding Research Council evaluated the effects of a number of industrial environments in producing *IGA* in austenitic stainless steels. The steels represented were 316, 316L, 317, and 318 (molybdenum bearing), and 302, 304, 304L, and 347 (nonmolybdenum bearing). The results are summarized in Table 7.5. The report itself gives a detailed account of the testing.

Table 7.5 Effect of Various Environments in Producing IGA on Several Austenite Stainless Steels[b]

Environments Producing Intergranular Attack
Acetic acid, Glacial, turbulent vapor, 293°F
Acetic acid (35%), + 1% formic acid, 267.5°F
Acetic anhydride + acetic acid, 212°F 230°F
Chlorinated kraft pulp, pH 2 to 2.8, 75°F
Cornstarch slurry, pH 1.5 @ HCl, 110 to 130°F
Monochlorobenzene (64%) & DDT (35%), 212°F
Monochlorobenzene, plus steam, HCl and air, 750 to 930°F[a]
Nitric acid (60%), plus chlorides and fluorides, 190°F
Nitric acid (50%), plus chlorides and fluorides, 180°F
Nitric acid (48%), plus chlorides and fluorides, 175°F
Nitric acid (47% to <1%, plus chlorides and fluorides, 176°F
Nitric acid (12% maximum), plus chorides and fluorides, 155°F
Nitric acid (20%) + 6 to 9% metal nitrates and 2% sulfates, 190°F maximum
Phosphoric acid, dilute, plus dilute nitric and sulfuric acids, pH 1.6, 160°F maximum
Phosphoric acid ("Phossy" Water), pH 6.2, 140°F maximum[a]
Phthalic anhydride (crude grade), 450°F average
Sulfuric acid (98%), 110°F[a]
Sulfuric acid (78%), ambient temperature[a]
Sulfuric acid, (4%), 190°F
Sulfuric acid, pH 2.0, 150°F
Sulfuric acid, (0.1%), plus 1% ammonium sulfate and 50 to 100 ppm chlorides, 194 to 220°F[a]
Starch, Milo, pH 1.6 due to sulfuric acid, 110 to 120°F
Acetic acid vapor, steam, air, methylene chloride and traces of hydrogen chloride. 260°F maximum
Boron trichloride (99%), trace HCl, 180°C maximum
Hydrochloric acid, Cl_2, H_3BO_3, chloride salts, 180°C maximum
Lactic acid (20%), corn steepwater liquor pH 3.5 to 4.5, 165°F
Maleic acid (0.5%), xylene (75%), water, 105°C
Maleic acid (2%), maleic anhydride (13%), xylene (8%), 105°C
Phenol, acetone, p-cresol, sodium sulfate, water, 60°C
Phthalic anhydride, phthalic acid, maleic anhydride, benzoic acid, water vapor, 400°F
Sodium chloride (25%), water, 180°F
Sodium sulfite (25%), water, 115°F
Sulfuric acid (0.1%), acetic acid (0.1%), acetaldehyde, 125°C maximum
Sulfuric acid (0.3%), pH 3. Ammonium sulfate, 158°F
Sulfuric acid (105%), 115°F
Sulfurous acid (5% total SO_2), ammonium bilsulfate, 75°F
Sulfurous acid (1% SO_2) pH 1.5 to 2. Calcium bisulfite liquor, 74°F
Sulfurous acid (<1% SO_2) pH 5.5. Washed sulfite pulp, weak calcium bisulfite liquor
Sulfurous acid (0.1% SO_2), pH 4.5. Ground corn slurry
Xylene (75%), water, maleic acid (0.5%), 105°C

[a] Some tests in this environment did not produce *IGA*.

[b] From Ref. 23.

7.7.3 Preventative Measures

Since the most common method of inducing *IGA* is through heat treatment in the sensitizing region, the problem may be eliminated by the careful control of heat treating measures. Where sensitization has occurred, or is suspected, a high temperature solution treatment, consisting of annealing at 2000°F ± 50°F, followed by water quenching, can alleviate the condition. It is extremely important that the quench rates be high since failure to cool rapidly through the critical range may aggravate, rather than relieve, the condition.

An additional method of control is to vary the alloy selected. Since the mechanism of *IGA* in austenitic stainless steels depends on the precipitation of chromium carbides, the use of low carbon stainless (304L or 316L) reduces the susceptibility. Carbide precipitation is held to low intensity, and protection from *IGA* is achieved during short heating cycles. The material may become susceptible during long term heating in the 900 to 1500°F range.

Stabilized grades containing titanium (321) or columbium (347, 318, or 309C) are effective in minimizing *IGA* because they have a stronger affinity for carbon than chromium. Therefore, chromium is left in solution to impart corrosion resistance. The concentration of the stabilizer necessary to insure protection is controlled by the carbon content.

The minimum value for titanium grades is five times the carbon content and for columbium grades, ten times the carbon content.

7.8 EXFOLIATION

Exfoliation or lamellar corrosion is a special form of intergranular attack, which primarily affects aluminum and magnesium alloys. It is markedly directional and is characterized by attack of the elongated grains on a plane parallel to the rolled or extruded surface. This results in a characteristic delamination or stratification of the surface structure. Grain boundary attack can occur in susceptible alloys without exfoliation, however, as Lifka and Sprowls[24] have clearly shown (Fig. 7.15).

7.8.1 Causative Factors

The overall susceptibility of an alloy appears to be related to both the cooling rate and compositional differences, although the exact relationship is not clear.

Certain environmental factors increase the tendency towards exfoliation, such as the presence of chloride or bromide ions in the environment, higher temperatures, an acidic condition, and intermittent wetting and drying. The latter condition presumably results in the creation of insoluble

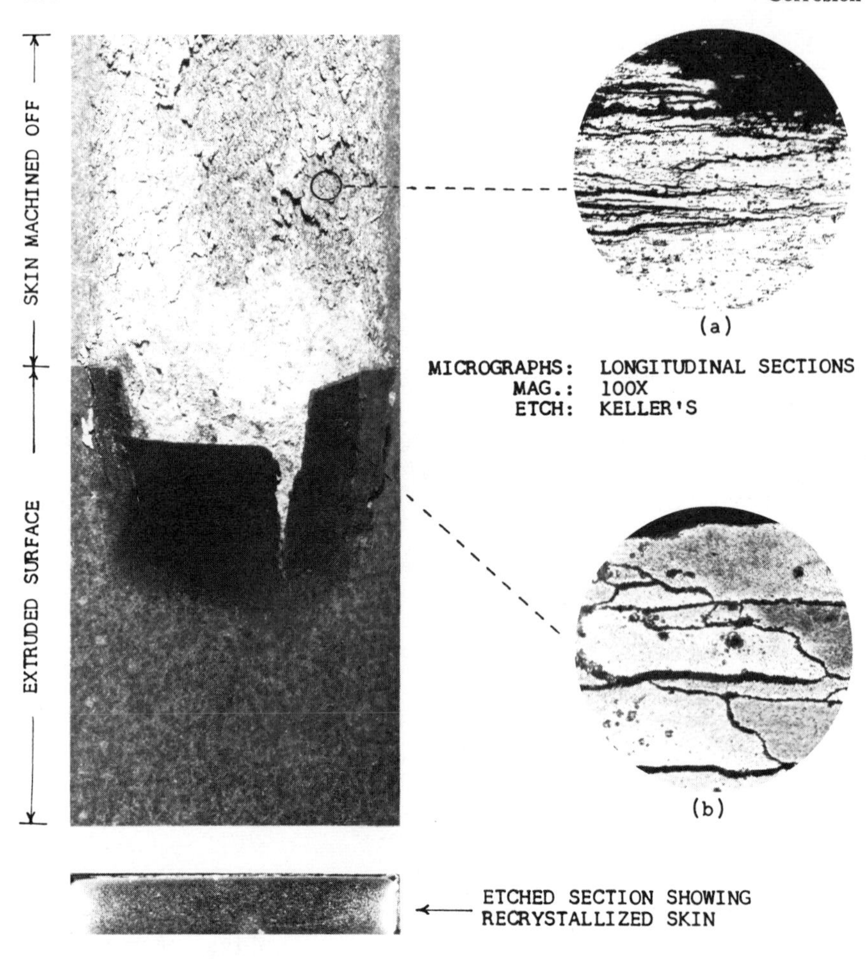

Fig. 7.15 Bar of 2024-T351 exposed for three years. (*a*) Fibrous bar interior exhibits exfoliation. (*b*) Recrystallized surface shows intergranular attack but no exfoliation.

corrosion products, which exert a stress normal to the surface. Lifka and Sprowls devised a test for exfoliation that utilizes all these factors to determine susceptibility. This test offers a rapid means for laboratory evaluation of alloys and appears to correlate well with natural environments.

7.8.2 Improving Exfoliation Resistance

Exfoliation can be alleviated by various methods. Bell and Campbell[25] suggested the use of extended aging cycles for an Al-Cu alloy. Organic and sprayed metal coatings[26] may also be utilized if they are complete and are

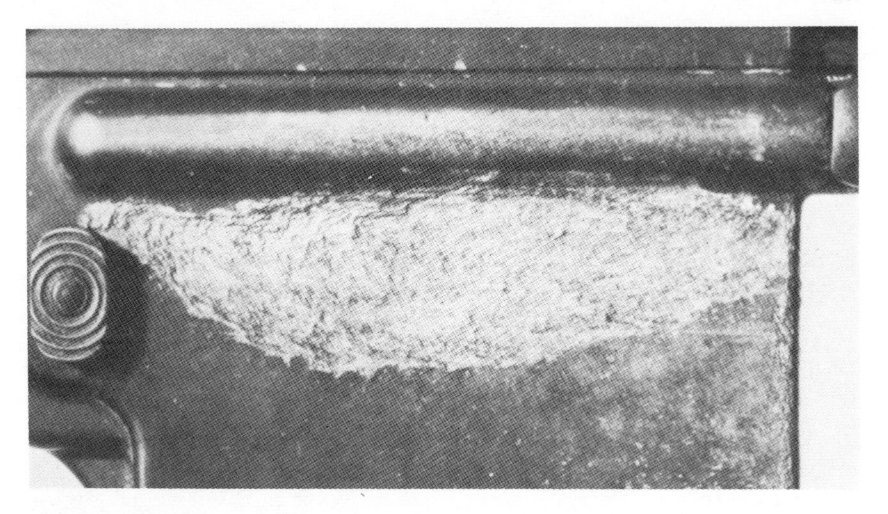

Fig. 7.16 Exfoliation resulting from the application of a graphite-bearing dry film lubricant to a 7075 component.

compatible. Figure 7.16 shows a 7075 aluminum component that was coated with a graphite-bearing dry film lubricant. The use of graphite-bearing lubricants promotes corrosion in a number of systems because it is cathodic and sets up a galvanic couple. Therefore, it would be extremely unwise to coat an exfoliation-sensitive alloy with any coating that would be cathodic to it.

Promoting an equiaxed grain structure on the surface or throughout the system would also decrease the suceptibility. This equiaxed surface layer, however, will not completely eliminate exfoliation, but will only retard it. Once the surface layer has been penetrated by corrosion, the banded layers below are vulnerable to attack, and the surface would be undermined and spalling of the surface would be expected. Conversely, overworking a susceptible alloy to improve the strength can be detrimental because it tends to create an anisotropic grain structure.

7.9 EROSION CORROSION

Erosion corrosion is exemplified by an increase in the corrosion rate caused by relative motion between the surface and the environment. Erosion corrosion has a distinctive appearance. The surface usually exhibits severe weight loss, and there are many scooped-out, rounded areas. The overall appearance, as shown in Fig. 7.17, presents a carved almost sculptured impression.

Fig. 7.17 Typical surface resulting from erosion corrosion.

Figure 7.18*a* illustrates the progressive thinning caused by erosion corrosion in a tubular section. The study also illustrates that a failure may have multiple causes. In this case, progressive thinning of the wall continued until the hoop stress generated by the internal pressure exceeded the yield strength of the material and the section failed by overstress, as shown in Fig. 7.18*b*.

7.9.1 Environmental Effects

Many alloys possess good corrosion resistance because insoluble surface films are created and maintained. Although increases in velocity occasionally decrease the rate of degradation because of kinetic factors, generally film maintenance is impaired and increased corrosion rates result. The effect of velocity is illustrated in Fig. 7.19, which resulted when high pressure gas, escaping past a seal, produced the damage shown. Similarly, turbulence can have an adverse effect on the protecting film. The work of Venzcel[27] indicates that erosion corrosion occurs when the Reynolds number required to produce turbulence is exceeded. The degree of degradation often depends on the alloy. According to LaQue,[28] iron modified cupro-nickel alloys have the highest resistance to turbulence; high zinc brasses and aluminum brass

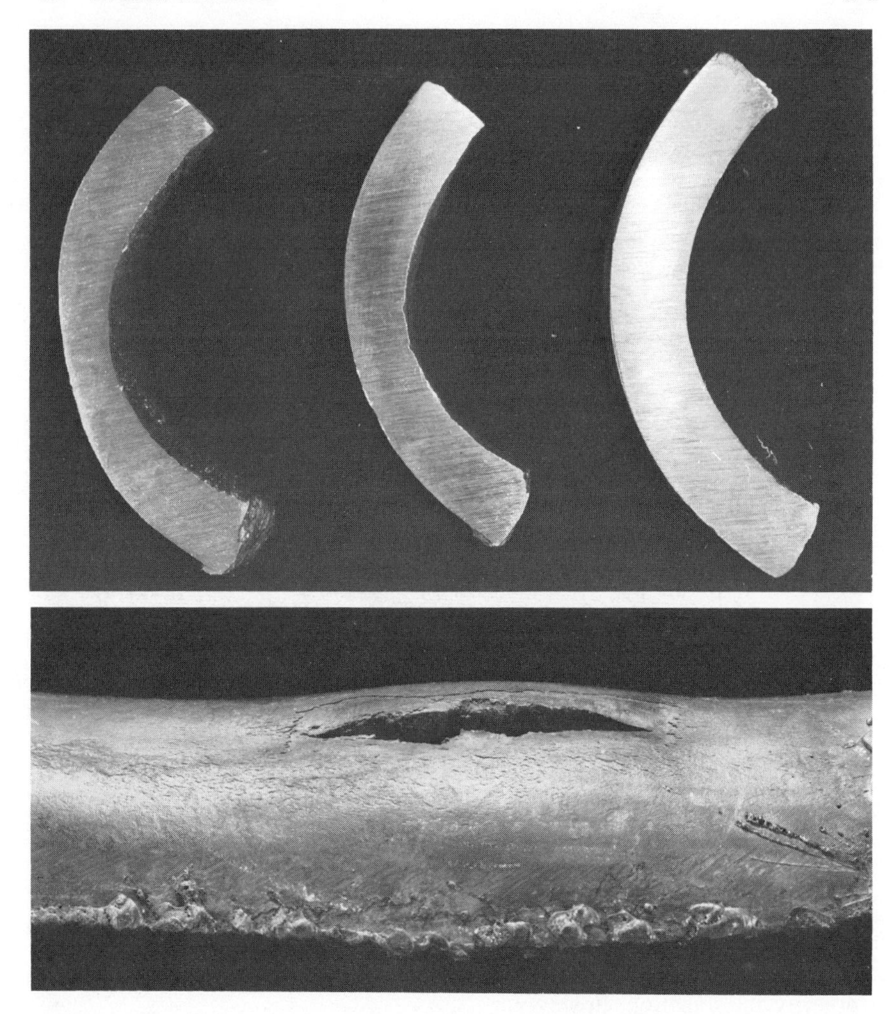

Fig. 7.18 Progressive thinning caused by erosion corrosion culminating in failure by overstress.

are less resistant; and copper, low zinc brasses, and silicon bronze show the least resistance.

Direct impingement can cause accelerated attack where the flow direction is normal to the metal surface, even when surfaces parallel to the flow direction are unaffected. One possible explanation is that the area stripped of the protective film by the impinging flow is anodic, relative to the remaining cathodic areas.

Fig. 7.19 Erosion damage caused by leakage of high pressure gas past a seal.

7.9.2 Prevention of Erosion Problems

The primary methods used to combat erosion corrosion are changing to a more resistive material or modifying the design to make it less severe. It is difficult to make absolute statements about the selection of materials for erosive application without testing them. However, materials that possess good corrosion resistance to the environment in question *and* high hardness will, in general, perform satisfactorily. Tests should always be performed to verify the selection, however.

With respect to design, the recommendations are not so vague. Changes made to reduce turbulence, such as the removal of protuberances, increased tube diameters, and reduced velocities, are beneficial. Modifications that reduce impingment effects, such as increased radii on elbows and angled baffle plates, are also helpful.

7.10 CAVITATION DAMAGE

Cavitation damage is a well-known type of degradation associated with rapid movement of liquid near the liquid metal interface. Stresses in the liquid environment induced by the rapid movement create large numbers of

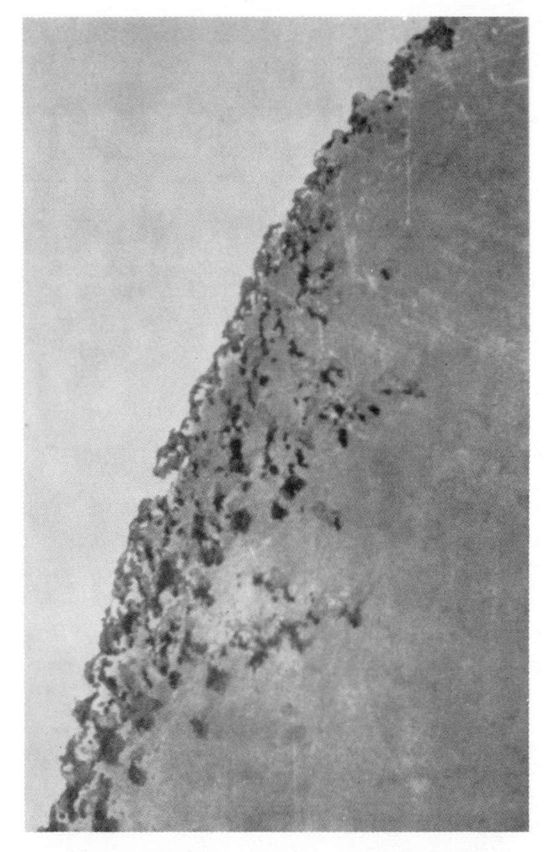

Fig. 7.20 Pitting and honeycombing as a result of cavitation damage.

momentary cavities. The duration of each cavity is short, and its collapse produces a shock wave that impinges on the metal surface. The shock wave, first produces a compressive stress on the surface and, upon reflection, a tensile stress normal to the surface. Because of the high rate of formation and decay of the cavities, damage is produced rapidly.

Cavitation damage superficially resembles pitting, but the surface appears considerably rougher and has many closely spaced pits. Microsections may show evidence of plastic deformation and honeycombing, as in Fig. 7.20.

7.10.1 Environmental Factors Affecting Cavitation

Cavitation damage in aqueous media can be intensified by a number of factors, such as the amount of entrained air and the presence of dust particles that act as nucleation sites for cavity formation. Temperature produces a mixed effect, with the maximum degradation occurring at approximately 125°F. Above this temperature, the rate of attack tends to decrease.

The corrosiveness of the media also plays an important role. Godfrey[29] states that the attack in sea water and brine is 50% greater than in distilled water. Unexpectedly, a sulphuric acid solution also only produces a 50% increase in attack. The degree of degradation may, therefore, be related to the pitting tendency of the environment, rather than to the general corrosivity. This pitting would cause maximum disruption of the streamlining and tend to create turbulence.

7.10.2 Control of Cavitation

The effects of cavitation damage can be minimized by the proper choice of materials, such as stainless steels instead of brasses. Hard facing of the surface with a resistant alloy is also beneficial.

In an opposite approach, rubber and other elastomeric coatings have been used to minimize cavitation. The elastic nature of the surface allows the reflection of the shock wave without intense damage.

Cathodic protection has also been applied successfully. Design changes that increase the pressure of the system or decrease the turbulence are also effective. Evans[30] describes a number of cases where marine propellers have been protected by this method.

7.11 STRESS CORROSION

7.11.1 Introduction

The problem of the premature failure of alloys by stress corrosion has long been recognized. At least fifty years ago, the ASTM held discussions on the season cracking of brass.[31]

The interaction of mechanical tensile stresses in a chemically aggressive environment causes cracking that is impossible to predict by independent measurements or corrosion testing. Four basic requirements are necessary to cause stress corrosion cracking: a susceptible alloy, an aggressive environment, applied or residual stresses, and time.

7.11.2 Theories of Stress Corrosion

Stress corrosion has been observed in many diverse materials and environments. Current literature contains numerous references to the fracture of titanium, brass, aluminum, magnesium, and various steel alloys.[32–40] With such a variety of distinctly different conditions of fracture, it is difficult to find common features. In work by Hoar and Hines,[41]

Hines,[42] and Parkins,[43] the following characteristics, consistent with this type of fracture, have been observed.

1. There must be simultaneous action of stress and corrosion; alternate application will not produce similar results. The fracture surfaces are macroscopically brittle, with little signs of ductile tearing.
2. The induction period necessary to produce crack initiation and pitting is relatively long, compared with actual crack propagation. Stress plays little part in the induction period; corrosion is the primary driving force. Embrittlement of this region surrounding the crack tip may also be a significant factor.
3. The mode of cracking can be intergranular or transgranular, but is predominantly one or the other. No generalizations can be made regarding the effect in heat treatment or compositions in the mode of cracking. However, one type of cracking occurs more readily for a given alloy.
4. The rate of attack is quite rapid at the crack tip and is much less rapid at the sides. The crack propagation process appears to be self-catalyzing; the much larger rate is sustained only at the advancing crack tip.
5. Conditions for cracking are specific as to alloy and environment. Specific ions are usually necessary to promote cracking conditions. Although many environments may produce similar corrosion rates, the susceptibility of different alloys to stress corrosion may be widely divergent. Certain corrodents that cause a relatively violent reaction with an alloy do not cause cracking. Thus the question about the contributing interaction of corrodent anions to failure is raised.

Tensile stress at the corroding surface is essential to stress corrosion cracking. These tensile stresses may be residual or applied. Both are detrimental to service life which is dependent on the magnitude of these stresses. Residual stresses may be more of a problem, however, since they are frequently concealed and therefore neglected when the safety factor of the component is designed.[44] These internally induced stresses arise from processing differences, such as nonuniform deformation during cold working, or unequal cooling rates from annealing temperatures. Other built-in stresses include those induced by press or shrink fits and other fastener deformations, such as rivets or bolts.

The many theories that have been advanced to explain stress corrosion cracking generally consider one of two mechanisms: (1) one in which electrochemical processes account for crack propagation by means of dissolution along preferential paths, and (2) an alternating step-fracture sequence where mechanical fracture is triggered by corrosion. To reach the characteristic high rate of propagation, it is necessary to satisfy either a type of corrosion process capable of high penetration rates in the region of the crack tip, or to find a means whereby corrosion can produce localized embrittlement and lead to intermittent mechanical fracture.

7.11.3 Preferential Path Theories

An early theory proposed by E. H. Dix[45] suggested that susceptible paths for corrosion follow under the action of tensile stresses. This mechanism allows corrosion to follow a path held open by stresses and explains integranular attack along grain boundaries, which are more reactive to corrosion. It is difficult to imagine, however, that the relatively high rates of crack propagation are caused solely by grain boundary differences.

The feature most common to recent theories of this type is the attempt to explain the very high and localized attack of metal at the tip of the crack. The difference in current density between the crack tip and the walls has been linked to the highly localized attack at the tip. The elastic strain energy and plastically strained material at the crack tip may account for some of the concentrated attack, if supplemented by other factors responsible for high current density.

Several dislocation theories were proposed to explain fracture observations. Swann and Pickering[46] deduced that moving dislocations transport solute atoms to the free surface of the metal. Movement of these atoms to active slip planes where a high density of dislocations is concentrated produces the chemical inhomogenity necessary to initiate and to propagate stress corrosion cracking. Observation of the serrated type stress-strain curve (Portevin-LaChatelier effect) associated with intermittent plastic flow illustrates that solute migration does occur with moving dislocations. Preferential corrosive attack[47] of components is a prerequisite, causing faults that fracture mechanically.

Varying alloy compositions have exhibited a marked difference in susceptibility to stress corrosion. Graf[48] advanced the theory that susceptibility should pass through a maximum as the concentration of a more noble alloying element is increased. Segregation of the alloying elements at grain boundaries, or separations from solution, establishes cathodic sites that accelerate attack at adjacent anodic areas. At higher alloying compositions, the process of dissolution does not occur, and continuous paths in the more active metal no longer exist. Of particular significance to stainless steels is recent work by Copson,[49] which studies the effects of nickel and chromium alloying. The matrix iron material is much more electrochemically active than either alloy, and suceptibility increases to a maximum at about 10% nickel, then decreases, as predicted by Graf. Copson also points out the fallacy of relying on "stable" protective films in steels.

When such films are ruptured by highly corrosive environments, such as ferric chloride solution, subsequent failure is quite rapid.

In apparent conflict with the previously mentioned electrochemical theory are studies of the wedging effect caused by buildup of corrosion products. Nielsen[50] obtained excellent photomicrographs of filled microcracks in stainless steels where the volume of corrosion product is twice that of the removed metal. Pickering[51] et al. studied stainless steels in dilute chloride solutions, at autoclave temperatures of 400°F. Pressures in the range of 4000 to 7000 psi have been generated by corrosion products that were identified as ferric and ferrous oxides, cubic chromium oxide, and complex chrome-iron oxides.

7.11.4 Mechanical Damage Theories

Keating[52] and Evans[53] advanced theories to explain high crack propagation rates in terms of alternating mechanical and electrochemical stages. Notches are produced by corrosion, it is theorized, then brittle fracture occurs along a mechanically weak path until an obstacle is reached. Crack propagation results from the repeated occurrence of this cycle. This theory is quite useful at grain boundaries, where it is known that precipitates are anodic to the maxtrix material and, therefore, are susceptible corrosion sites.[54] Logan[55,56] theorized that local yielding disrupts the polarized condition of surface oxide films to the extent that the crack sides remain passive and the advancing crack tip becomes active. The crack propagates if the strain at the tip ruptures the protective film faster than it can be repaired. Hoar and Hines[44] further speculated that this local yielding assisted in the actual removal of cations from the metal lattice.

Hines[57,58] discussed the conditions necessary for electrochemical attack at a crack tip. He postulated that an ohmic drop exists within the crack, whereby the crack tip and surface exhibit markedly different potentials and, therefore, different corrosion rates.

Mechanical mechanisms, including embrittlement and stacking fault theories, have been advanced to explain the cracking of normally ductile materials. Barnett and Troiano[59] described the steps to achieve hydrogen-induced static fatigue of high strength AISI 4340 steel. Under sustained load, the process is one of crack initiation and slow crack growth, followed by cataclysmic brittle fracture. The stages were identified and followed, using photographic and electrical resistance measurements. Kim and Loginow[60] recently reported measurements of hydrogen intake and permeability in a Ni-Cr-Mo steel. A concept of hydrogen trapping was proposed to explain the higher concentration in steels of increasing yield strength of the same composition. Evans[61] advanced an embrittlement theory in which hydrogen formed near the crack tip could diffuse into the highly strained crack tip and cause propagation of brittle fracture. Reduction of hydrogen ions at the crack tip is likely, in light of the lower potential and higher acidity exhibited. Brown[62] reported pH and metal ion changes at the crack tip, using liquid nitrogen to freeze the solution in a cracked specimen. These data are used to support the hydrogen embrittlement theory for high strength steels. It must be noted, however, that hydrogen cannot be responsible in all cases, since metal systems that are more electrochemically noble than the hydrogen potential also suffer stress cracking.

Further evidence supporting the alternating electrochemical-mechanical theory of crack propagation has recently been added by McEvily and Bond[63] and Logan et al.[64] The former work, on brass, contains excellent replicas of electron micrographs, clearly showing discontinuous step-wise crack propagation. Rupture of the protective tarnish has been shown to concentrate the effects of environment. Subsequent work with aluminum, zinc, and magnesium alloys showed similar results. Logan's work with 321 stainless steels has produced electron micrographs showing the discontinuous process of fracture. It was thought that the tendency of austenitic stainless steel to work harden produces fracture increments too small to be detected by simple elongation measurements.

7.11.5 Application of Fracture Mechanics to Stress Corrosion Cracking

In most alloys, stress corrosion begins with some form of localized attack, such as pitting or intergranular attack, which raises the effective stress and initiates a crack.

Since one of the prerequisites to stress corrosion cracking is the presence of tensile stresses, and since there is now an existing crack, the fracture mechanics equations developed in Chapter 2, can be applied. By using these equations, we can quantitatively describe the stress state at the crack tip by the stress intensity factor, K.

Tiffany,[65] Johnson,[66] and Brown[67] were among the first to apply the concepts of fracture mechanics to stress corrosion studies. Tiffany and Johnson used tensile specimens and Brown used a precracked cantilever beam specimen, which required the relatively simple apparatus shown in Fig. 7.21. Using this approach, a precracked specimen is loaded to some initial value of stress intensity, K_{Ii}, which can be calculated from the equation

$$K_{Ii} = \frac{4.12M(1/\alpha^3 - \alpha^3)^{1/2}}{BW^{3/2}}$$

where

$$\alpha = 1 - \frac{a}{W}$$

$$M = \text{moment}$$
$$a = \text{crack depth}$$
$$W = \text{specimen depth}$$
$$B = \text{specimen width}$$

For any single specimen, as the crack length increases by stress corrosion cracking, the stress intensity at the crack tip increases until the critical value K_{Ic} is reached, and final failure occurs by brittle fracture, as shown in Fig. 7.22.

If a number of specimens are loaded to various initial levels of K_{Ii} and the respective times to failure are recorded, a threshold value of stress intensity, K_{Iscc}, is observed, below which no crack growth is noted. This behavior is illustrated in Fig. 7.23.

The significance of this threshold value becomes apparent when one considers that when K values higher than this value exist in a component, subcritical crack growth ocurs. Therefore, the lower the K_{Iscc} value is the more susceptible the material.

The use of precracked cantilever beam, or other similar specimens, permits the analysis of test results in terms of fracture mechanics, if the yield strength and fracture toughness are such that plane strain conditions can be achieved. This type of analysis is desirable for two reasons, First, it separates the crack initiation and crack propagation stages of stress corrosion. Titanium alloys were formerly believed to be immune to stress corrosion; however, the use of precracked specimens has shown them to

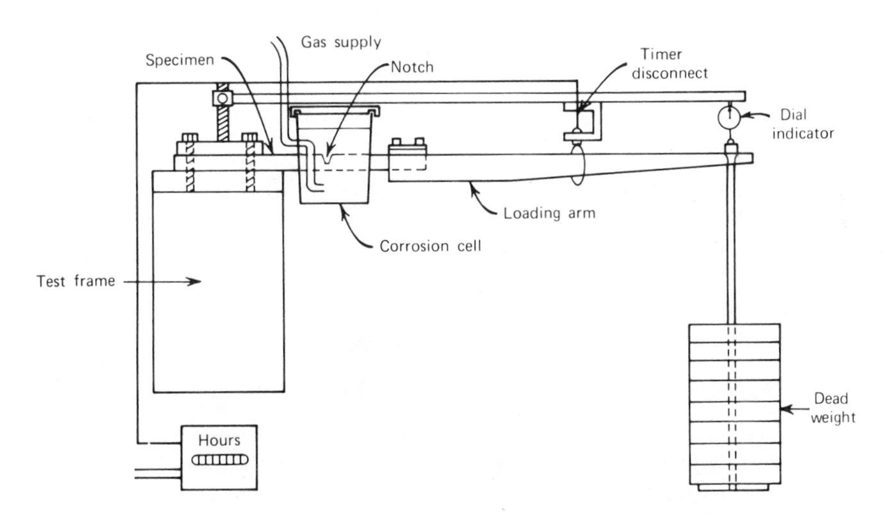

Fig. 7.21 Schematic drawing a fatigue-crack cantilever-beam test apparatus.

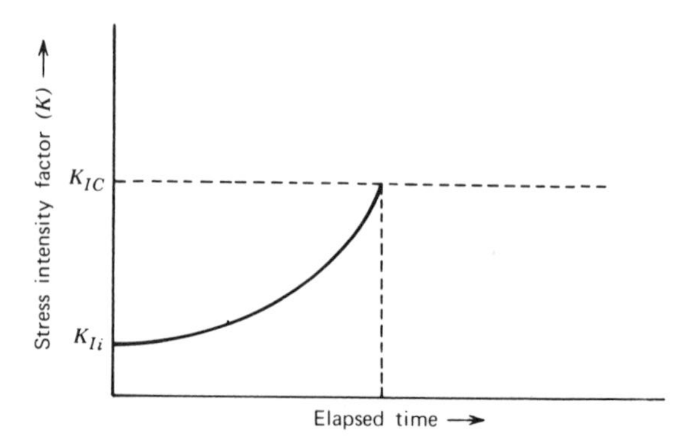

Fig. 7.22 Increase in stress intensity factor (K) at the crack tip to a critical value.

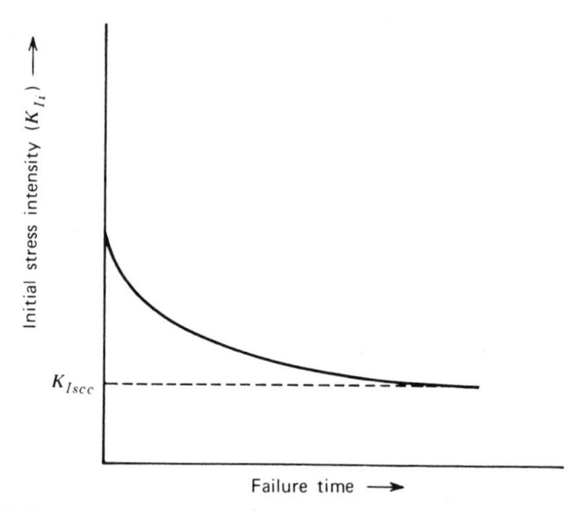

Fig. 7.23 Typical stress corrosion curve showing fracture time as a function of initial stress intensity factor.

be susceptible. Such analysis also allows one to observe the effects of stress in terms of behavior at the crack tip, rather than on the body as a whole.

7.11.6 Appearance of Stress Corrosion Cracks

Stress corrosion failures generally exhibit little ductility and have the macroscopic appearance of a brittle fracture. There may be multiple cracks originating from the surface, but failure usually results from the progression of a single crack on a plane normal to the main tensile stress.

In austenitic stainless steels, cracks are usually transgranular and are frequently associated with a specific crystallographic plane. These alloys also exhibit intergranular cracking on certain media, notably caustic solutions and highly oxygenated chloride solutions.

Intergranular cracking is the predominant mode of failure for martensitic stainless steels. However, transgranular cracking has been observed when these alloys were tempered below 850°F.

In high strength steels, the crack path is intergranular, as shown in Fig. 7.24. It is widely believed that stress corrosion cracking in these materials is related to a hydrogen embrittlement mechanism, as a result of hydrogen being generated at the crack tip as a corrosion product. From a metallographic viewpoint, there is a great deal of similarity between the two. Note

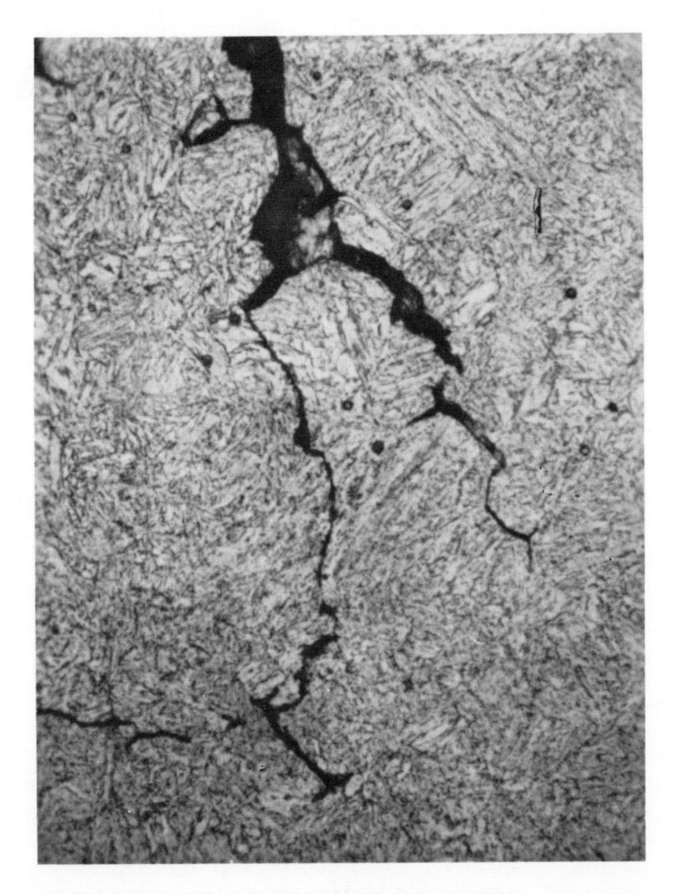

Fig. 7.24 Photomicrograph of crack tip showing intergranular crack path of stress corrosion failure.

Fig. 7.25 Electron microscope fractograph of a stress corrosion failure illustrating the intergranular nature of the crack. High strength steel (AMS 6427) (4150×).

the resemblance between Figs. 7.24 and 9.6. This is also true on a fractographic basis. Figure 7.25 is an electron fractograph of a stress corrosion fracture surface. There is a striking similarity between it and the fractograph shown in Fig. 9.8.

Cracks in brasses may be either intergranular or transgranular, depending on several variables. Gilbert described the crack morphology in several media for α and β brasses and for aluminum bronzes.

Stress corrosion cracking in aluminum alloys is characteristically intergranular. Al-Cu, Al-Zn-Mg, and Al-Mg alloys are most often affected.

Of the magnesium alloys, Mg-Al and Mg-Al-Zn are suceptible to stress corrosion damage; attack is usually transgranular.

7.11.7 Prevention

Since stress corrosion simultaneously requires the application of stress, the presence of a susceptible alloy, and a specific environment, stress corrosion problems would be minimized by changing any or all of these factors.

1. *Reduction of the stresses.* This can be accomplished in a variety of ways. Applied tensile stresses can be reduced by increasing the thickness of the section so that the net section stress is decreased, or by reducing the

load. If the weight of the component is critical, the effective tensile stresses can be reduced by induced compressive elastic stresses at the surface, such as those created by shot peening, glass bead peening, or various other surface treatments.

Residual processing stresses can be eliminated by stress relief annealing. The precise temperature at which this should be conducted depends on the alloy.

2. *Modification of the environment.* The elimination of trace chloride ions and oxygen is very effective with austenitic stainless steels, for example. Inhibitors can be added to a closed system if they are added in sufficient concentration to be effective.

3. *Substitution of one alloy for another.* This is often the most expedient solution, and may also be the only solution when neither the environment nor the stress profile can be modified. This can usually be accomplished if some trade-off in properties is permitted.

The primary material attribute to be considered in material selection is a high K_{Iscc} value. This places the primary emphasis on propagation rather than initiation but this is undoubtedly correct since the propagative stage can be so rapid.

4. *Anodic or cathodic protection.* These can be used with some restrictions, to protect a structure against stress corrosion cracking. Cathodic protection should not be applied to high strength steels or to other alloys susceptible to hydrogen embrittlement, since the application of cathodic currents increases the rate of hydrogen evolution. The existence of cracks or crevices in the structure also limits the efficacy of external polarization, since the applied currents are rapidly attentuated within the crack. With large crevices, the required level of protection may not be attainable.

REFERENCES

1. M. G. Fontana and N. D. Greene, *Corrosion Engineering*, McGraw-Hill, New York, 1967, p. 2.
2. O. Bauer and O. Vogel, *Mitt. Deut. Materialprufungsanst.*, Vol. 6, 114 (1918).
3. W. G. Whitman and R. P. Russell, *Ind. Eng. Chem.*, Vol. 16, 276 (1924).
4. M. G. Fontana and N. D. Greene, *op. cit.*, p. 36.
5. G. Schikorr, *Korrosion*, Vol. 16, 43 (1963).
6. M. G. Fontana and N. D. Greene, *op. cit.*, p. 50.
7. N. D. Greene and M. G. Fontana, *Corrosion*, Vol. 15, 25t (1959).
8. A. P. Bond, G. F. Bolling, and H. A. Domian, *J. Electrochem. Soc.*, Vol. 112, 178c (1965).
9. A. P. Bond, G. F. Bolling, and H. A. Domian, *J. Electrochem. Soc.*, Vol. 113, 773 (1966).
10. M. G. Fontana and N. D. Greene, *op. cit.*, p. 57.

11. W. D. France and N. D. Greene, *Corrosion*, Vol. 24, No. 8, 247–251 (August 1968).
12. G. F. Bush, *Corrosion of Automobiles*, ASM Failure Analysis Conference, Philadelphia, June 1968.
13. U. R. Evans, *The Corrosion and Oxidation of Metals*, Arnold, London, 1960, p. 107.
14. G. J. Shafer, J. R. Gabriel, and P. K. Foster, *J. Electrochem. Soc.*, Vol. 107, 1002 (1960).
15. H. L. Espy, *Met. Progr.*, September, 109–115 (1964).
16. H. T. Shirley, *Corrosion*, L. L. Schreir, Ed., Wiley, New York, 1963, p. 362.
17. G. M. Ugiansky and G. A. Ellinger, *Corrosion*, Vol. 24, No. 5, 134 (1968).
18. W. D. Clark, *J. Inst. Met.*, Vol. 73, 263 (1947).
19. W. B. Brooks, *Corrosion*, Vol. 24, No. 6, 171 (June 1968).
20. N. D. Greene and N. Ahmed, to be published.
21. U. R. Evans, *op. cit.*
22. S. Alm and R. Kiessling, *J. Inst. Met.*, Vol. 91, 190 (1963).
23. Welding Research Council, *Bull. No. 93*, January 1964.
24. B. W. Lifka and D. O. Sprowls, *Corrosion*, Vol. 22, 7–15, (January 1966).
25. W. A. Bell and H. S. Campbell, *J. Inst. Met.*, Vol. 89, 464–471 (August 1961).
26. V. E. Carter and H. S. Campbell, *J. Inst. Met.*, Vol. 89, 472–475 (August 1961).
27. J. Venzcel, L. Knutsson, and G. Wranglen, *Corr. Sci.*, Vol. 4, 1 (1964).
28. F. L. La Que, *Failure Analysis*, J. A. Fellows, Ed., ASM, Ohio, 1969, pp. 229–326.
29. D. J. Godfrey, *Corrosion*, Vol. 1, L. L. Schreir, Ed., Wiley, New York, 1963, p. 8, 97.
30. U. R. Evans, *The Corrosion and Oxidation of Metals*, 1st Supplement, St. Martins, New York, 1968.
31. "Topical Discussion on Season and Corrosion Cracking of Brass," *Proc. Am. Soc., Test. Mater.*, Vol. XVIII, Part II, 147–219 (1918).
32. R. Newcomer, H. C. Tourkakis, and H. C. Turner, *Corrosion*, Vol. 21, No. 10, 307–315 (October 1965).
33. M. H. Peterson, B. F. Brown, R. L. Newbegin, and R. E. Groover, *Corrosion*, Vol. 23, No. 5, 142–148 (May 1967).
34. H. H. Johnson and J. Leja, *Corrosion*, Vol. 22, No. 6, 178–189 (June 1966).
35. F. A. Champion, *The Assessment of the Susceptibility of Aluminum Alloys to Stress Corrosion*, Symposium on Stress Corrosion Cracking of Alloys, (STP64) ASTM, 358–378 (1944).
36. S. W. Dean and H. R. Copson, *Corrosion*, Vol. 21, No. 3, 95–101 (March 1965).
37. H. H. Johnson and A. M. Willner, *Appl. Mater. Res.*, Vol. 4, No. 1, 34–40 (January 1965).
38. W. A. Van Der Sluys, *Trans. ASME; J. Basic Eng.*, 28–34, March (1967).
39. M. Henthorne and R. N. Parkins, *Corros. Sci.*, Vol. 6, 357–369 (1966).
40. G. G. Hancock and H. H. Johnson, *Mater. Res. Stand.*, Vol. 6, No. 9, 431–435 (September 1966).
41. T. P. Hoar and J. G. Hines, *J. Iron Steel Inst.*, Vol. 182, 124 (February 1956); Vol. 182, 166 (October 1956) (second part).
42. J. G. Hines, *Corrosion*, L. L. Shreir, Ed., Vol. 1, Wiley, New York, 1963, p. 83.
43. R. N. Parkins, *Stress Corrosion Cracking and Embrittlement*, W. D. Robertson, Ed., Wiley, New York, 1966, pp. 140–157.
44. R. H. Copson, "Effect of Mechanical Factors on Corrosion," *The Corrosion Handbook*. H. H. Uhlig, Ed., Wiley, New York, 1948, p. 569.
45. E. H. Dix, Jr., *Trans., Am. Inst. Min. and Metal. Eng. Inst. Met. Div.*, Vol. 137, 11 (1940).

46. P. R. Swann and H. W. Pickering, *Corrosion*, Vol. 19, No. 11, 369t–372t (November 1963).
47. A. J. Forty, *Physical Metallurgy of Stress Corrosion Fracture*, T. H. Rhodin, Ed., Wiley-Interscience, New York, 1959, pp. 99–116.
48. L. Graf, *Stress Corrosion Cracking and Embrittlement*, W. D. Robertson, Ed., Wiley, New York, 1966, pp. 48–60.
49. H. R. Copson, *Physical Metallurgy of Stress Corrosion Fracture*, T. H. Rhodin, Ed., Wiley-Interscience, New York, 1959, pp. 247–273.
50. N. A. Nielson, *op. cit.*, pp. 121–154.
51. H. W. Pickering, F. H. Beck, and M. G. Fontana, *Corrosion*, Vol. 18, 230t–239t (June 1962).
52. F. H. Keating, *Internal Stresses in Metals and Alloys*, Institute of Metals, London, 1948, p 311.
53. U. R. Evans, *Corrosion*, Vol. 7, No. 238–244 (1951).
54. R. N. Parkins, *Chem. Ind.*, February 28, 1953, 180–184.
55. H. L. Logan, *J. Res. Nat. Bur. Stand.*, Vol 48, No 2, 99–105 (February 1952)
56. H. L. Logan, *Physcial Metallurgy of Stress Corrosion Fracture*, T. H. Rhodin, Ed , Wiley-Interscience, New York, 1959, p. 295.
57. J. G. Hines, *Corros. Sci.*, Vol. 1, 2–20, (1961).
58. J. G. Hines, *Corros. Sci.*, Vol. 1, 21–48 (1961).
59. W. J. Barnett and A. R. Troiano, *Trans. AIME, J. Met.*, April, 486–494 (1957).
60. C. D. Kim and A. W. Loginow, *Corrosion*, Vol. 24, No. 10, 313–318 (October 1968).
61. U. R. Evans, *Stress Corrosion Cracking and Embrittlement*, W. D. Robertson, Ed., Wiley, New York, 1966, pp. 158–162.
62. B. F. Brown, *The Iron Age*, November, 21, 1968, 28–29.
63. A. J. McEvily, Jr. and A. P. Bond, *J. Elect. Soc.*, Vol. 112, 131–138 (February 1965).
64. H. L. Logan, M. J. McBee, and D. J. Kakan, *Corros. Sci.*, Vol 5, 729–730 (1965).
65. C. F. Tiffany, *Mater. Res. Stud.*, Vol. 4, No. 3, 107 (1964).
66. H. H. Johnson and A. M. Willner, *Appl. Mater. Res.*, Vol. 4, 34 (1965).
67. B. F. Brown and C. D. Beachem, *Corros. Sci.*, Vol. 5, 745–750 (1965).

8

Wear

8.1 GENERAL

Wear is probably the most important factor in the deterioration of machinery with moving components, often limiting both the life and the performance of such equipment. Therefore, the economic consequences of wear-induced failure are of major concern.

Wear is the loss of material from the surface by transfer to another surface or the creation of wear debris. However, the rearrangement of the material is equally important since surface matter is sometimes simply displaced to another area, but serious defects are, nevertheless, created in the surface. We are concerned with these surface defects because they may act as foci for other types of damage, such as fatigue or stress corrosion.

Rabinowicz[1] states that modern research has established four primary wear mechanisms: adhesive wear, plastic deformation (abrasive wear), spalling or pitting, and chemical or corrosive action. The question of classification is more than an academic convenience. The correct solution to a troublesome wear problem may depend strongly on the identification of a specific basic mechanism. Applying a thin film boundary lubricant has little beneficial effect if the damage is caused by the presence of abrasive particles, but is extremely effective if the damage is caused by adhesion.

Wear is affected by a variety of conditions, such as the type of lubrication, loading, speed, temperature, materials, surface finish, and hardness. Consequently, the visible manifestations of wear may be the result of a combination of primary forms. Dorinson[2] indicates that it may be difficult to recognize the various damages types. Usually, however, one type of damage is predominant.

8.2 ADHESIVE WEAR

Adhesive wear, also referred to as galling or seizing, is the most common type of wear and since expected, contributes little to the occurrence of sudden failures. Adhesion results from the welding together and subsequent shearing of asperities, which occurs when two metal surfaces are slid past each other. Theoretically, the load (W) between two bodies in sliding contact is divided among a number of asperities. The total area (A) of these asperities is related to the yield strength of the softer material (p) and the applied load (W) as follows:

$$A = \frac{W}{p} \tag{1}$$

The frictional force (F) is then equivalent to the force required to shear these adherent areas or

$$F = As \tag{2}$$

where s is the shear strength of the bond.
Rearranging equation 1, and dividing by 2

$$f = \frac{F}{W} = \frac{s}{p} \tag{3}$$

where f = coefficient of friction
This reaffirms the classic observation that the coefficient of friction is independent of the load or the *apparent* area of contact. This relationship must remain approximate since some assumptions have been made. First, a small plowing contribution has been neglected. Second, the terms s and p do not reflect the shear strength and yield strength, respectively, of actual bulk material since the material in the surface is subject to considerable local pressure and heating, which may alter the properties markedly.

Failures that do occur through adhesive wear can usually be traced to the improper selection of materials or a malfunction of the lubrication system.

8.2.1 Characteristics of Adhesive Wear

When two metal surfaces come in contact, the high local pressures caused by interfering asperities causes rupturing of the interfering surface films. As the clean metal surfaces contact each other, there is a natural tendency towards adhesion, which is a result of electrostatic imbalances on the surface. If the surfaces are in motion, relative to one another, the adhesions formed will be broken. The plane of fracture may be the original interface, but it often occurs in the weaker of the base metals.

Rabinowicz and Tabor,[3] in an autoradiographic study of wear and lubrication, found that if the surfaces are not lubricated, the transfer of material increases linearly with the load applied normal to the surface. The presence of a lubricant reduces the coefficient of friction as well as the

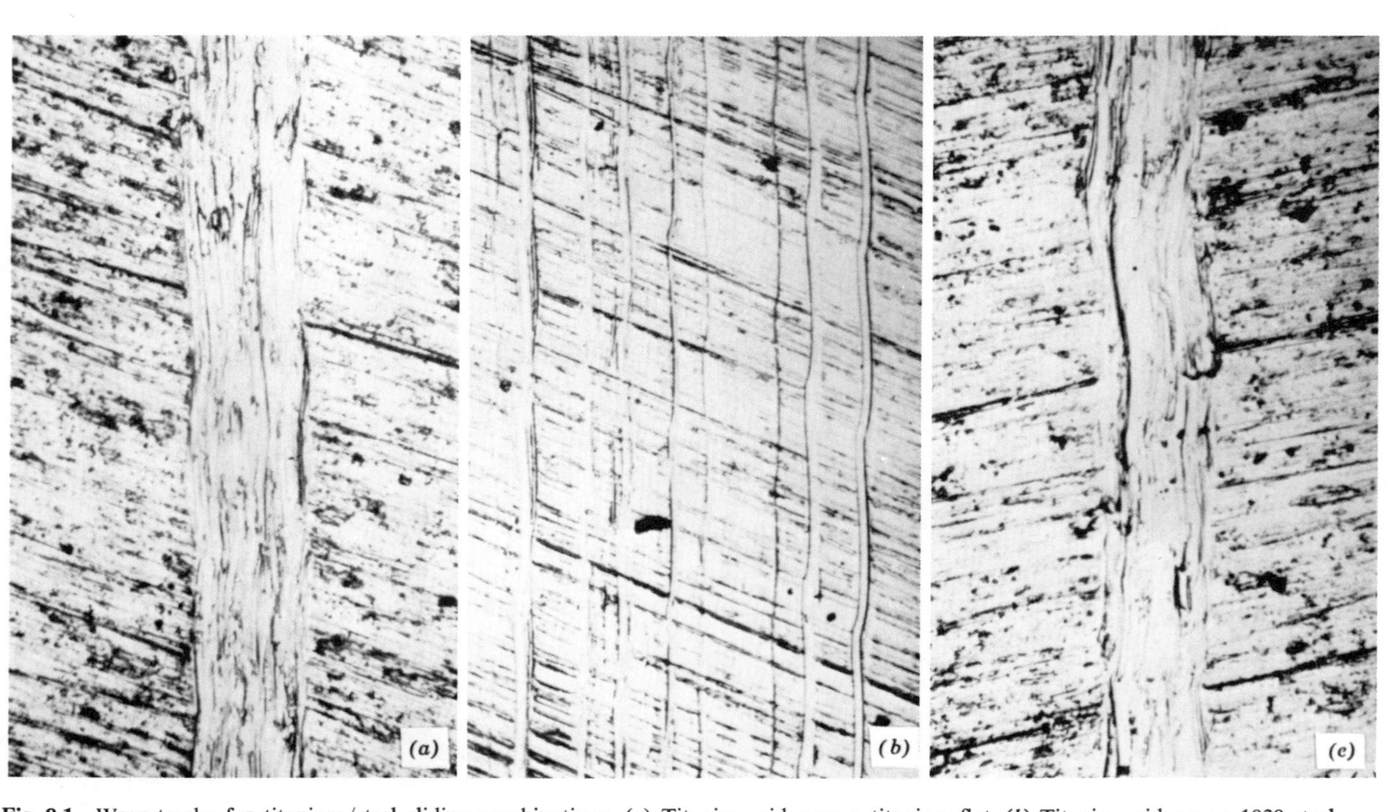

Fig. 8.1 Wear tracks for titanium/steel sliding combinations. (*a*) Titanium rider on a titanium flat. (*b*) Titanium rider on a 1020 steel flat. (*c*) A 1020 steel rider on a titanium flat (200X).

amount of transferred material. The reduction in the amount of transferred material is undoubtedly the result of reducing the number and/or size of the adhesions, since Sikorski[4] showed that the coefficient of adhesion is directly related to the coefficient of friction for many metals.

It is possible, based upon a number of such studies, to make some general statements regarding adhesive wear.

1. The amount of wear is proportional to the normal load.
2. The amount of wear is proportional to the distance slid.
3. The amount of wear is proportional to the hardness of the surface being worn.

These observations suggest that the solutions to many wear problems may be found by altering these parameters.

Examining metallographic specimens that have been subjected to conditions promoting adhesive wear shows that there is a definite transport of material, usually from the softer material to the harder one.

Figure 8.1 shows photomicrographs of the wear track of various steel/titanium sliding combinations. The photomicrograph of the titanium rider on a titanium flat (*a*) looks remarkably like (*c*), which illustrates the steel

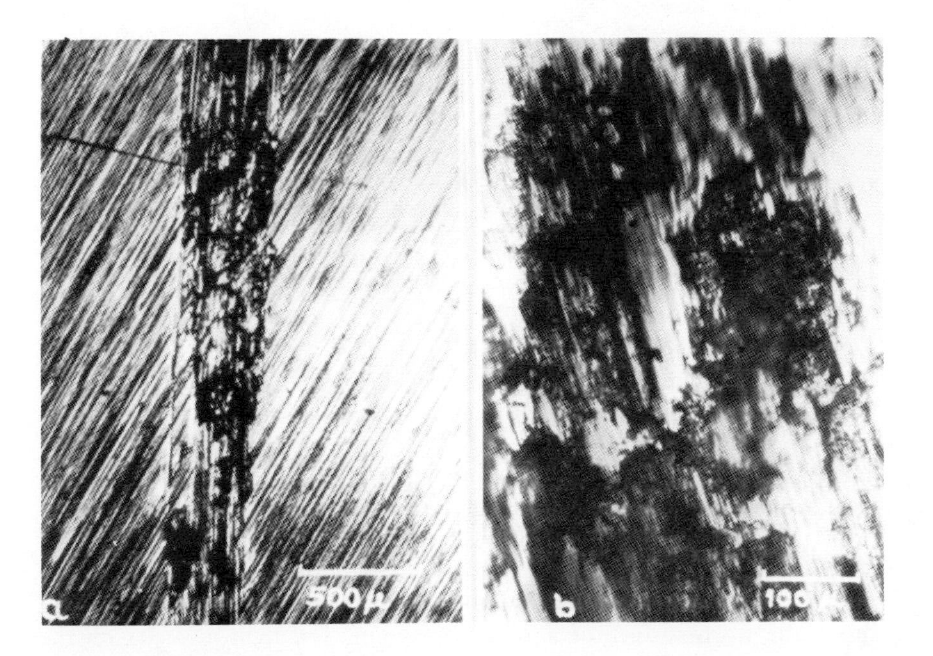

Fig. 8.2 Massive adhesions caused by a rider after 12 traverses on a steel surface (from Ref. 2). (*a*) 50X. (*b*) 500X.

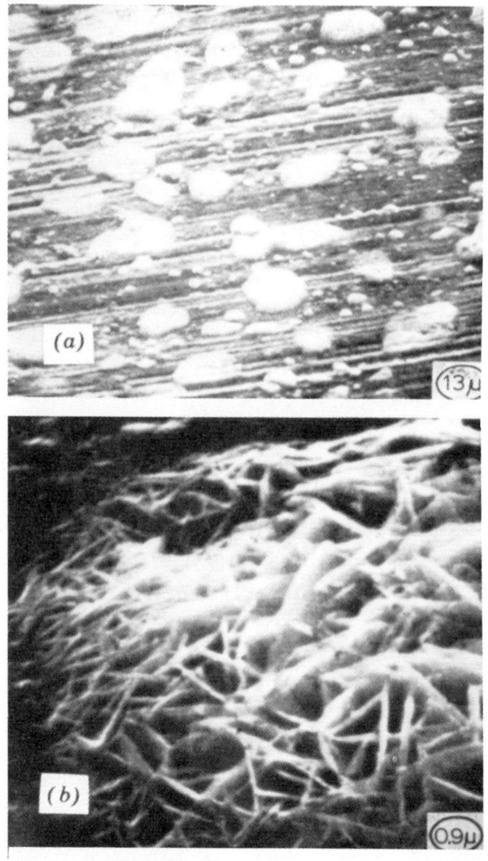

Fig. 8.3 Photographs taken with Scanning Electron Microscope (SEM). Worn surface of 70–30 brass pin (from Ref. 5). (*a*) 550X. (*b*) 800 X.

rider on a titanium flat. This occurs because the surface of the steel rider becomes covered with adhering titanium particles, producing a titanium on titanium situation.

Figure 8.2 illustrates the substantial nature of the adhesions that can occur. Bhattacharyya[5] suggested that the scanning electron microscope (SEM) would be a valuable tool for examining wear surfaces. The greater depth of focus with the SEM permits the ready differentiation of surface features and the resolution of minute details. The photograph in Fig. 8.3 shows the worn surface of a 70–30 brass pin. The light particles on the surface are the ZnO crystals which resulted from the oxidation of zinc during frictional heating. In an enlarged view the acicular appearance of the ZnO crystals is evident.

8.2.2 Minimixing Adhesion Effects

To minimize the effects of adhesive wear, there are a number of factors to consider. Assuming that the lubrication system is satisfactory, the selection of materials should be studied. Research and experience suggest the application of the following rules to minimize adhesive wear.

1. Materials forming the couple should have low solid solubility. Roach, Goodzeit, and Hunnicut[6] consider alloying ability to be the most important factor affecting strength in adhesive bonds.
2. Materials should be as hard as possible, within the limits set by the other engineering requirements. In general, hard materials are more resistant to plastic deformation and are usually characterized by a low coefficient of adhesion.

As Rabinowicz points out, silica is the primary contaminant apt to be introduced into a sliding system. Its hardness is about 800 kg/mm^2 ($\sim$ 60 R_c). Therefore, it is best to have the material hardness at or above this level.

Sikorski[4] summarized the factors influencing the adhesion of metals; this summary is reproduced in Table 8.1.

Table 8.1 Factors Influencing the Adhesion of Metals[a]

Property	Kind or Magnitude	Coefficient of Adhesion
1. Surface contamination	High	Low
2. Crystal structure	Cubic	High
	Hexagonal	Low
3. Work-hardening coefficient	High	High
4. Purity	High	[b]
5. Hardness	High	Low
6. Elastic modulus	High	Low
7. Melting point	High	Low
8. Recrystallization temperature	High	Low
9. Atomic radius	Small	Low
10. Surface energy	High	[c]

[a] From Ref. 8.

[b] Preliminary results obtained on zinc (Zn), which has a hexagonal close-packed structure, indicate that purity does not seem to affect the coefficient of adhesion to any large extent. In contrast, a large effect of purity was observed in the case of copper (Cu), which has a face-centered cubic structure.

[c] From the physico-chemical point of view, a high surface energy would suggest a high adhesion strength. From the mechanical point of view, a high surface energy often implies a high hardness, which, generally, is accompanied by a low coefficient of adhesion.

Examining the data shows that the factors tending to decrease adhesion are those that minimize the degree of plastic deformation occurring at the interface. Sikorski[8] discussed this aspect of adhesion in detail. Surface contamination is also important; if the surface is clean, adhesion is relatively high. If adherent oxides, or other films that prevent intimate metal/metal contact are present, adhesion is low. Consequently, metals that form tough adherent surface oxides tend to have lower adhesion.

8.3 ABRASIVE WEAR

Although adhesive wear is the most common form of wear damage, abrasive wear is more dangerous. It may occur suddenly, with the introduction of a contaminant, for example, thereby producing high wear rates and catastrophic failure of a system.

Abrasive wear occurs when two surfaces, one of which is harder and rougher than the other, are in sliding contact. Similar damage may also result when hard, abrasive particles are imbedded in a softer matrix. The damage occurs because of a plowing action; the harder particles or asperities create grooves or furrows in the softer material. An example of such plowing is shown in Fig. 8.4.

The material formerly contained in the furrows is transformed into wear particles that are usually loose and nonadhering. This loss of material from the surface explains the high wear rates typical of abrasive wear.

8.3.1 Types of Abrasive Wear

There are three predominant types of abrasive wear: abrasion caused by gouging, by grinding, and by erosion. Although they possess similar overall characteristics, each is sufficiently different from the other to warrant individual discussion.

Gouging, as the name implies, usually results in massive physical deformation of the surface. It is the result of an abrasive of relatively large diameter, which is driven in and along the surface under heaving loading. The nature of the loading requires that the abrasive be well supported. An example of gouging on a 316 stainless steel surgical plate is shown in Fig. 8.5. This was caused by contact with surgical tools during the insertion of the plate.

Abrasive wear due to grinding occurs when two surfaces are in sliding contact, and abrasive grains are present between them. The grains may be fixed to one surface or to an integral part of that surface (two-body wear), or loosely held between the surfaces (three-body wear). According to Rabinowicz,[9] two-body wear does not occur if the hard sliding surface is smooth and three-body wear does not occur if the particle is small or

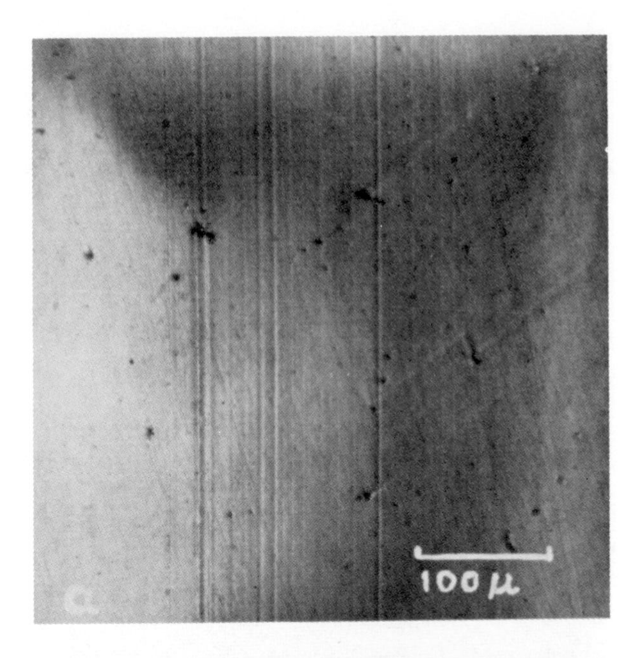

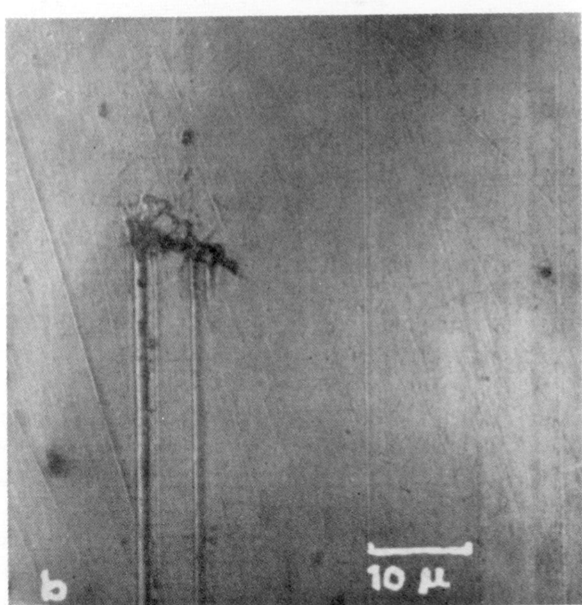

Fig. 8.4 (*a*) Plowing of a polished steel disk by a rider (10X). (*b*) Enlarged view shows particle trapped by rider (90 ×) (from Ref. 2).

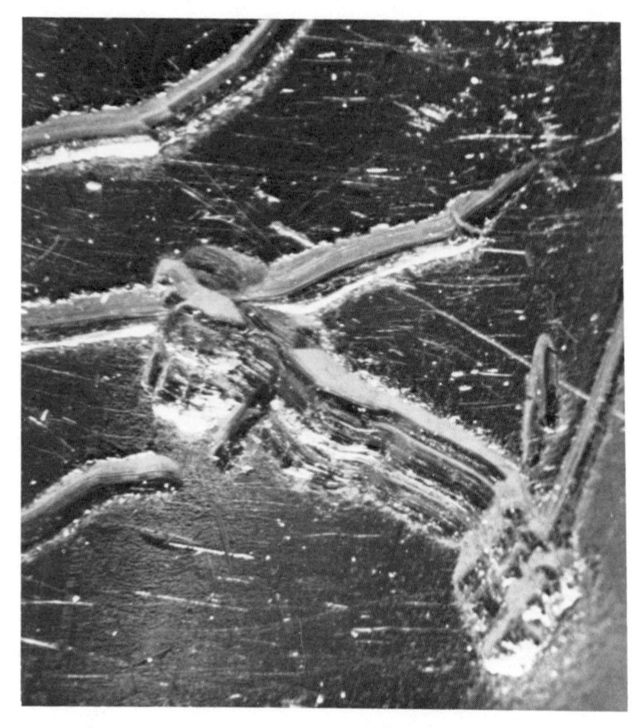

Fig. 8.5 Gouging on the surface of a stainless steel surgical plate (20X).

softer than the sliding members of the couple. Abrasive wear occurs when the abrasive grains are fractured under load producing sharply faceted fragments which remove material by plowing and scratching the surface.

Erosive wear occurs as a result of the impingement of abrasive grains suspended in a fluid, either gaseous or liquid. Each contact produces a small scar in the metal surface. The wear rates are relatively low under normal circumstances, but can reach high values at elevated temperatures where the yield strength of the material is reduced or at extreme velocities. In some situations, the medium itself may be corrosive and may contribute to the overall degradation of the surface. Chapter 7 discusses corrosion effects in erosion.

8.3.2 Minimizing Abrasive Wear

Typical wear rates, in applications where abrasive wear is observed, are shown in Table 8.2. Note that applications involving gouging are characterized by extremely high wear rates. The other mechanisms do not generate such high wear rates, except for those processes involving high velocity erosion, such as sandblasting.

Table 8.2 Typical Wear Rates in Processes Involving Abrasion[a]

Mechanism	Wear Rate (mils/hr)
1. Austenitic manganese steel in gouging abrasion	
(a) Hammers in impact pulverizers	5 to 1000
(b) Shovel dipper teeth	5 to 500
(c) Wearing blades on coarse ore scrapers	5 to 100
(d) Ball mill scoop lips	4 to 15
(e) Crusher liners for crushing siliceous ores	2 to 20
(f) Chute liners handling coarse siliceous ores	0.1 to 10
2. Low alloy high carbon steel in high-stress (grinding) abrasion	
(a) Rod and ball mill liners in siliceous ores	0.5 to 5.0
(b) Grinding balls in wet grinding siliceous ores	0.15 to 0.45
(c) Grinding balls in wet grinding raw cement slurries	0.05 to 0.15
(d) Grinding balls in dry grinding cement clinker	0.005 to 0.015
3. Pearlitic white iron in low-stress abrasion (erosion)	
(a) Sandblast nozzles	100 to 1000
(b) Sandslinger liners	50 to 250
(c) Pump runner vanes pumping abrasive mineral slurries	0.1 to 5.0
(d) Agitator and flotation impellers in abrasive mineral slurries	0.05 to 1.00
(e) Screw type classifier wear shoes in sand slurries	0.05 to 0.20

[a] from Ref. 17.

For one material to be significantly abraded by another, it must be the softer of the two. To minimize abrasion, therefore, the hardness of the surface must be greater than that of the abrasive. This can be accomplished by various methods: (a) the alloy can be changed so that it is inherently harder; (b) the heat treatment can be changed to effect different hardness levels; (c) the surface can be modified by creating or applying a hard surface layer, such as those attained by anodizing, electroplating, flame spraying, nitriding, or a number of other methods. Most surface treatments are limited in use to light to moderate wear situations since wear resistance is severely impaired once the layer is degraded.

8.4 PITTING AND SPALLING

8.4.1 General

Pitting, considered from a wear viewpoint, is the result of the fatigue failure of the surface metal. Repeated application of relatively low stresses may result in numerous pit-like cavities in the metal surface. The characteristics of surface fatigue damage are somewhat different from those of

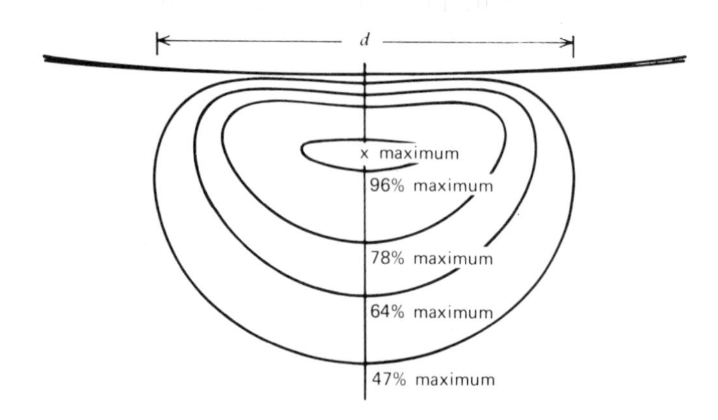

Fig. 8.6 Location of maximum elastic stress relative to the surface (from Ref. 1).

ordinary fatigue. One primary difference between bulk and surface fatigue is that no apparent endurance limit exists; that is, there is no stress level below which the material remains unaffected by surface fatigue damage. Test data is also subject to wider scatter than that obtained in ordinary fatigue tests. Consequently, it is exceedingly difficult to design a highly stressed bearing component, on the basis of test results, with the positive knowledge that surface fatigue has been eliminated.

There is some question as to whether cracking in pitting initiates at the surface or at a subsurface location. There is evidence that surface damage may initiate at either location, depending on the specific circumstances.

Cracks may initiate at a subsurface location when the maximum Hertz stresses, which occur slightly below the surface, are coinicident with an inclusion site, as shown in Fig. 8.6. Current improvements in melting technology, such as vacuum arc remelt, vacuum induction melting, and electron beam refining, have resulted in materials with significantly reduced inclusion counts. In these materials, the importance of inclusions relative to crack initiation has been minimized, and study should be directed to the effects of segregation, retained austenite, and the role of banded or fibered structures on crack initiation.

The type of surface movement, rolling or sliding, also helps to determine whether the crack will initiate at the surface or subsurface. With pure rolling, the maximum shear stress occurs below the surface, as previously indicated; when a sliding component is added, however, the maximum shear stresses may occur at the surface.

8.4.2 Environmental Factors Affecting Pitting

Environment can have a marked effect on surface fatigue. It is known that lubricant characteristics play an important role in determining whether surface pitting occurs. With mineral oils, pitting occurs more readily with those of lower viscosity. A common theory is that the oil enters an open surface crack. The oncoming rolling surface then seals the crack orifice, which places the trapped oil under pressure and creates a wedge effect, thereby propagating the crack. The premise supposes that higher viscosity oils will not enter a crack as readily as those of lower viscosity. However, viscosity alone is not the only factor, since with silicone fluids life is independent of viscosity.[10]

The reactivity of the lubricant is also important; surface active lubricants tend to lower the life of the component,[11] as compared to straight mineral

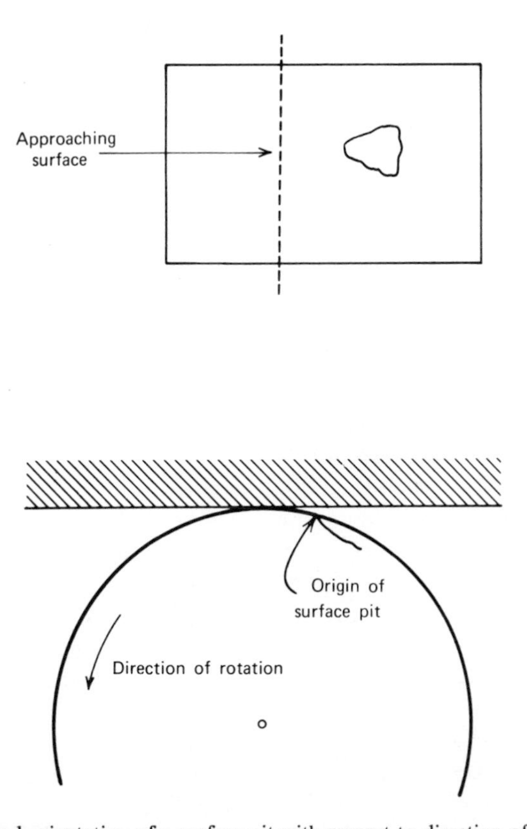

Fig. 8.7 Shape and orientation of a surface pit with respect to direction of motion.

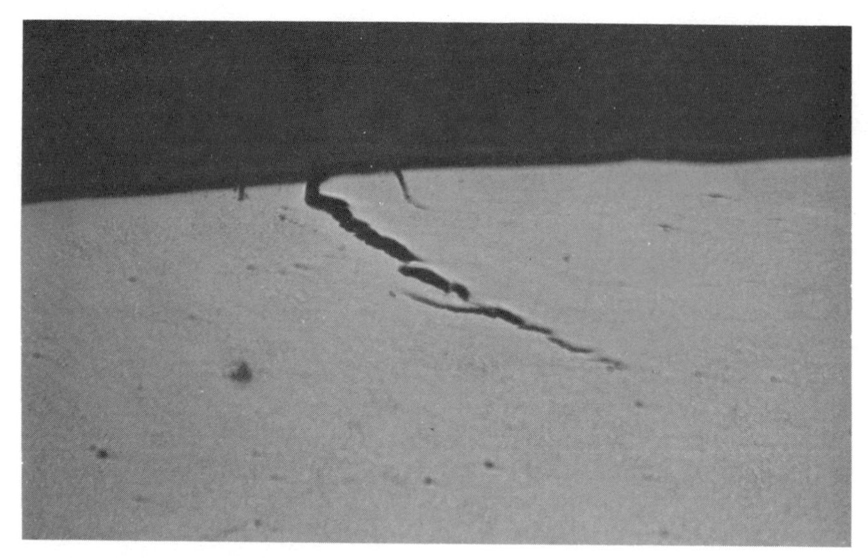

Fig. 8.8 Photomicrograph of a hardened steel wear surface showing initial stages of pit formation—unetched (150X) (from Ref. 13).

oils. Water contaminated lubricants can also accelerate the propagation of surface cracks. The mechanism is probably a form of corrosion fatigue and there is also some evidence that hydrogen embrittlement is a factor.[12]

8.4.3 Pitting Appearance

Pits are generally triangular or fan shaped, with the apex of the triangle pointing towards the rotation of the surface, as shown schematically in Fig. 8.7. A photomicrograph of an actual surface pit in the initial stages is shown in Fig. 8.8. Note the gradual incline to the surface near the origin.

Electron microscope fractographic techniques have been utilized to study pitting and spalling[13,14] under rolling contact fatigue conditions. Figure 8.9 shows a scanning electron photomicrograph of a spall cavity in a 51100 ball. Closeups of the cavity at higher magnifications (Fig. 8.10) show that there was a change in topography as the crack extended from the leading to the trailing edge. The fracture surface of the leading edge was comprised primarily of equiaxed and elongated cavities, typical of fracture by the ductile growth and coalescence of holes. This area extends approximately one third the distance to the trailing edge. The remainder of the surface exhibited the parabolic tongues shown in Figs. 8.10*A*, *B*, and *C*. Syniuta and Corrow[13] suggest that the initial crack resulted from fatigue at the leading edge and upon some extension becomes fast growing and completes

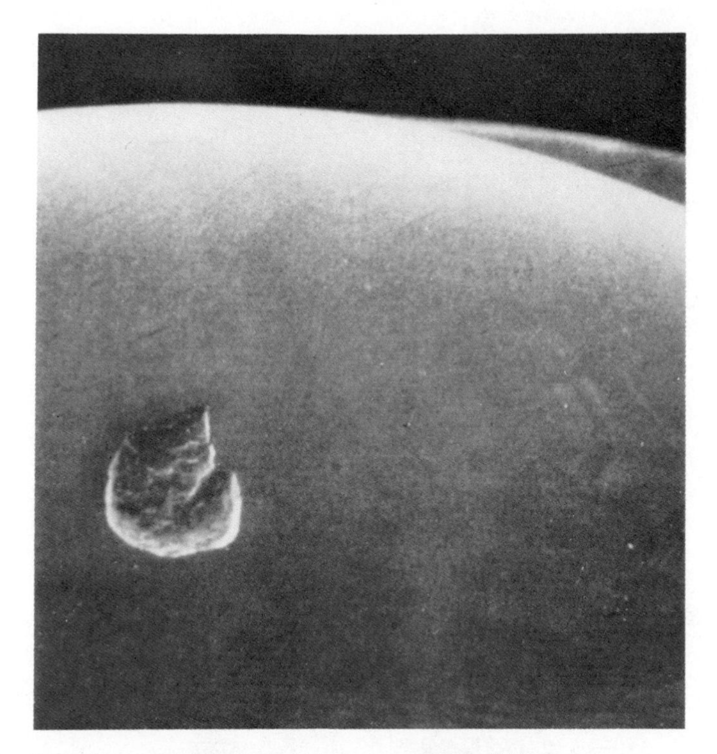

Fig. 8.9 SEM photograph of a surface pit in a rotating ball (50X).

the fracture. This final fracture area is believed to result in the changed fracture appearance in the trailing edge area. A schematic cross-section through the spall is shown in Fig. 8.11.

Spalling is generally considered to be a special type of pitting, which results when several pits join or when a crack runs parallel to the surface, rather than running to it, for some distance. Consequently, spalling defects are typically rather large. The latter type of spalling is often associated with the surface hardened parts and occurs near the core-case interface. This type of wear occurs because the shear strength of the subsurface material is inadequate to withstand the shear stress to which it is subjected. For this reason, a spalling condition is often rectified by increasing the case depth or the core hardness.

8.5 FRETTING

Fretting is the most common form of corrosion-assisted wear. Fretting, or fretting corrosion, is due to a slight oscillatory motion between two

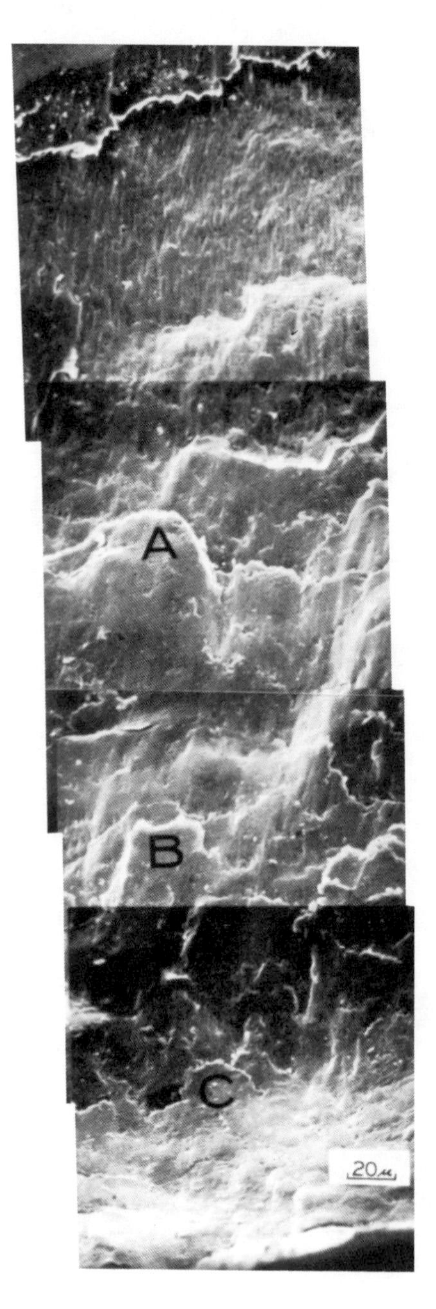

Fig. 8.10 Magnified view of fracture surface of pit in Fig. 8.9, illustrating change in fracture appearance from the area of initiation to the area of fast fracture (300X).

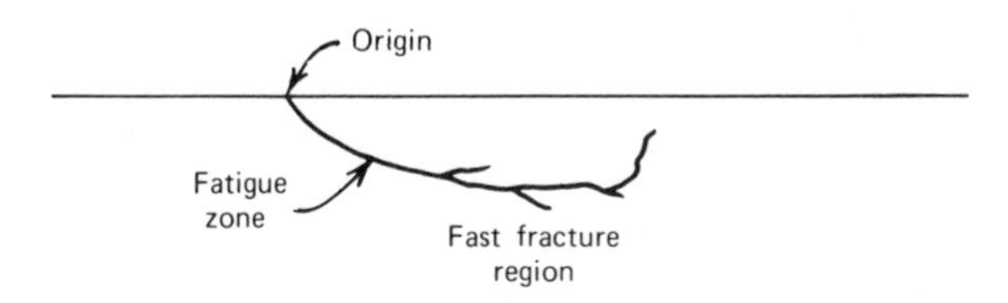

Fig. 8.11 Schematic view of cross-section through a spall.

Fig. 8.12 Fretting damage of inner race of bearing assembly (from Ref. 16).

mating surfaces under load, and manifests itself as pits in the surface surrounded by oxidation debris. An example of fretting damage is shown in Fig. 8.12. The wear-oxidation theory is one of the mechanisms proposed to explain fretting. When two metal surfaces are placed in contact, under load, adhesion between contacting asperities occurs. Upon displacement of

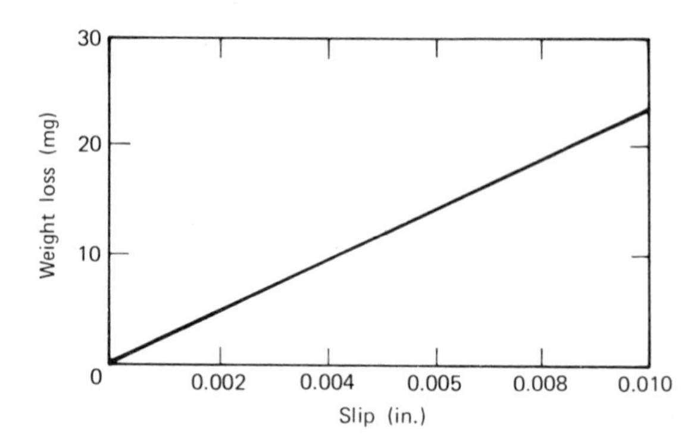

Fig. 8.13 Effect of displacement on total weight loss resulting from fretting (from Ref. 15).

the surfaces, the welded areas rupture and wear fragments are produced. These fragments, which are small and subject to high frictional temperatures, are readily oxidized. These metallic oxides are not removed from the area because of the small displacements. Since they are abrasive in nature, they contribute heavily to the wear rate.

The relative motion required to produce damage may be quite small. Displacements of 10^{-6} cm are sufficient to cause damage, but the amplitudes usually seen in service are in the order of a few thousandths of an inch. Figure 8.13 shows the effect of displacement amplitude on total weight loss for a mild steel.[15]

The total amount of fretting damage increases with increasing cycles, as expected. The increase is essentially linear, except for the initial portion when little abrasive material is available to cause damage. Frequency effects are observed, but tend to be small. Uhlig[15] indicates that the wear rate is higher at low frequencies but decreases to a constant value as frequency increases.

Fretting damage alone is sufficient to cause malfunction in many close tolerance designs of modern industry. The problem is intensified, however, since fatigue failures frequently initiate from fretting pits.[16] When the initiation phase of fatigue is eliminated, crack propagation and final fracture can be quite rapid, resulting in catastrophic failure.

REFERENCES

1. E. Rabinowicz, *Friction and Wear of Materials*, Wiley, New York, 1965, p. 113.
2. A. Dorinson, *Wear*, Vol. 11, 29–40, (January 1968).
3. E. Rabinowicz and D. Tabor, *Proc. Roy. Soc.*, Vol. 208A, 455 (1951).

4. M. E. Sikorski, *Wear*, Vol. 7, 144–162 (1964).
5. S. Bhattacharyya, *Wear*, Vol. 12, 131–134 (August 1968).
6. A. E. Roach, C. L. Goodzeit, and R. P. Hunnicutt, *Trans. ASME*, Vol. 78, 1659 (1956).
7. E. Rabinowicz, *op. cit.*, p. 165.
8. M. E. Sikorski, *J. Basic Eng.*, Vol. 85, 279 (1963).
9. E. Rabinowicz, *op. cit.*, p. 167.
10. D. Scott, NEL Report LDR 44/60, National Engineering Laboratories, East Kilbride, U. K., 1960.
11. G. S. Reichenbach, *Introduction to the Mechanical Behavior of Materials*, F. A. McClintock and A. S. Argon, Eds., MIT, Cambridge, Mass., 1962.
12. L. Grunberg, D. T. Jamieson, D. Scott, and R. A. Lloyd, *Nature*, Vol. 188, 1182–1183 (1960).
13. W. D. Syniuta and C. J. Corrow, *Wear*, Vol. 15, 187–199 (1970).
14. S. Borgese, ASME Paper No. 67–WA/CF–3, 1967.
15. H. H. Uhlig, I. M. Feng, W. D. Tierney, and A. McClellan, *Fundamental Investigation of Fretting Corrosion*, NACA Technical Note No. 3029, (December 1953).
16. R. L. Widner and J. O. Wolfe, *Met. Prog.*, April, 79–86 (1968).
17. H. S. Avery, *Surface Protection Against Wear and Corrosion*, American Society for Metals, 1954.

9

Hydrogen Degradation

9.1 INTRODUCTION

Damage resulting from hydrogen entrapment has been a problem for a number of years. It is primarily associated with body-centered cubic (*BCC*) materials, particularly high strength steels, which for economic reasons comprise the bulk of the affected materials.

The damage may present itself in a number of forms, such as flakes or "fish eyes", surface cracks, or a single crack caused by hydrogen, which results in premature failure. The general characteristics of hydrogen embrittled materials are:

1. A loss of tensile strength and ductility, which diminishes with low test temperatures and high strain rates.
2. Increased hydrogen contents, which generally result in an increased loss of mechanical properties.
3. Increasing the tensile strength of the material causes increased susceptibility, as manifested by a decrease in failure time at a particular stress level.
4. Increased notch acuity lowers the applied stress required to cause failure.

9.2 THEORIES OF HYDROGEN DAMAGE

Hydrogen degradation manifests itself in various forms, and has been studied for a number of years. Therefore, it is not surprising that a number of theories have been advanced, the most important of which are described in this chapter.

The hydrogen pressure theory advanced by Zapffe and Sims[1] was one of the earliest attempts to explain the unusual temperature and strain rate effects observed with hydrogen embrittlement. The theory proposed that atomic hydrogen diffused through the metal lattice to preexisting voids, and recombined there to form molecular hydrogen, according to the reaction:

$$2H \rightarrow H_2$$

This molecular hydrogen generates exceedingly high pressures, which facilitate cracking. The unusual temperature and strain rate dependency of hydrogen embrittlement could be explained in terms of the diffusion rate of hydrogen. Low temperatures or high strain rates would not permit adequate diffusion of hydrogen to the voids.

Petch and Staples[2] did not agree that internal pressure was primarily responsible for hydrogen cracking. Instead, they felt that a reduction in surface free energy (γ_s), caused by the adsorption of hydrogen on the walls of microcracks, facilitated the extension of these cracks by reducing the energy requirements. The strain rate and temperature dependency was explained on the basis that hydrogen could not diffuse into rapidly running cracks fast enough to cause embrittlement.

After investigating this premise, Bilby and Hewitt,[3] concluded that the surface work term, although operative, was rather small when compared to the plastic work term, γ_p. According to their calculations, the crack propagates when

$$(\sigma + P)nb > 2(\gamma_s + \gamma_p)$$

where

σ = applied stress
P = hydrogen pressure in voids
n = number of dislocations comprising crack
b = Burgers vector

or when the energy input due to the combined action of the applied stress and the internal pressure exceeds the energy required to produce two new surfaces, plus the associated plastic work where the plastic work portion was considerably greater than the surface energy consideration.

In a recent theory Troiano et al.[4] proposed that under the influence of an applied stress, hydrogen would diffuse toward regions in a triaxial stress-state. Furthermore, these regions of triaxiality would result from a network of internal voids acting as internal stress raisers.

The hydrogen, which exists in the stressed region of the lattice near the voids, may donate its electron to the metal, thereby decreasing the binding energy or cohesion of the lattice in this region. Hydrogen uniformly distributed throughout a metal lattice is nondamaging because its concentration is so small. But when hydrogen is segregated in highly stressed regions by stress-induced diffusion, the result is a lowering of the cohesive strength of the metal locally, with subsequent crack propagation and embrittlement.

Venett and Ansell,[5] in a study of the effects of high pressure hydrogen on the mechanical properties of 304 stainless steel, observed that the degradation noted was directly related to martensite transformation resulting from plastic deformation. As a result, they concluded that lattice hydrogen rather than adsorbed hydrogen, was by far the most damaging. They further suggested that hydrogen damage might result from a combination of mechanisms, rather than from a single specific mechanism, depending on the material and the circumstances.

9.3 EVIDENCE OF HYDROGEN DAMAGE

Hydrogen damage in steels may manifest itself as flakes, blisters, and general embrittlement. Flakes are small internal fissures, which occur in

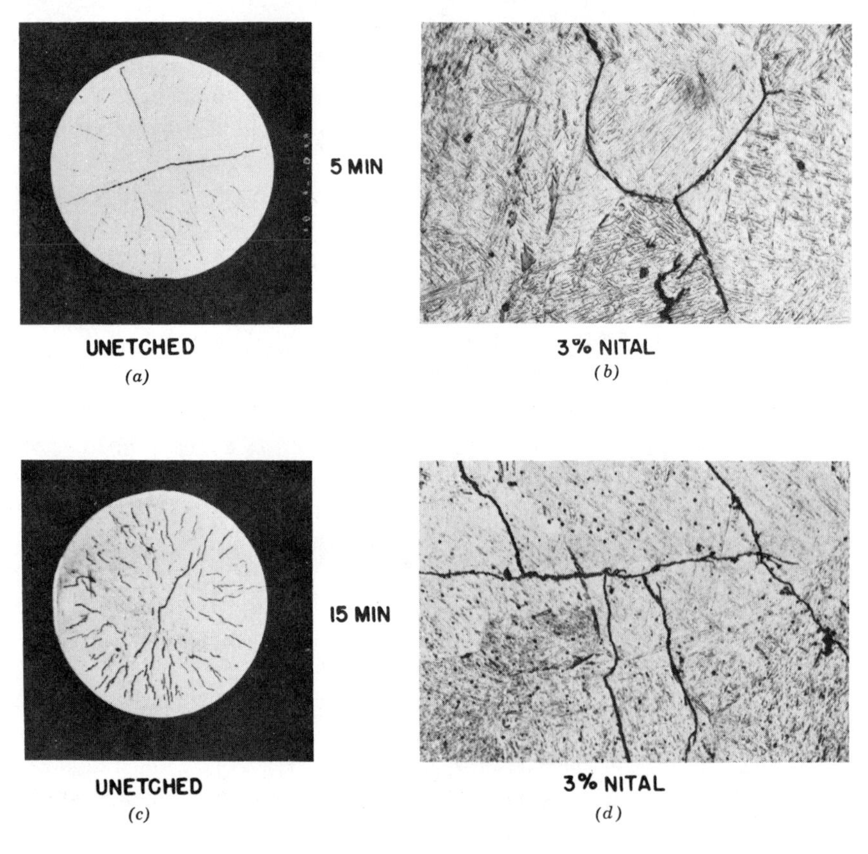

Fig. 9.1 Flake patterns of transverse specimens of SAE 5145 isothermally transformed at 1200°F for indicated times, after austenization at 2050°F in hydrogen atmosphere. (*a*) 1×. (*b*) 240×. (*c*) 1×. (*d*) 240×.

Fig. 9.2 Macroscopic view of hydrogen flakes on a fracture surface (3×).

the interior of a forging in a plane parallel to the forging direction. These cracks may be detected by ultrasonic inspection, or destructively, by etching transverse sections, as shown in Fig. 9.1.

On a fracture surface, these flakes appear as bright, highly reflective spots. Typical representations are shown in Fig. 9.2 and 9.3*a*. On a microscopic scale, the flakes contain some flat areas combined with other areas which appear to be contoured grain boundaries, as shown in Fig. 9.3*b*. According to Phillips and Kerlins,[6] the striations visible in Fig. 9.3 are characteristic of hydrogen flakes.

In some cases, the entrapped hydrogen may produce blisters because of the pressure generated by the gas (Fig. 9.4). This generally occurs with materials, such as ferrite, that have considerable inherent ductility.[7] Evans[8] indicated that the blisters are nucleated at inclusions situated slightly below the metal surface. Edwards[9] made a thorough study of the cause of blistering in mild steel in 1924, and concluded that the generation of gas in voids near the surface caused delamination of the heavily worked steel plate by separating the various strata near the surface. The form of the blisters shown in Fig. 9.5 supports his premise.

In ordinary rolled sheet or plate, banded structures containing elongated and flattened inclusions are common. Therefore, it is quite possible that the

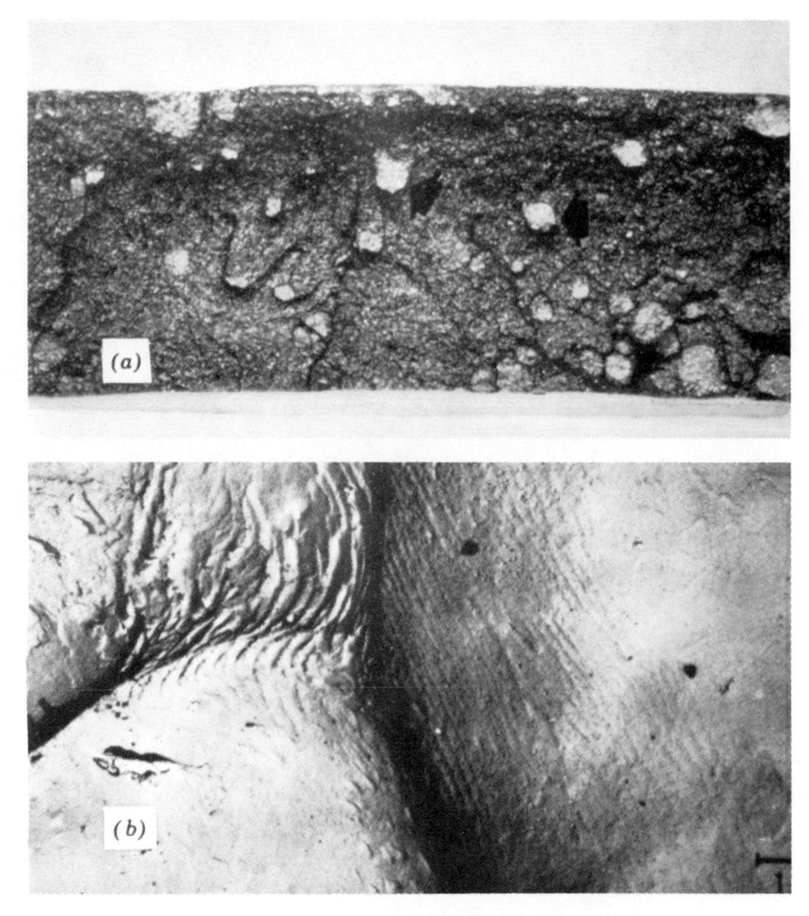

Fig. 9.3 (*a*) Macroscopic and microscopic appearance of hydrogen flakes on fracture surface of 4340 plate Macroscopic (1×). (*b*) Electron fractograph (6000×) (from Ref. 6.).

delaminations visible in Fig. 9.5 directly result from the interaction of the accumulation of hydrogen at the inclusion/matrix interface.

As previously mentioned, the principal effects of hydrogen embrittlement in steels are a decrease in tensile strength and ductility, when tested under static loads or low strain rates. According to Tetelman,[10] impact tests do not indicate susceptibility and, for this reason, are not recommended to determine whether embrittlement exists. On a fractographic basis, the visual features vary. Microcracks initiate internally, often near inclusions or other interfaces, and propagate intergranularly for an indeterminate distance. They may also originate from electroplated surfaces, as shown in

Fig. 9.4 Blisters formed by hydrogen in a carbon steel plate surface.

Fig. 9.6. The regions between the adjacent microcracks fail in a ductile manner and present evidence of dimpling, as shown in Fig. 9.7. When the cross-section has been sufficiently reduced, and the load bearing capability sufficiently impaired, final failure occurs from overstress.

There has been some question[11–13] as to whether what has been reported as the stress corrosion failure of high strength steel, has not in fact, been the result hydrogen embrittlement, which results from hydrogen generated as a corrosion product. From a fractographic standpoint, they appear very similar. Whiteson et al.[14] initiated a study to discern the differences in appearance between failures caused by the two mechanisms. They concluded that the differences were very subtle and consisted of the following:

1. Stress corrosion failures generally began at the surface, whereas hydrogen embrittlement failures began internally.
2. With stress corrosion cracking (SCC), intergranular regions show considerably more secondary cracking.
3. There is a greater amount of corrosion attack at the origin with stress corrosion than with hydrogen embrittlement.
4. Fine indications, such as hairlines, are less pronounced with SCC.

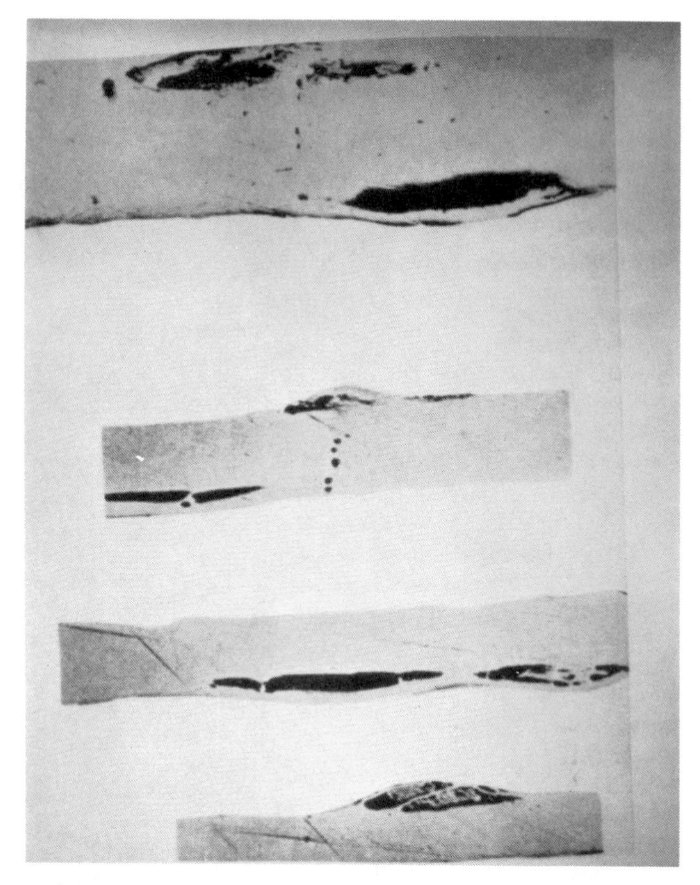

Fig 9.5 Transverse sections through blisters caused by pickling of mild steel plates (from Ref. 9).

However, hydrogen cracks originating from the surface have been observed, particularly with electroplated components, and (3) and (4) may well be the result of the intrusion of the corrosion medium on the crack surface after the fracture has occurred. The fractures resulting from the two processes are extremely similar. Figure 9.8 shows fractures in the same alloy.

Much of the information on hydrogen damage has been the result of work in aqueous systems. Recently, a number of studies have indicated that gaseous hydrogen at high pressures may also cause embrittlement. Dodge, along with VanNess[15] and Perlmutter,[16] investigated the effect of high pressure hydrogen on a number of iron and nickel base alloys. In general, only the nickel base and high chromium iron base alloys proved

Fig. 9.6 Intergranular crack resulting from hydrogen embrittlement during chromium plating (transverse section—35×).

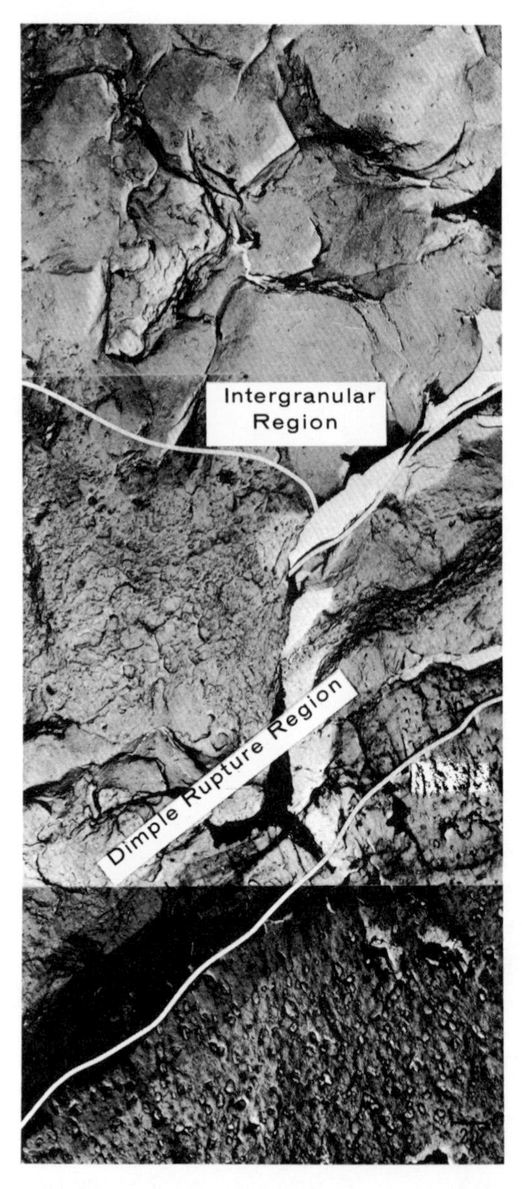

Fig. 9.7 Transition from intergranular fracture to ductile rupture (2500×) (from Ref. 14).

resistant. It was concluded that the austenitic stainless steels were the most suitable for use with hydrogen at high pressures and elevated temperatures. Ansell and Venett[17] recently tested a number of face-centered cubic (FCC) materials, 304 and 310 stainless steel, OFHC copper, and aluminum, at hydrogen pressures up to 10,000 psi and various exposure times. The aluminum and an aluminum alloy were unaffected. The copper and 304 stainless steel lost some strength and ductility, which was associated with the occurrence of plastic deformation. The damage was independent of exposure time, but was related to hydrogen pressure.

The fractographic appearance of the 304 stainless steel changed from a totally dimpled surface to one that was predominantly dimpled (60 to 70%), but which exhibited large areas of quasi-cleavage and intergranular fracture (Fig. 9.9).

9.4 SOURCES OF HYDROGEN

The origin of the embrittling hydrogen is varied. A major source of hydrogen in steels results from water vapor reacting at high temperatures with the liquid iron. The water vapor may come from the scrap used to charge the furnace, the slag ingredients or from the refractory materials lining the furnace. The resulting hydrogen may become entrapped during solidification as solubilities decrease.

Hydrogen may also become available during acid pickling or plating operations. Exposure in service to process fluids bearing hydrogen, as in catalytic cracking, can also cause embrittlement. Similarly, hydrogen may be generated as a corrosion product in certain environments, and thereby become available to cause embrittlement. In welding, the principal source of hydrogen is moisture in the electrode coating.

9.5 EMBRITTLEMENT OF OTHER METALS

9.5.1 Titanium and Zirconium Alloys

The α alloys of both titanium and zirconium possess a hexagonal close-packed (HCP) structure at room temperature, and both exhibit a marked sensitivity to hydrogen embrittlement. Unlike the iron base alloys, there is a pronounced tendency to form a stable hydride when the equilibrium solubility limit is exceeded. As the hydrogen concentration increases above this limit, the amount of hydride increases. The size of the hydride particles is directly related to the kinetics of the nucleation process. Slow cooling from higher charging temperatures and high supersaturation tends to generate large particles, whereas the reverse is true for rapid cooling and low supersaturation.

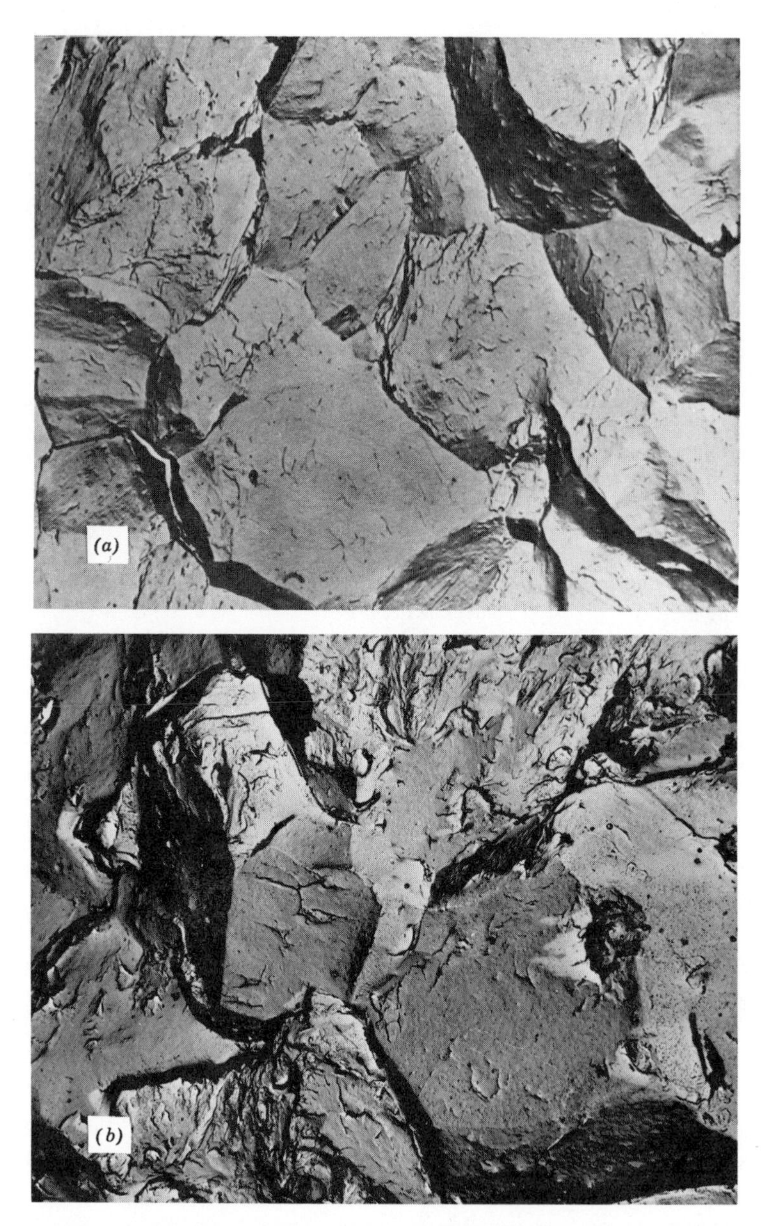

Fig. 9.8 Electron fractographs illustrating the similarity between stress corrosion cracking and hydrogen embrittlement of same alloy (from Ref. 13). (*a*) Stress corrosion cracking. (*b*) Hydrogen embrittlement.

Fig. 9.9 Electron fractograph illustrating quasi-cleavage in 304 stainless steel exposed and tested in high pressure hydrogen (6750×) (from Ref. 5).

The brittleness is associated with the fracture of the hydride particle or its interface, and is not the result of hydrogen in solution.

The embrittlement of the α alloys is visible through a marked decrease in the impact strength and tensile ductility, when tested at high strain rates[18] and low temperatures.[19]

The two-phase $\alpha\beta$ alloys may also become embrittled by exposure to hydrogen at elevated temperatures; however, in contrast to the α alloys, the embrittlement is more pronounced when tests are conducted at low strain rates. The cracking in $\alpha\beta$ alloys occurs intergranularly along the grain boundary between the α and β phases.[20]

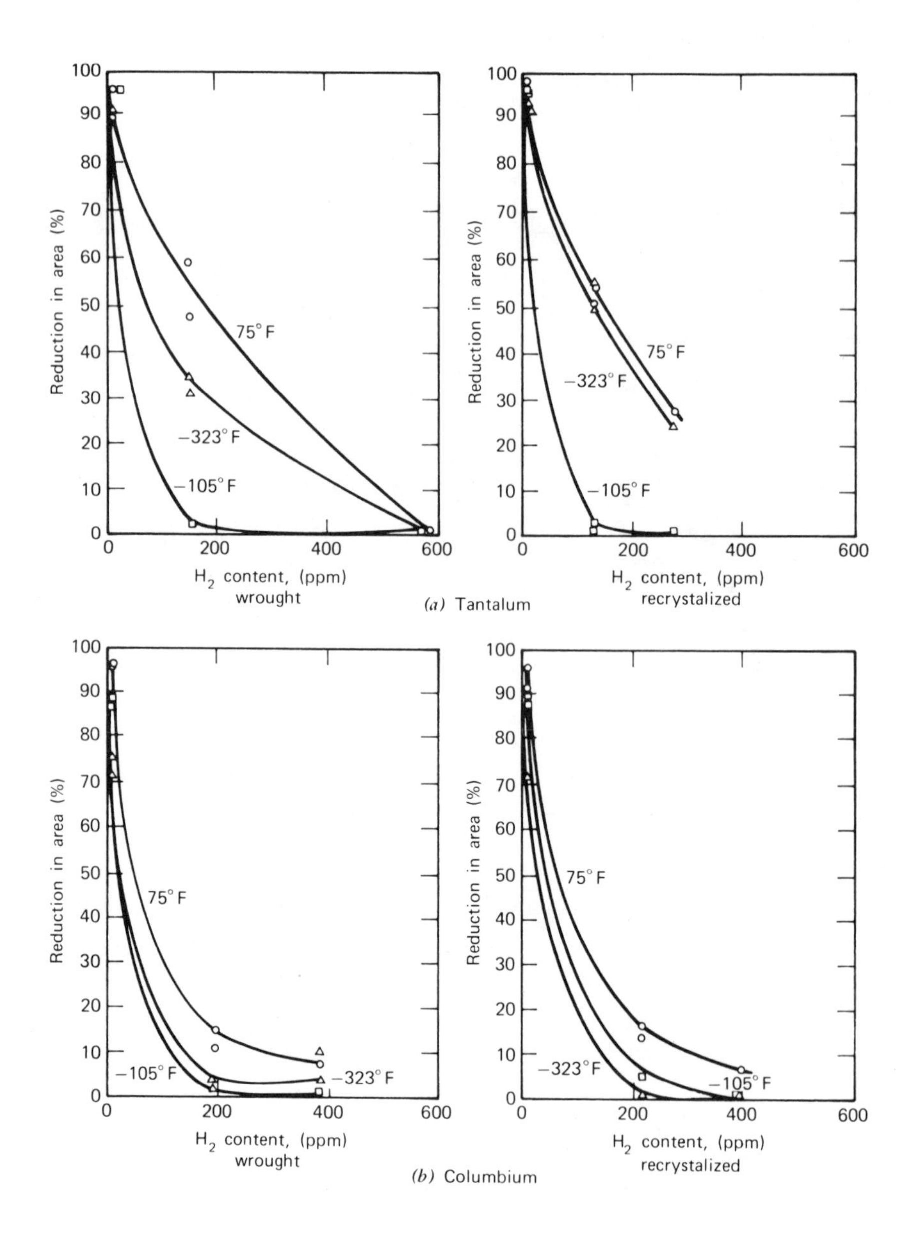

Fig. 9.10 Effect of hydrogen content on the ductility of (*a*) tantalum and (*b*) columbium.

9.5.2 Refractory Metals

Hydrogen exerts a damaging effect on the refractory metal group consisting of tungsten, vanadium, tantalum, and columbium. The evidence indicates that the hydrogen exists in a supersaturated solution and that stable hydrides are not present.[21] The effect of hydrogen on the ductility of tantalum and columbium, as measured by reduction in area, is shown in Fig. 9.10. Examination of the curves shows that even low concentrations are extremely detrimental. Impact strength is also affected. The general result of increased hydrogen contents, up to about 400 ppm, is an increase in the transition temperature.[22]

Hydrogen, even in relatively low concentrations and at room temperature, promotes brittle cleavage fractures in the refractory metals. Fractographic examinations reveal that the fracture surface is composed of cleavage facets, each of which represents the transgranular fracture of an individual grain. Bakish[23] indicated that cleavage occurred in the (100) and (110) planes in tantalum that was charged with hydrogen. Whether the fracture initiates within these planes or at the grain boundaries has yet to be determined.

9.5.3 Other Metals

Stoloff[24] reported that zirconium bearing magnesium alloys may become embrittled by heat treatment in a hydrogen atmosphere. He also indicated that other alloys bearing hafnium or thorium may exhibit similar behavior.

REFERENCES

1. C. A. Zapffe and C. E. Sims, *Trans. AIME*, Vol. 145, 225–259 (1941).
2. N. J. Petch and P. Staples, *Nature*, Vol. 169, 842 (1952).
3. B. A. Bilby and J. Hewitt, *Acta Met.*, Vol. 10, 587 (1962).
4. A. R. Troiano, *Campbell Memorial Lecture*, *Trans. ASM*, Vol. 52, 54 (1960).
5. R. M. Venett and G. S. Ansell, *Trans. ASM*, Vol. 60, No. 2, 242–251 (June 1967).
6. A. Phillips and V. Kerlins, *Met. Prog.*, Vol. 95, 81–85, (May 1969).
7. A. S. Tetelman, Ph.D. Thesis, Yale University, 1971.
8. U. R. Evans, *The Corrosion and Oxidation of Metals*, St. Martin's, New York, 1968, p. 427.
9. C. A. Edwards, *J. Iron, Steel Inst.*, Vol. 11, 110, (1924).
10. A. S. Tetelman and A. J. McEvily, *Fracture of Structural Materials*, New York, 1967, p. 456.
11. G. L. Hanna, A. R. Troiano, and E. A. Steigerwald, *Trans. ASM*, Vol. 57, 658–671, (September 1964).
12. H. P. Leckie and A. W. Loganow, *Corrosion*, Vol. 24, 291–297 (1968).
13. V. J. Colangelo and M. S. Ferguson, Vol. 25, 509–514 (1969).
14. B. V. Whiteson, A. Phillips, R. A. Rawe, and V. Kerlins, *Special Fractographic Techniques for Failure Analysis*, Annual Meeting of ASTM, Boston, Mass., June 1967.

15. H. C. Van Ness and B. F. Dodge, *Chem. Eng.* Prog., Vol. 51, 266 (1955).
16. D. D. Perlmutter B. F. Dodge, *Ind. Eng. Chem.*, Vol. 48, No. 5, 885 (1958).
17. G. S. Ansell and R. M. Venett, *Hydrogen Embrittlement*, Final Report, Contract #AT (30–1)–3479, R.P.I., June 1968.
18. R. Haynes, *J. Inst. Met.*, Vol. 88, 509 (1959).
19. C. J. Beevers, *Trans. AIME*, Vol. 233, 780 (1965).
20. P. Cotterill, *Hydrogen Embrittlement of Metals*, Pergamon, 1965, p. 201.
21. A. Lawley, W. Liebmann, and R. Maddin, *Acta Met.*, Vol. 9, 841 (1961).
22. A. G. Imgram, Technical Report ASD 61–474, Battelle Memorial Institute, August 1961, pp. 36–37.
23. R. Bakish, *J. Electrochem. Soc.*, October, 574 (1958).
24. N. S. Stoloff, *The Effects of Solutes on Fracture Behavior of Metals*, Fourth Annual Symposium on Fundamental Phenomena, Boston, Mass., February 1966.

10

Metalworking Defects

10.1 INTRODUCTION

The majority of parts in use today are produced by metalworking techniques that involve plastic deformation. These include rolling, forging, swaging, explosive forming, and a variety of others.

Most metalworking is done hot, primarily because of the lower strength (requiring less working force) and the higher ductility (allowing greater reductions) at higher temperatures. Cold working is employed for finishing operations to attain better surface finishes and tolerances, or to utilize the work hardening characteristics of some metals.

Because plastic deformation is involved, several factors are important.

1. Residual stresses can be built into a part.
2. The microstructure is altered because of the working.
3. There is a limit to the amount of deformation tolerable, which, if exceeded, would cause the part to crack.

The defects and characteristics of worked metal are considered in three categories: stresses, microstructure, and suface flaws.

Although various metalworking processes differ in the type of loading and the amount and type of reduction involved, the effect of a surface crack is the same whether it is the result of forging, rolling, or swaging. Because the processes are different, the likelihood of encountering the various defects is not the same for each.

10.2 STRESSES

Working involves high stresses. If the stresses are sufficiently high and tensile, cracking can result while the piece is being formed. This can occur if the metal is worked too far for its inherent ductility. Figure 10.1 shows

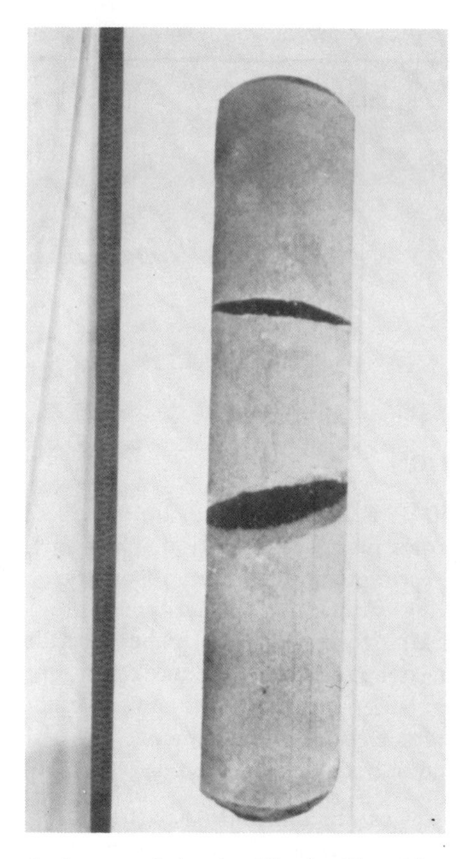

Fig. 10.1 Forging cracks due to underheating. The ductility of the metals was insufficient at the working temperature to accommodate the plastic deformation imposed (Reproduced by permission, from Metals Handbook, 1955 Supplement, American Society for Metals, 1955).

large cracks in a forging due to underheating,[1] that is, the temperature of the stock was too low for the necessary low strength and high ductility needed to permit forging. This problem can easily be uncovered during inspection, thus it is not usually a concern of the failure analyst. However, the cracks may be too small to identify by nondestructive testing, or they may be internal (Fig. 10.2). If so, there is a potential for service failure.[2] For example, Fig. 10.3 shows a fatigue failure of a crank pin, initiated at a longitudinal pipe,[3] which acted as a stress concentrator. The pipe was caused by shrinkage during solidification of the initial ingot, which was elongated during working.

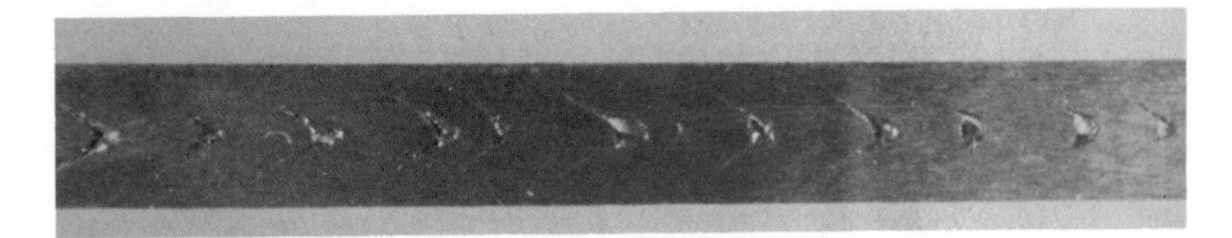

Fig. 10.2 Center burst in a drawn 1024 rod (Reproduced by permission, from Metal Progress, American Society for Metals, 1969).

Residual stresses can also be built into a worked piece. They can be beneficial, if compressive, or detrimental, if tensile. Worked parts invariably contain residual stresses, whose magnitude increases as the amount of deformation increases and the working temperature decreases. This accounts for the fact that most forgings with high length/diameter ratios must be straightened after working.

Residual stresses also develop because of the lack of uniform working through the thickness of a part. The stress pattern varies from the surface

Fig. 10.3 Fracture of a crank pin in fatigue, caused by longitudinal pipe cavity (Reproduced by permission, Society for Experimental Stress Analysis).

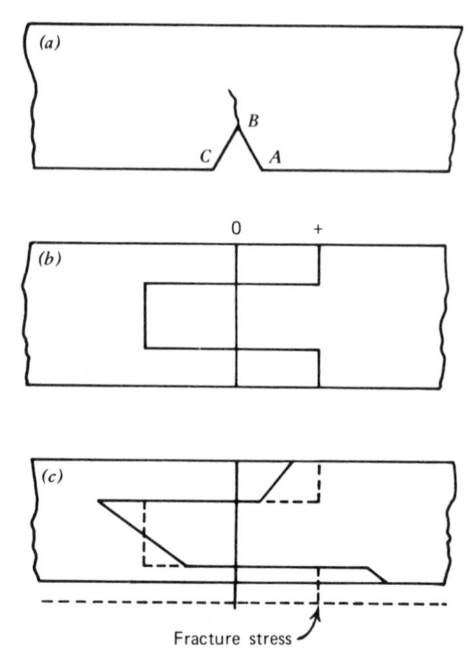

Fig. 10.4 Cracking during machining of the notch in a Charpy impact bar, caused by the relaxation of residual stresses (Reproduced by permission, from Metals Handbook, 1955 Supplement, American Society for Metals, 1955).

to the interior, which causes a variation in residual stress through the thickness.

Residual stresses can cause a variety of effects including warping, which occur particularly during subsequent machining, or fracturing. Figure 10.4 shows a bar that fractured because of the relief of residual stresses during milling of the notch in the Charpy bar.[4] The relieved residual stresses exceeded the fracture stress of the material, and caused cracking, as shown in Fig. 10.4*c*.

Many parts are heat treated after forming; thus any residual stress problem can be compounded, particularly if a quenching operation is involved. Two failures traced to residual stresses are shown in Figs. 10.5 and 10.6. Figure 10.5 shows a fractured connecting rod from a diesel engine[5] that failed because of residual stresses developed during straightening, and Fig. 10.6 shows a piston rod from a diesel engine that failed because of the relief of residual stresses developed during forming.[6]

Localized residual stresses can cause alteration of microstructure,[7] as

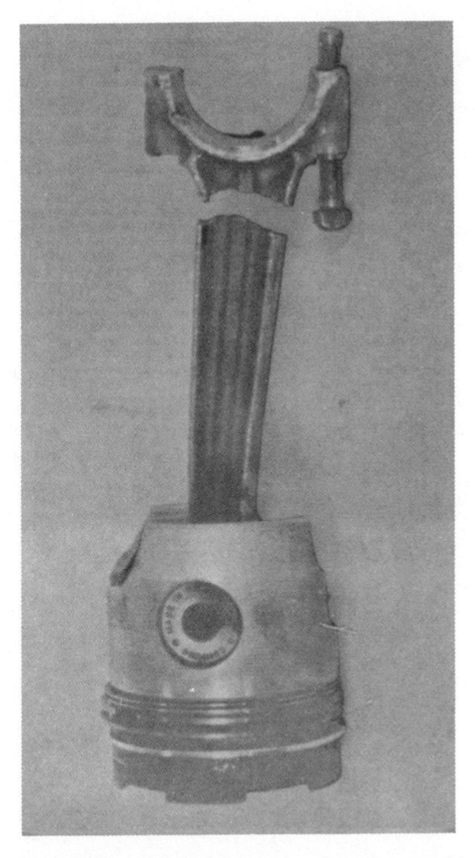

Fig. 10.5 4137 steel connecting rod that failed due to stresses caused by cold straightening (Reproduced by permission, from Metal Progress, American Society for Metals, 1969).

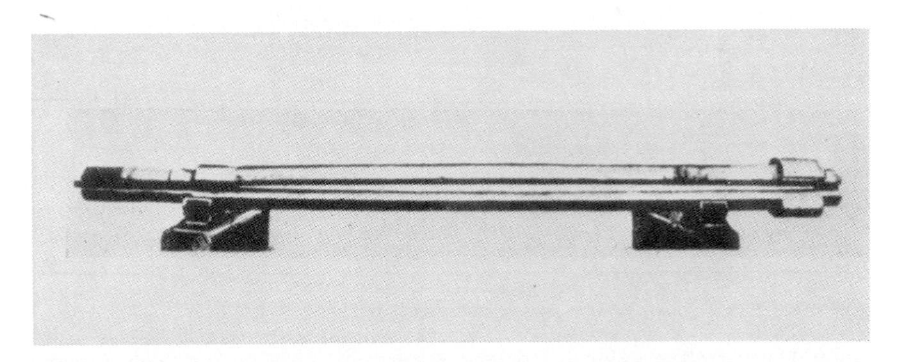

Fig. 10.6 Failure in a marine diesel engine piston rod resulting from residual stresses (Reproduced by permission, from Practical Metallurgy, American Society for Metals, 1948).

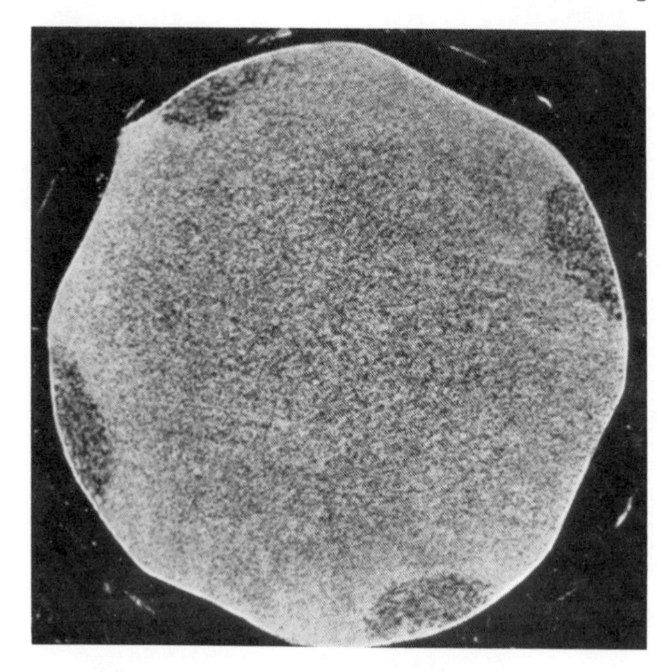

Fig. 10.7 Exaggerated grain growth due to localized residual stresses in swaged tungsten rod (Reproduced by permission, Elsevier Scientific Publishing Co.).

evidenced by the excessive grain growth shown in Fig. 10.7. In this respect, it is similar to overheating. However, this type of local grain growth may also be caused by a critical amount of deformation. Whatever the cause, since grain growth usually involves a decrease in strength, these areas can act as fracture initiation sites.

Since stresses are additive, residual tensile stresses are deleterious because they increase the effect of load on a component by adding to the applied load. There are also several phenomena that require prolonged loading, such as hydrogen embrittlement or stress corrosion. In these cases, residual stresses can supply the necessary loading conditions. Therefore, if the service conditions are such that hydrogen embrittlement or stress corrosion are possible, it is important that residual stresses be prevented or eliminated.

10.3 MICROSTRUCTURE

When a metal is deformed, two types of preferred orientation are possible, crystallographic texturing and mechanical fibering. The latter is more important in commercial components. Crystallographic texturing causes the alignment of crystallographic planes and directions, with respect to the working planes and directions. Segregation, inclusions, and microconstituents are aligned and elongated in directions determined by the metal flow resulting in mechanical fibering. The occurrence and importance of texturing and fibering vary with metal systems in use and the deformation conditions. Either condition can result in anisotropic mechanical properties.

One benefit of plastically working a metal is that the dendritic pattern associated with the original casting is refined, since many of the inherent interdentritic voids are welded shut. Underworked parts may exhibit a residual dendritic pattern. The degree of refinement increases with the amount of working. For example, the macrostructure shown in Fig. 10.8[8]

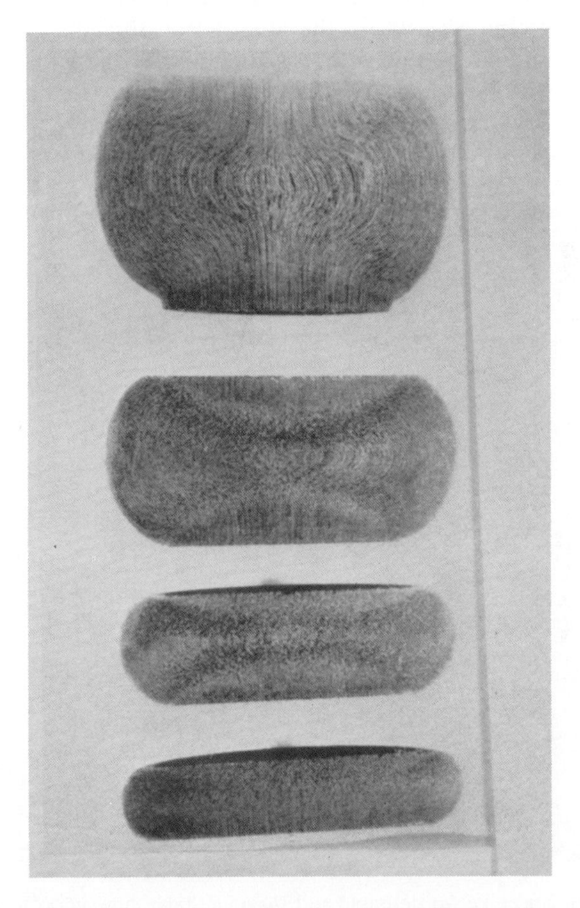

Fig. 10.8 Effect of increased plastic working on mechanical fibering (Reproduced by permission from Metal Progress, American Society for Metals, 1966).

became increasingly finer as the amount of working increased. This figure also illustrates the macroscopic effect of mechanical fibering. Originally, the material was formed into bars, aligning the inclusions, grains, and segregated areas. The dark lines in the top sample are flow lines caused by the aligned inclusions and segregated areas. Subsequently, the samples were upset, that is, worked normal to the original working direction. This caused the realignment of the fibering, which is shown as the bulging of the flow lines. With sufficient working, both the cast structure and the fibering were refined.

However, if there is sufficient segregation in the initial ingot, it may remain in the final shaped part.

Flow lines associated with mechanical fibering can be shown by differences in the etching reaction of the aligned constituents. In components, the flow lines should be aligned in the direction of the maximum stresses, since they are strongest in that direction.[9] In this case, the greatest stresses act along the flow lines, rather than across them.

Fig. 10.9 Comparison of (*a*) desirable and (*b*) undesirable flow lines. In the latter, the flow lines are too concentrated at the critical section (Reproduced by permission, Elsevier Scientific Publishing Co.).

Proper and improper flow patterns are shown in Fig. 10.9. In Fig. 10.9*b*, the improper case, the stresses act across the flow lines, which is the material's weakest direction.

Since orientation is so critical in formed parts, the origin and importance of the flow lines, or the mechanical fibering, should be considered. Alignment of microconstituents can occur in most metal systems[10] and in all forming processes, as shown by the aligned microstructure of the cupro-nickel in Fig. 10.10.

The most graphic example of the alignment of microconstituents is seen in banded mild steel, Fig. 10.11, which consists of alternating layers of pearlite and ferrite.[11]

Since the grains are elongated in the direction of metal movement, the weak grain boundaries are also aligned, which contributes to the transverse weakness. However, the alignment of inclusions is sometimes considered more important than the alignment of microconstituents.

If the inclusions are ductile, they may deform with the matrix during working; if brittle, they may crack and cause internal fissures. In most cases, the inclusions are more brittle than the matrix. They contribute to both the weakness of the metal normal to the fibering direction and to the strengthening in the fibering direction.

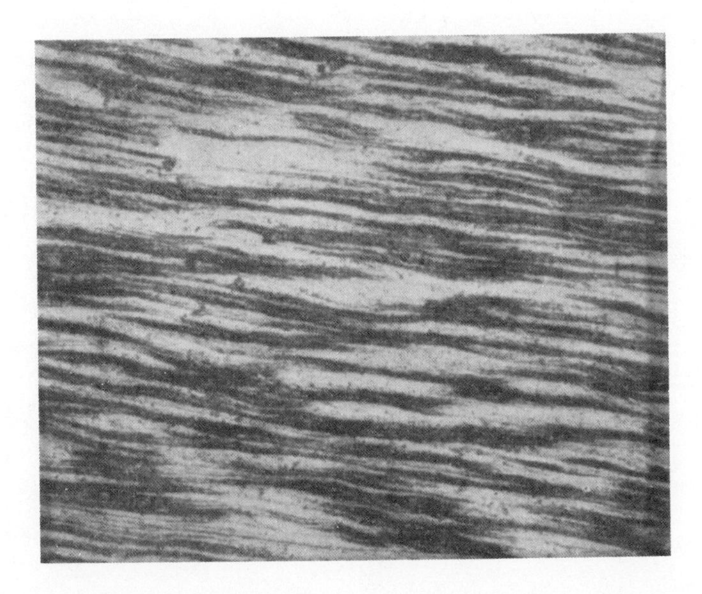

Fig. 10.10 Longitudinal microstructure in worked cupro-nickel (Reproduced by permission, Elsevier Scientific Publishing Co.).

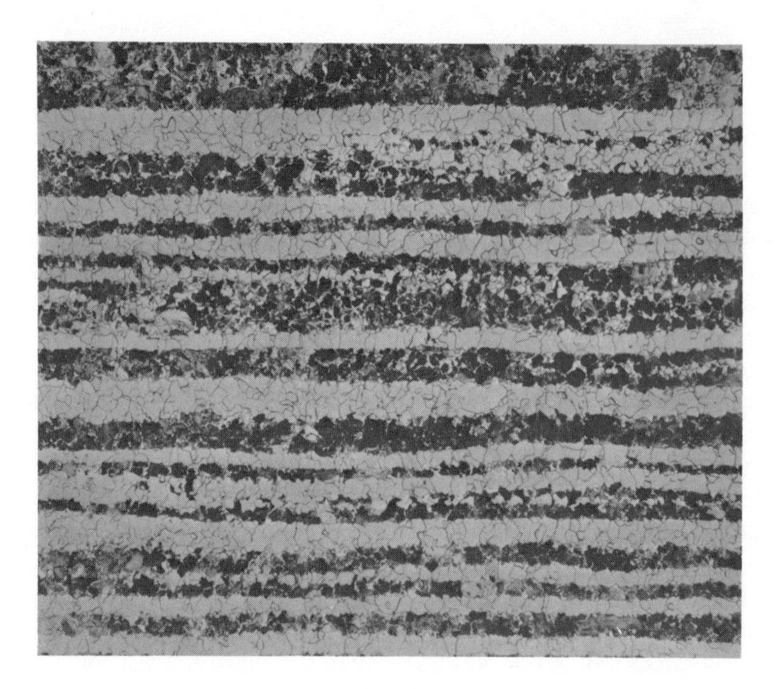

Fig. 10.11 Banding in low carbon steel plate showing alternating layers of ferrite and pearlite.

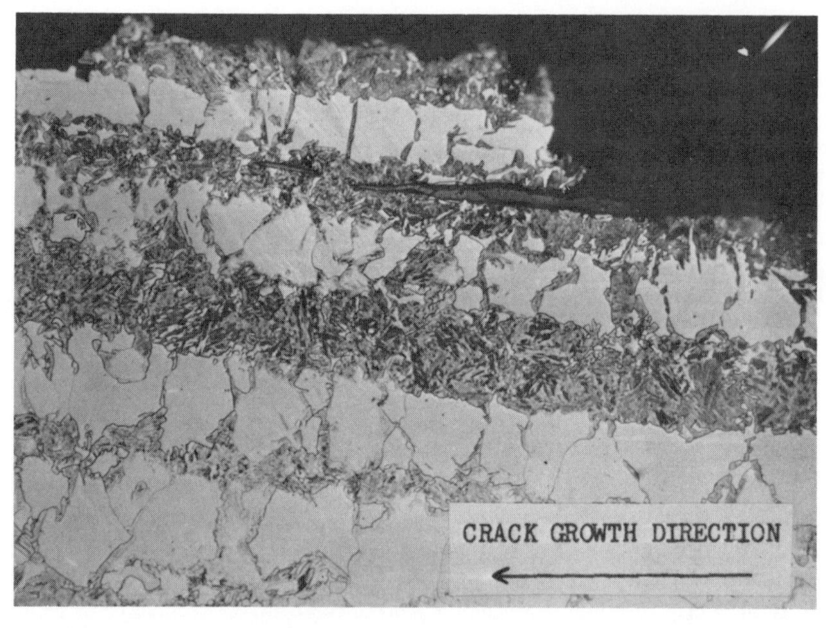

Fig. 10.12 Cracking at an inclusion-matrix interface in low carbon steel (Reproduced by permission, Elsevier Scientific Publishing Co.).

Applied stresses may cause cracking either of the inclusion or at the inclusion-matrix interface, Fig. 10.12, resulting in internal fissures and stress concentration.[11] These can also cause decreased fatigue resistance by acting as nuclei for fatigue failures, as shown in Fig. 10.13. This illustrates a service failure in a crank shaft, which was associated with the striated aligned inclusion.[3]

Failures resulting directly from weaknesses due to the aligned microstructure and inclusions have been reported. Figure 10.14 shows splitting of grains in copper due to weak grain boundaries,[12] and Fig. 10.15 shows a fracture along grains during drawing in a mild steel. In this case, the crack propagated in the direction of the flow of metal, that is, in the direction of the tensile load rather than normal to it, as anticipated. Figure 10.16 shows splitting along fibers in tungsten wire.[13] These three examples illustrate the significance of mechanical fibering irrespective of the metal system.

If the original ingot material is defective, for example, if it contains gross segregation or cavities (pipe) in the ingot, working accentuates the defect, and contributes to failure. Figures 10.17 and 10.18 show pipe in a section of rail, and cracks at seams in a high speed steel cutter, respec-

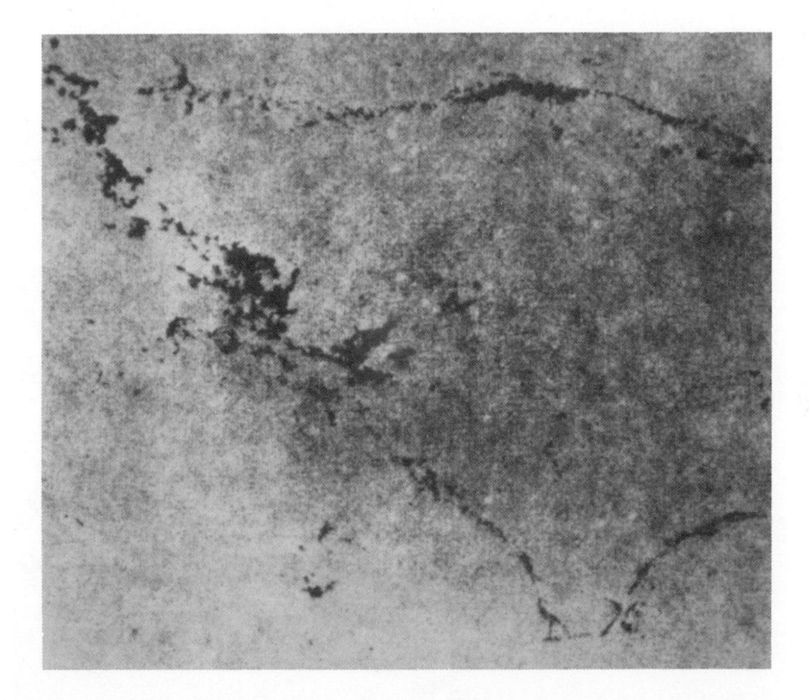

Fig. 10.13 Cross-section near the fracture origin in a crankshaft showing striated nature of inclusions (Reproduced by permission, Society for Experimental Stress Analysis).

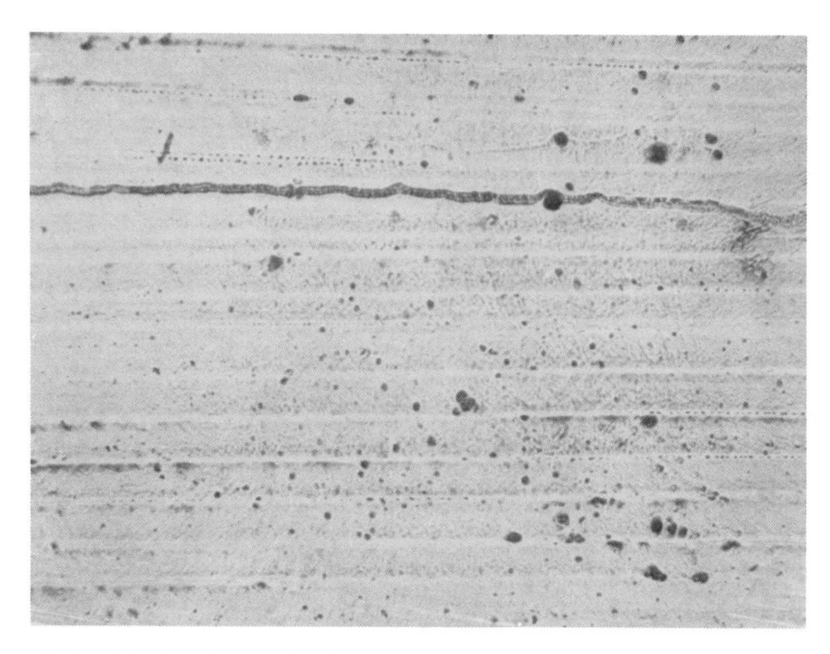

Fig. 10.14 Splitting along grains in cast copper (Reproduced by permission, Elsevier Scientific Publishing Co.).

Fig. 10.15 Fracture along grains in drawn steel (Reproduced by permission, Elsevier Scientific Publishing Co.).

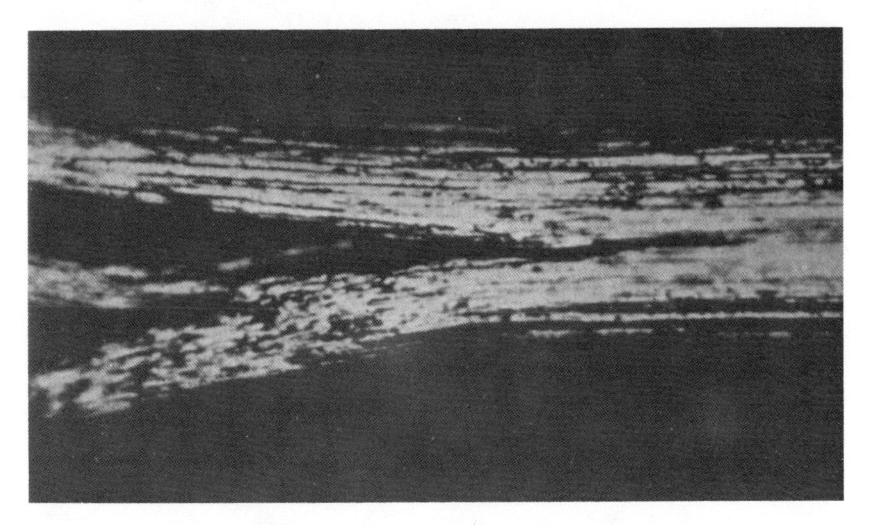

Fig. 10.16 Split along fibers in tungsten wire (Reproduced by permission, Elsevier Scientific Publishing Co.).

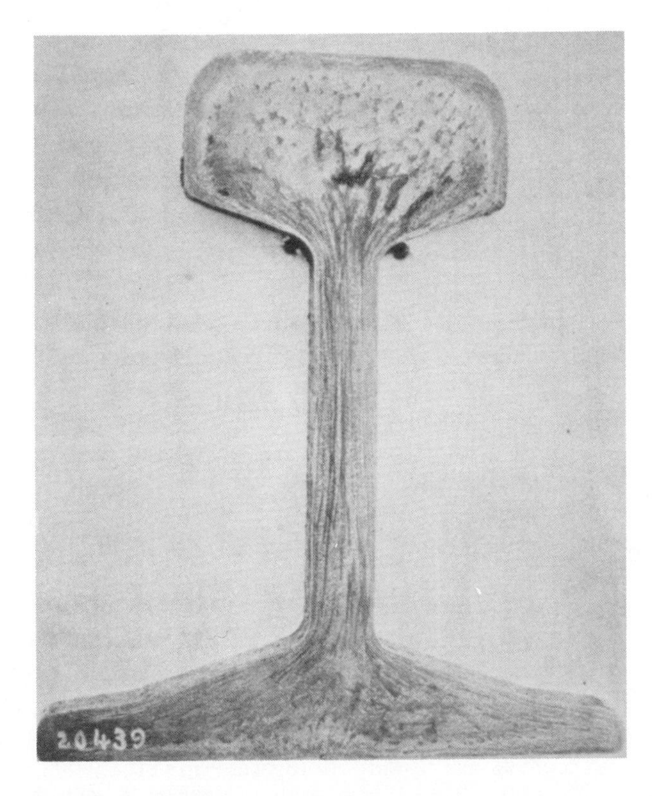

Fig. 10.17 Elongation of pipe in the original ingot during the rolling of the rail.

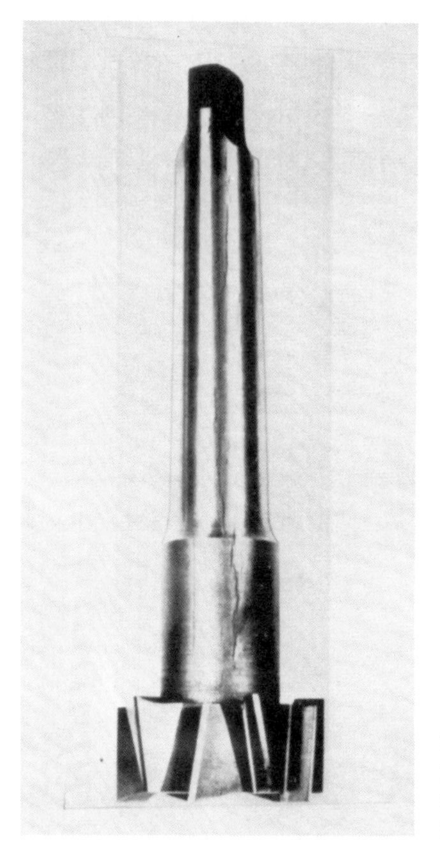

Fig. 10.18 Cracks at seams in a high speed tool cutter due to segregation (Reproduced by permission from Metals Handbook, 1955 Supplement, American Society for Metals, 1955).

tively.[14,15] In both cases, the effect of working was similar. A defect was present in the original ingot, and subsequent working extended the defective material.

A final failure is shown in Fig. 10.19. This fatigue failure, in a 1050 crankshaft, began at a large inclusion.[16] The explanation for this failure concluded that a large inclusion originally existed in unsound material in the interior of the metal. During forging, it was extruded to the outside, where it acted as a stress concentrator and crack initiator.

10.4 SURFACE DEFECTS

Various types of surface defects are found in formed parts. Some are important from the standpoint of performance, such as cracks; others are primarily aesthetically displeasing, such as stretcher strains[17] in steel (Fig. 10.20). Surface defects are usually more deleterious than internal defects

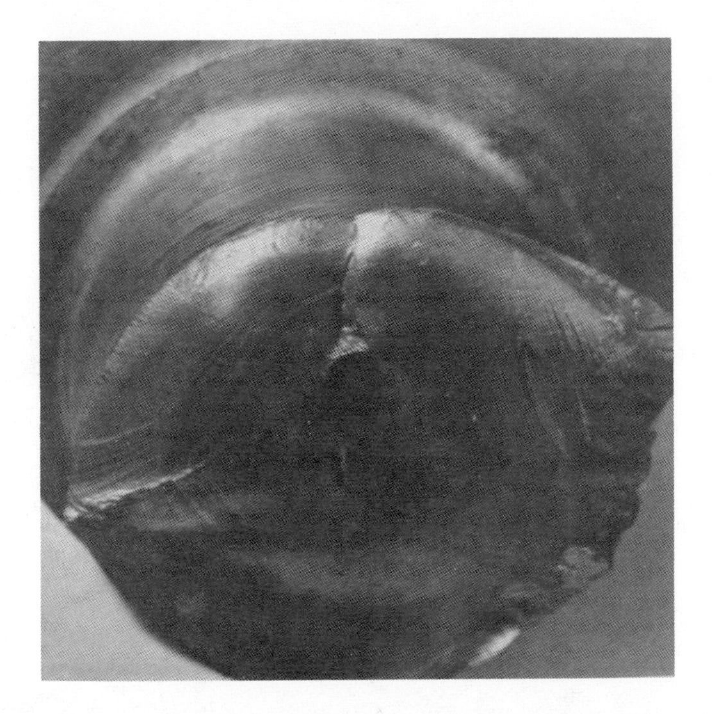

Fig. 10.19 Crankshaft fatigue that started at an inclusion related to unsound center material (Reproduced by permission from Metal Progress, American Society for Metals, 1969).

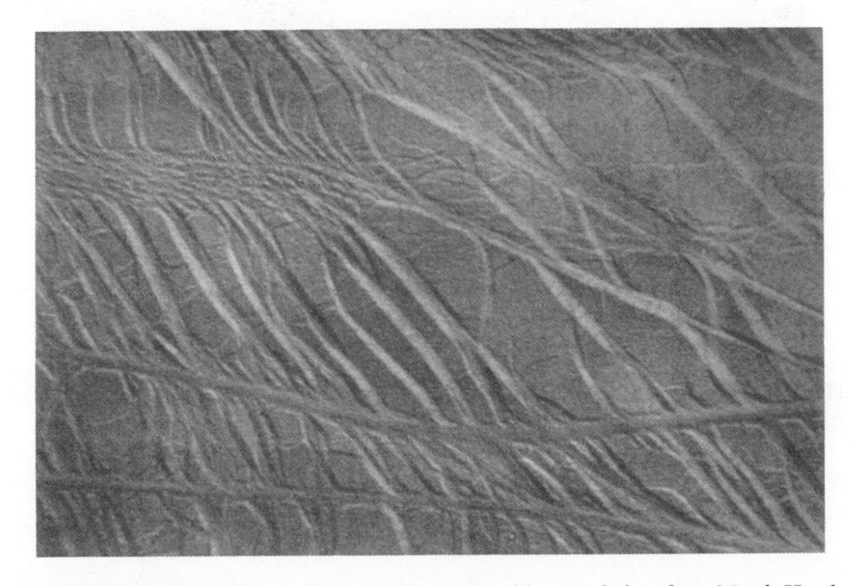

Fig. 10.20 Stretcher strains in sheet steel (Reproduced by permission, from Metals Handbook, American Society for Metals, 1955).

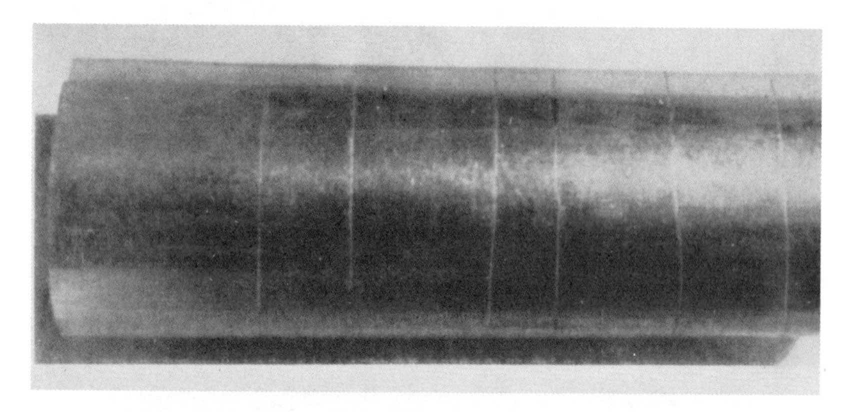

Fig. 10.21 Cracks in a hot forged steel mandrel (Reproduced by permission, Elsevier Scientific Publishing Co.).

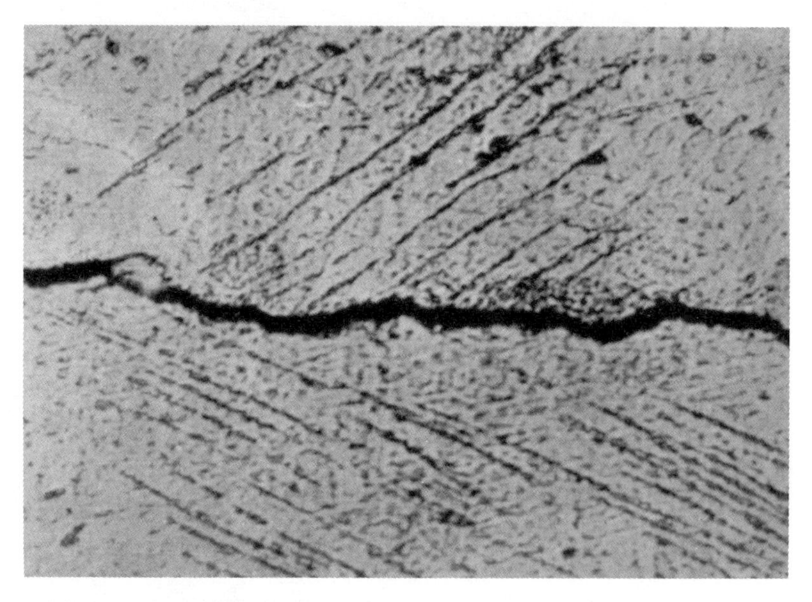

Fig. 10.22 Crack caused by cold working deformation (Reproduced by permission, Elsevier Scientific Publishing Co.).

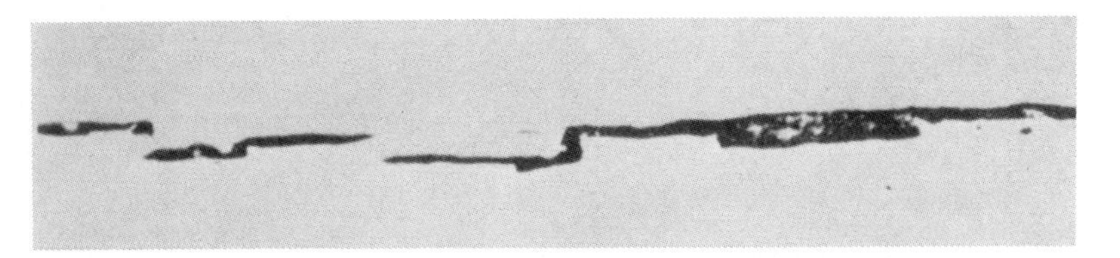

Fig. 10.23 Crack in patented steel wire (Reproduced by permission, Elsevier Scientific Publishing Co.).

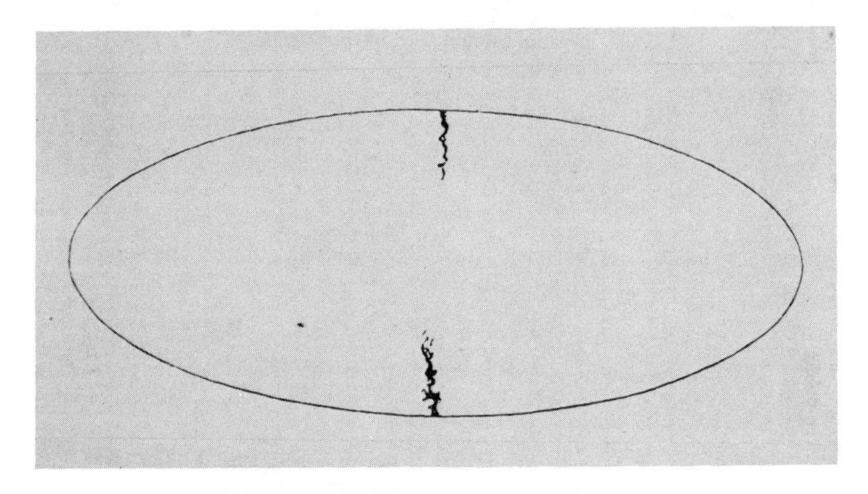

Fig. 10.24 Cracks caused during rolling by crushing between the rolls (Reproduced by permission, Elsevier Scientific Publishing Co.).

because applied stresses are higher at the surface in most applications.

Cracking may occur during any hot or cold working operation. Cracks in a hot forged steel mandrel are shown in Fig. 10.21,[18] and Fig. 10.22 shows a crack in cold worked steel.[19] Figures 10.23 and 10.24 illustrate typical cracks caused during patenting[20] and rolling,[21] respectively.

As shown previously, surface cracking occurs if the metal is underheated, since its ductility is too low for the amount of straining. In some instances, it can occur because of overheating. Any low melting point constituents or segregated areas that are present may melt at the temperature used to preheat the metal for working, which causes a weakness known as "hot shortness." This can occur in a steel with a high sulfur content. Another distinct potential for cracking exists in those metals where the working causes phase precipitation, which can result in increased yield strength and reduced ductility.

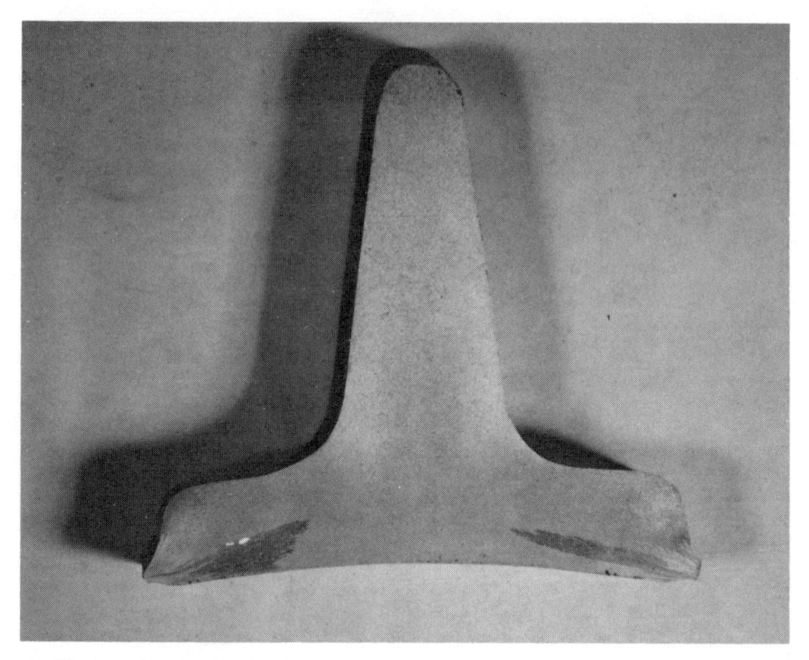

Fig. 10.25 Large grains in aluminum forging due to excessive deformation at the flash line.

The problem with overheating is not confined to precipitation or melting. Excess grain growth may occur, which results in a weakening of the metal. Excess heating during the forming operation can also cause large grains. In Fig. 10.25 overheating, which is associated with very large deformation, caused large grains in the aluminum forging. Such conditions are not discernible by nondestructive testing. There would not have been a deleterious effect in this illustration, since the load on the section with the large grains was small. However, in the presence of a high stress, the weaker large grains acted as crack initiators.

Surface flaws, as well as internal ingot flaws, can be magnified by working. Any defect on the surface of the original ingot will exist on the final piece. Therefore, extensive programs to remove surface flaws by grinding or scarfing are necessary. Despite this precaution, problems still exist. The formation of laps[22,23] is another common problem. Figure 10.26 and 10.27 show the formation of a lap, and the result of this process, respectively. These occur when the metal on the surface of the piece being worked folds over. Because of oxidation, it is not possible for them to weld together, which results in a lap. Laps can also occur when a crack fills in during

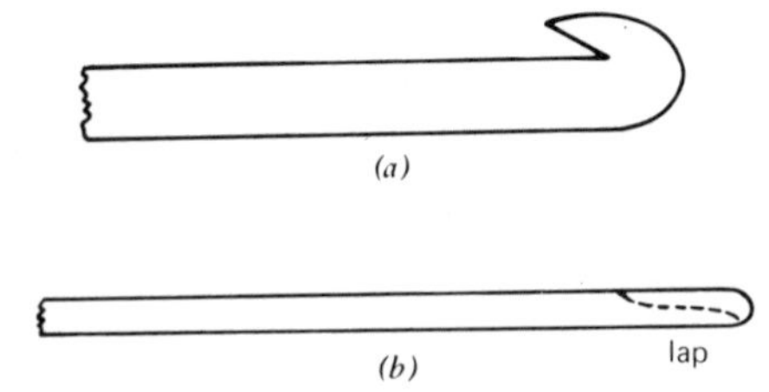

Fig. 10.26 Schematic showing the formation of a forging lap (Reproduced by permission, Elsevier Scientific Publishing Co.).

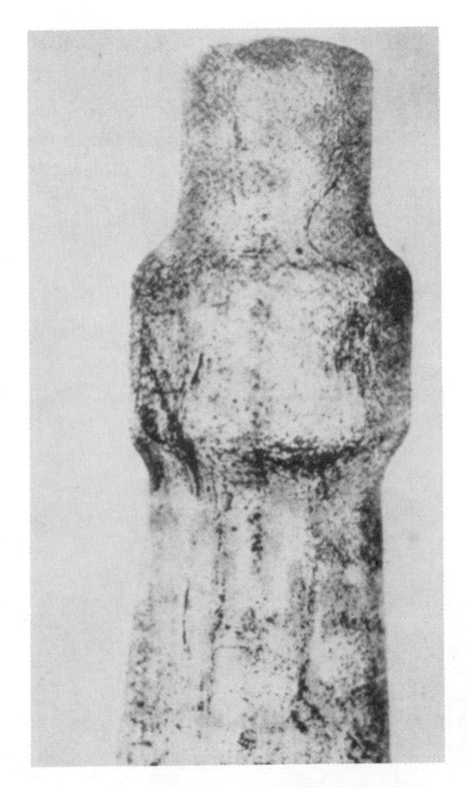

Fig. 10.27 Example of a forging lap (Reproduced by permission, Elsevier Scientific Publishing Co.).

Fig. 10.28 Fracture surface of a crane hook showing that the origin of failure is a lap. Note how the macroscopic pattern points toward the lap (arrow) (Reproduced by permission, Society for Experimental Stress Analysis).

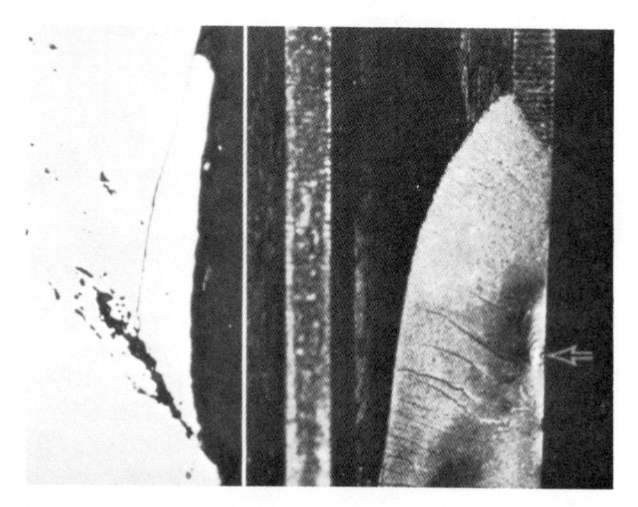

Fig. 10.29 Failure of a gear tooth originating in a forging lap (Reproduced by permission from How Components Fail, American Society for Metals, 1966).

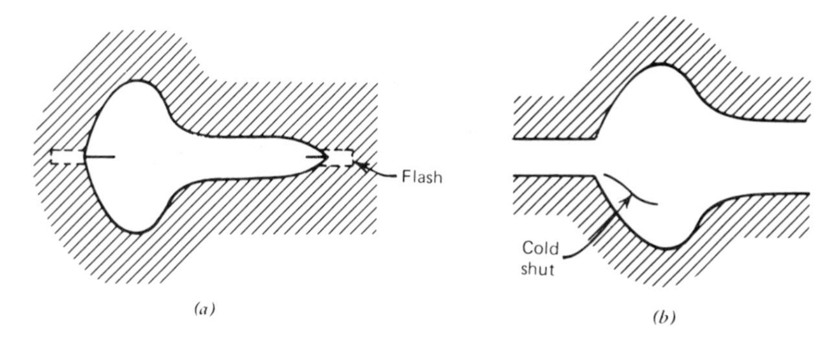

Fig. 10.30 Schematic of typical forging defects. (*a*) Cracking at the flash line. (*b*) Cold shut or fold.

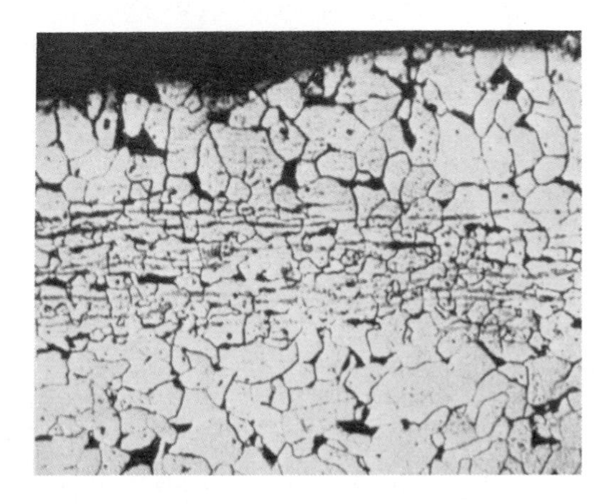

Fig. 10.31 Folding due to cold working in a boiler plate failure (Reproduced by permission, Elsevier Scientific Publishing Co.).

rolling. Two service failures directly associated with laps are shown in Figs. 10.28, the failure of a crane hook,[3] and 10.29, the failure of a gear tooth.[24] In both cases, the lap initiated a crack.

Cracks sometimes occur at flash lines, particularly if the flash is thin, as illustrated in Fig. 10.30. Another defect, cold shuts, is also illustrated in this figure. They are similar to a lap, except that there is no overlay of material, as shown in Fig. 10.31.[25]

If a surface defect, or crack, has been the cause of a failure, it is usually apparent. It is possible to locate the source of the failure by following chevron marking or beach marks, as described in Chapter 4. Large grains can be seen by the metallographic examination of a section. Folds or laps can be identified by the presence of an oxide scale on internal surfaces. Once the cause of a failure has been determined, it is possible to correct the material or forming practice to prevent its recurrence. However, this requires a fair understanding of the metallurgy of the alloy system and the forming practice that is used. Although many general principles are applicable to most metals, each system has its own critical temperatures and deformation limitations, which control the capabilities of the working process.

REFERENCES

1. *Metals Handbook*, 1955 Supplement, August 15, 1955, p. 153.
2. Z. Zimerman and B. Avitzur, *Met. Prog.*, February 1969, p. 103.
3. T. Dolan, *W. M. Murray Lecture*, SESA, 1969.
4. *Metals Handbook*, 1955 Supplement, *op. cit.*, p. 94.
5. R. C. Tittel, *Met. Prog.*, April, 1969, p. 76.
6. G. Sachs and K. Van Horn, *Practical Metallurgy*, ASM, 1948, p. 172.
7. E. P. Polushkin, *Defects and Failures of Metals*, Elsevier, Amsterdam, 1956, p. 136.
8. J. Schey, F. Shunk, and P. Wallace, *Met. Prog.*, November, 1966, p. 93.
9. E. P. Polushkin, *op. cit.*, p. 10.
10. E. P. Polushkin, *op. cit.*, p. 12.
11. F. Heiser, *Anisotropy of Fatigue Crack Propagation in Hot Rolled Banded Steel Plate*, Ph.D. Thesis, Lehigh University, 1969.
12. E. P. Polushkin, *op. cit.*, p. 96.
13. E. P. Polushkin, *op. cit.*, p. 265.
14. E. P. Polushkin, *op. cit.*, p. 73.
15. *Metals Handbook*, 1955 Supplement, *op. cit.*, p. 155.
16. R. C. Tittel, *op. cit.*, p. 77.
17. *Metals Handbook*, 1955 Supplement, *op. cit.*, p. 7.
18. E. P. Polushkin, *op. cit.*, p. 260.
19. E. P. Polushkin, *op. cit.*, p. 262.
20. E. P. Polushkin, *op. cit.*, p. 228.
21. E. P. Polushkin, *op. cit.*, p. 270.
22. E. P. Polushkin, *op. cit.*, p. 299.
23. E. P. Polushkin, *op. cit.*, p. 300.
24. D. Wulpi, *How Components Fail*, ASM, p. 18.
25. E. P. Polushkin, *op. cit.*, p. 229.

11

Casting Defects

Many types of defects, such as blows, misruns, sags, and buckles, can arise during the production of a casting. Some of these are surface defects, others are gross in nature. Most are readily discernible. This chapter discusses those which occur subsurface or internally, since these latent defects are usually responsible for an unexpected failure.

11.1 SEGREGATION

11.1.1 General

The distribution of chemical elements in an alloy is seldom uniform. Even "pure" metals contain random amounts of various impurities, in the from of tramp elements or dissolved gases. Therefore, the metal or alloy in castings, ingots, welds, and even wrought structures is usually comprised of regions of varying composition. This deviation from the mean composition is termed segregation. In the liquid state the atoms possess sufficient mobility, and there is generally sufficient mechanical mixing to render the melt uniform. The problem of segregation, therefore, arises upon solidification, and any consideration of segregation must involve the topic of solidification.

11.1.2 Pure Metals

Since completely pure metals do not exist, the discussion of solidification for a pure metal represents the limiting boundary condition. According to Gibbs' phase rule, the degrees of freedom at constant pressure may be calculated from

$$F = C - P + 1$$

where

C = number of components
P = number of phases in equilibrium

The idealized pure metal is composed of a single component and there are only two phases present (liquid and solid); therefore, there are zero degrees of freedom. That is, the liquid and solid have the same concentration, and the temperature of the interface, which is invariant, is the melting point of the pure metal.

In solidifying, the atoms in the liquid must give up a part of the energy they possess and assume positions in the crystal structure of the pure metal characteristic of the solid state. The difference in energy between the solid and liquid states corresponds to the latent heat of fusion.

The purest common structural metals, however, usually contain sufficient impurities to allow them to be considered alloys, from the solidification standpoint.

11.1.3 Alloys

Alloys, by definition, must be multicomponent systems, and solidification usually occurs (with few exceptions) over a range of temperatures. The mechanics of solidification for alloys differs from that of pure metals in two ways.

1. Solidification occurs over a range of temperatures, rather than at a constant temperature.
2. The solidification process requires a redistribution of the solute atoms and the dissipation of the latent heat of solidification.

11.1.4 Redistribution of Solute During Solidification

It is advantageous to begin a discussion of segregation by considering variations in the solidification process since it is intimately related to the form and degree of the segregation. Upon segregation, a redistribution of solute occurs, which depends on the boundary conditions. Three cases may be postulated.

1. Equilibrium is maintained in the system at all times.
2. Negligible diffusion in the solid with sufficient mechanical mixing in the liquid to render is completely homogeneous at all times.
3. Negligible diffusion in the solid and no mechanical mixing in the liquid; diffusion is the only method for modifying the composition of the liquid phase.

EQUILIBRIUM IS MAINTAINED IN THE SYSTEM AT ALL TIMES

As the melt of mean composition, C_0, loses heat, the temperature falls, and upon reaching the liquidus, solidification begins. The first solid to form is of composition C_1 at temperature T_i (Fig. 11.1). Since for the case shown the composition of the solid, C_1, is lower than that of the melt, C_0,

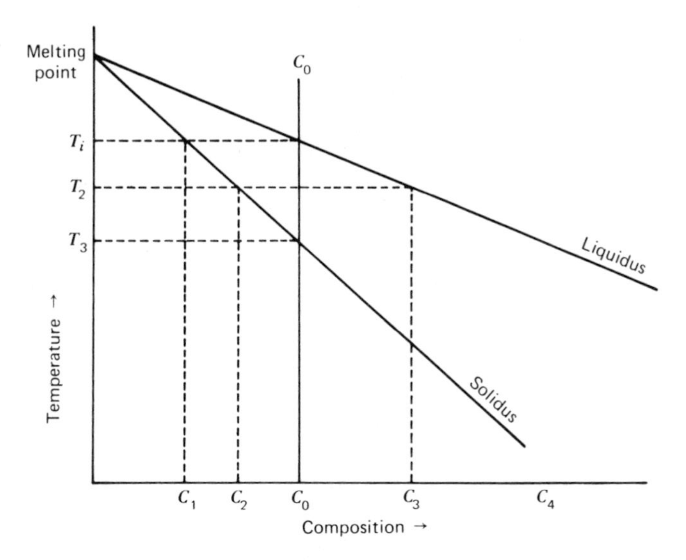

Fig. 11.1 Equilibrium solidification.

in terms of solute concentration, the formation of a significant amount of solid, therefore, must be accompanied by rejection of the solute and a corresponding increase in the solute concentration in the liquid. However, since we have assumed the system to be in equilibrium, compositional adjustments must occur in the solid and in the liquid. For example, at any temperature T_2, the concentration of the solute in the solid has been made uniform throughout, by diffusion, and is given by composition C_2. At this same temperature, the concentration of solute in the liquid has been made uniform throughout, by mixing and diffusion, and is given by C_3.

The percentage of the solid and liquid in equilibrium at any temperature can be calculated from the lever law.

At T_2

$$\text{percent of liquid} = \frac{C_0 - C_2}{C_3 - C_2} \times 100$$

$$\text{percent of solid} = \frac{C_3 - C_0}{C_3 - C_2} \times 100$$

As the temperature is slowly lowered, solidification continues. The volume of the solid increases while the volume of liquid decreases, and the homogenizing processes (diffusion and mixing) continue to operate to keep each at the equilibrium composition for the temperature in question.

At T_3 the last liquid has disappeared and solidification is complete. The excess solute in the last infinitesimal quantity of liquid is uniformly distributed throughout the solid phase by diffusion, and the composition of the solid at this point is C_0, the nominal compostional of the alloy.

NEGLIGIBLE DIFFUSION IN THE SOLID WITH SUFFICIENT MECHANICAL MIXING IN THE LIQUID TO RENDER IT HOMOGENEOUS

The nominal composition of the heat is C_0. Upon solidification, the first solid to form is C_1 at temperature T_i as shown in Fig. 11.2. At any temperature T_2 below T_i, the composition is given by the intersection with the solidus, C_2. Since by definition homogenization cannot occur, there is now a distinction from the previous case. A concentration gradient exists between the last solid to form and the first, and each infinitely small layer of solid is a slightly different composition than the previous. The *mean* composition C_m of the entire solid at any temperature T_x may be expressed by $(C_x - C_1)/2$. Therefore, a line AC may be constructed so that $AB = CD$. This line describes the locus of points representing the mean composition of solid phase.

The composition of the liquid at temperature T_x is given by C_l. Complete mixing is assumed, and no concentration gradient exists in the liquid.

The percentage of solid and liquid in equilibrium at any temperature T_x,

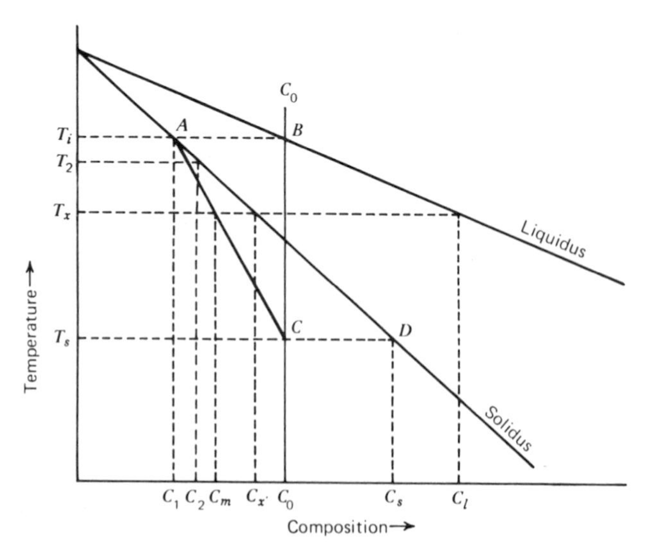

Fig. 11.2 Solidification with negligible diffusion in solid, with mechanical mixing in liquid.

may again be calculated from the lever law. However, the mean composition line AC must be used, rather than the solidus.

At T_x

$$\text{percent of solid} = \frac{C_l - C_0}{C_l - C_m} \times 100$$

$$\text{percent of liquid} = \frac{C_0 - C_m}{C_l - C_m} \times 100$$

At T_s solidification is complete. The temperature is defined by the coincidence of the mean solid composition line AC with the nominal composition C_0. At this temperature the last remnants of liquid solidify, forming a solid of composition C_s, with the mean composition of the entire solid being the nominal composition C_0.

NEGLIGIBLE DIFFUSION IN THE SOLID AND NO MECHANICAL MIXING IN THE LIQUID, DIFFUSION IS THE ONLY METHOD FOR MODIFYING THE COMPOSITION OF THE LIQUID PHASE

Given a melt of nominal composition C_0, upon cooling to T_i the composition of the first solid to form is C_i as shown in Fig. 11.3. Using the distribution coefficient, the composition $C_i = K_0 C_L$, therefore, $C_i = K_0 C_0$.

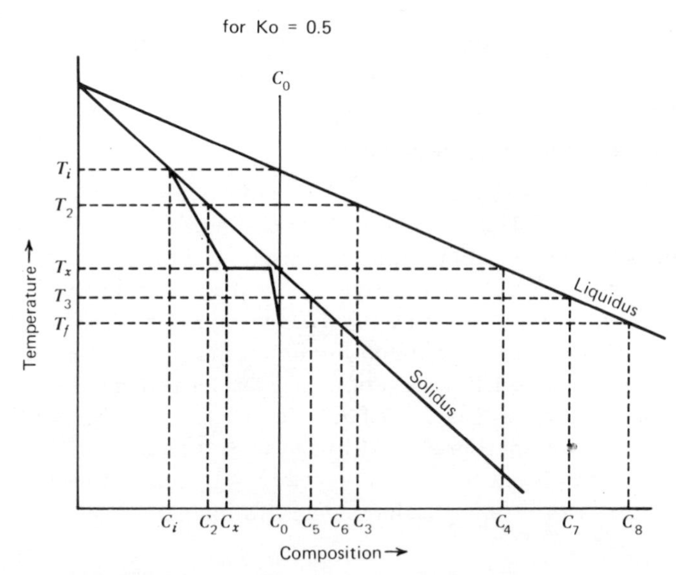

Fig. 11.3 Solidification with negligible diffusion in solid, with no mechanical mixing in liquid.

The composition of this solid is significantly lower than the melt; therefore, its formation must be accompanied by the rejection of solute, thereby enriching the liquid. Since the concentration of solute in the liquid can only change by diffusion, a concentration gradient is established in the liquid and the next solid to form is higher in solute than the previous one. Consequently, a concentration gradient is created in the solid as well. Solidification proceeds until the temperature falls to T_x, which marks the onset of dynamic equilibrium. As in the previous case, the mean or average composition of the solid up to this point may be calculated by $(C_i + C_0)/2$.

At T_x the composition of the solid forming at the interface is equal to C_0, and that of the liquid is equal to C_4. This composition is considerably richer in solute than the bulk of the liquid. A steep concentration gradient exists on the liquid side of the interface due to the lack of mechanical mixing. The mean composition shifts along the composition axis in the direction of the nominal composition. The steady state stage is limited, however, by the physical confines of the system. As the volume of the solid formed increases, there is, of course, a corresponding decrease in the liquid. When this liquid diminishes to the point where only the enriched volume of liquid material preceding the interface is available, the terminal stage of solidification begins. This stage is marked by increased concentrations of solute in the solid since the liquid from which solidification proceeds is heavily enriched. For example, at T_3, the composition of the solid formed is C_5, from liquid of composition C_7. Similarly, at T_f the composition of the final solid to form is C_6, from liquid of composition C_8. Upon total solidification, the mean composition is identical to the nominal composition, C_0.

11.1.5 Classification of Segregation Types

Segregation may be simply defined as a deviation from the nominal composition at a particular location in the casting. In general, segregation is the result of solute rejection at the solidification interface. In the preceding discussions, simple binary systems have been considered; however, solidification in real systems may involve the rejection of more than one solute. When the distances between minimum and maximum solute concentration are small, such segregation may be termed microsegregation. When these distances are large, relative to the casting, the term macrosegregation is more appropriate. Although these are general terms, other terms are also employed to describe types of segregation that vary in the specific degree or direction of solute concentration.

NORMAL SEGREGATION

The distribution of solute along a line normal to the solidification front characterizes this type of segregation. When such segregation results, the exterior or surface zones of the casting exhibit a higher concentration of the higher melting point components. Conversely, the central regions of the casting are richer in the lower melting components. This may be explained by the second case in Section 11.1.4 which is a simple approximation of nonequilibrium solidification occurring in many metallurgical processes. Figure 11.2 shows that the initial material to solidify had a composition of C_1 and the final solid was of composition C_8, a concentration much higher in solute than the initial melt. Also, the temperature range of solidification has almost doubled, and a linear concentration gradient has resulted, even though the average composition is still C_0.

DENDRITIC SEGREGATION

Segregation is usually determined with respect to the casting as a whole. However, segregation on a finer scale also occurs within the grain, in the form of dendritic segregation. According to Chalmers,[3] dendritic solidification is characterized by the formation of dendrites, a linear branched structure with arms parallel to specific crystallographic directions and

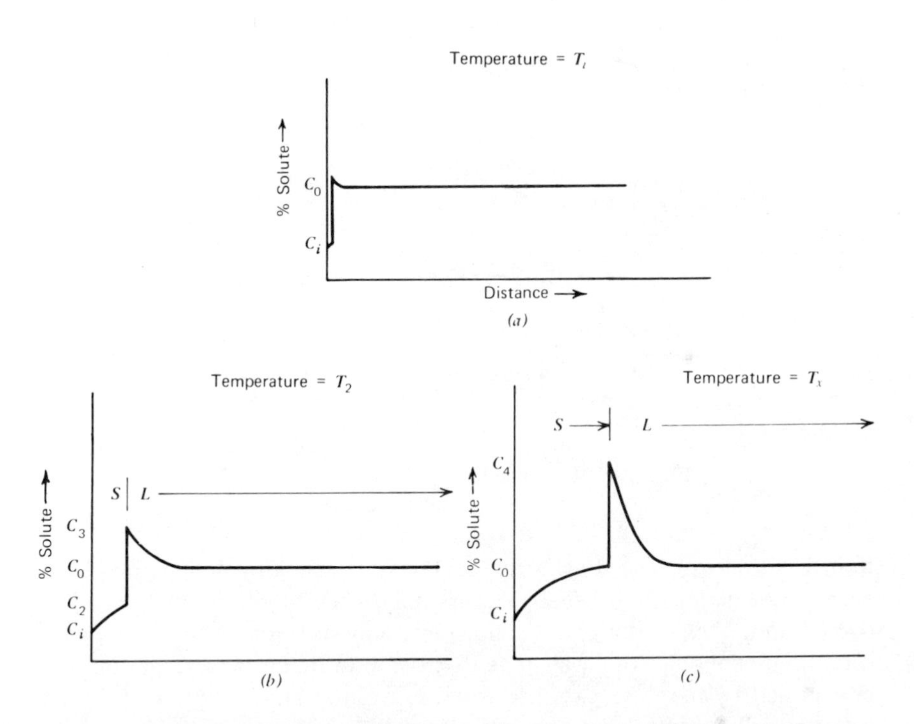

Fig. 11.4 Schematic showing the distribution of solute during dendritic solidification. (*a*) Temperature = T_i. (*b*) Temperature = T_2. (*c*) Temperature = T_x. (*d*) Temperature = T_3. (*e*) Temperature = T_f.

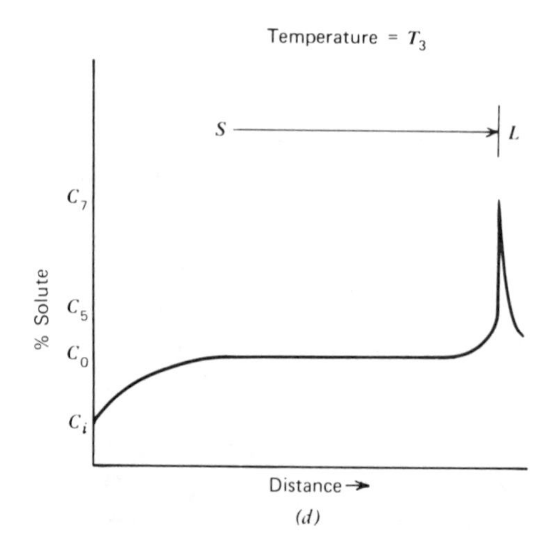

(*d*)

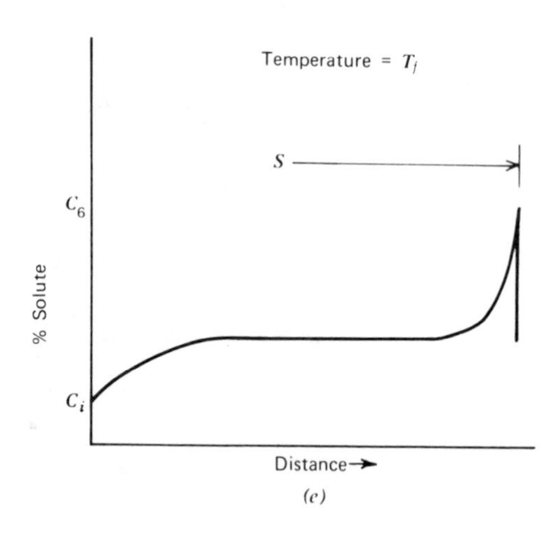

(*e*)

Fig. 11.4 (Continued)

branches that are spaced at fairly regular intervals. All solidification, however, is not dendritic. It usually occurs only when the liquid melt is supercooled and, even under these conditions, only a minor proportion of the total liquid solidifies this way. The discussion of dendritic segregation may be clarified by the third case presented in Section 11.1.4. Figure 11.4 illustrates the situation for dendrite solidification.

The temperatures shown refer back to Fig. 11.3. In Figure 11.4*a* solidification begins at temperature T_i, initiating at the dendrite nucleus. The initial solid is lower in solute and the liquid is slightly enriched. At temperature T_2, the composition of the forming solid is C_2, which is higher in solute content; however, the solute concentration in the liquid at the interface is also increasing. Figure 11.4*c* shows the conditions that exist when dynamic equilibrium commences. Tiller[1] indicated that the composition of the solid at any point along the curve from C_i to C_0, called the initial transient, is approximately equal to

$$C_s = C_0 \left\{(1 - K_0)\left[1 - \exp\left(K_0 \frac{R}{D} X\right)\right] + K_0\right\}$$

where

K_0 = distribution coefficient
D = diffusion coefficient in Liquid
R = rate of growth of interface

It should be noted that the rate of growth is now a consideration, and can appreciably affect the distribution of solute.

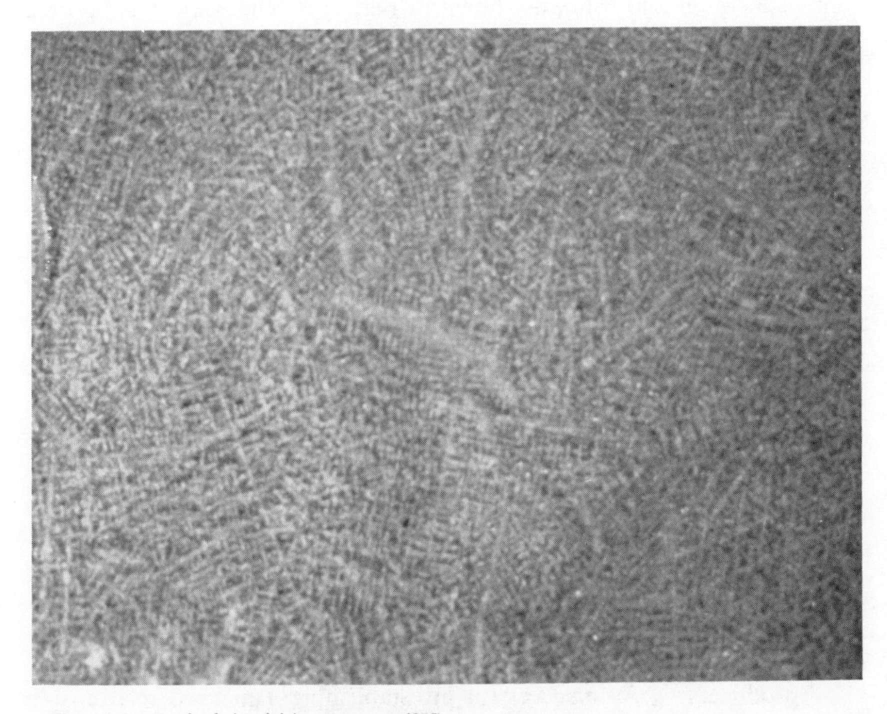

Fig. 11.5 Typical dendritic structure (2X).

Upon completion of the initial transient, the temperature remains constant at T_x and the solid of composition, C_0, forms from the liquid of composition, C_4, as long as such liquid is available. As the liquid interface proceeds ahead of the growing dendrite, it approaches the interfaces of competing dendrites. When the two concentration spikes impinge on one another, the solute content in the solid at the interface begins to increase, as shown in Fig. 11.4*d*. Finally, as the two interfaces meet, the conditions shown in Fig. 11.4*e* exist. The last liquid of composition, C_8 reacts to form the solid of composition, C_8. The photograph in Fig. 11.5 illustrates a typical dendritic structure.

INVERSE SEGREGATION

Inverse segregation is the type of segregation where the solute rejected during solidification is found in higher concentrations in early solidification zones than in later ones. This is explained by the fact that this situation is generally accompanied by shrinkage. Solidification stresses acting on the initial solid create shrinkage cavities that are in contact with the enriched liquid. The liquid permeates these cavities, thereby increasing the solute content of the initial solid, as shown in Fig. 11.6. This view is supported by the experiments of Winegard.[2] As additional proof, Chalmers[3] cites the observation of Adams[4] that inverse segregation does not occur in alloys that expand upon solidification.

GRAVITY SEGREGATION

Gravity segregation occasionally occurs in large systems where a marked density difference is created by the rejected solute. The net effect is a migration of the denser material to the lower zones of the liquid region.

FRECKLING

Freckling is a type of segregation commonly associated with vacuum-consumable electrode-melted alloys. The segregation is revealed upon macroetching as darkly etched spots, as shown in Fig. 11.7. The freckles are composed of a particle type segregate rich in various elements, such as Ni, Ti, and S. The composition and distribution of the freckles varies with the alloy. The example shown is a section from an 11 in. square billet of A286 alloy that was forged from a 26 in. diameter, consumable electrode ingot.

The effect of freckling is generally subtle; however, it has an adverse effect on transverse ductility. Repeated efforts, conducted by many superalloy producers, were unsuccessful in establishing a marked relationship between the degree of freckling and the gross mechanical properties.

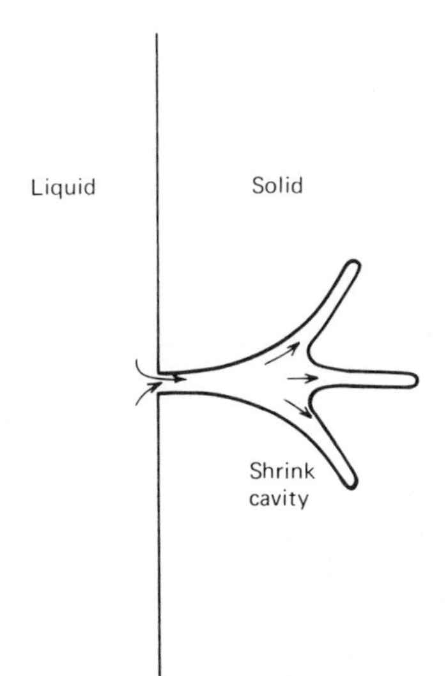

Fig. 11.6 Schematic illustration of inverse segregation.

Fig. 11.7 Freckling on a macroetch disk (1/4X).

11.1.6 Detection of Segregation

MACROETCHING

The most common method used to determine the existence of segregation is macroetching. Transverse or longitudinal sections, as desired, ground smooth but not polished, are cleaned and immersed in etching tanks. Segregation is revealed by the presence of patterns on the metal surface caused by differential etching of the segregated area. The intensity of the etch roughly corresponds to the degree of segregation. Examples of such patterns are shown in Fig. 11.8.

The composition and immersion parameters of the etch vary with the alloy system and also depend on the type of segregation to be revealed.

Kehl[5] prepared a comprehensive list of macroetches for general use. Additional information may also be found in the ASM *Metals Handbook*.[6]

SULPHUR PRINTING

Sulphur printing of steel is a specialized use of the macroetching technique. The method is fairly widespread, and is used specifically for the detection of sulphur segregation. A sheet of photographic printing paper is

Fig. 11.8 Ring pattern on a macroetch disk (1/4X).

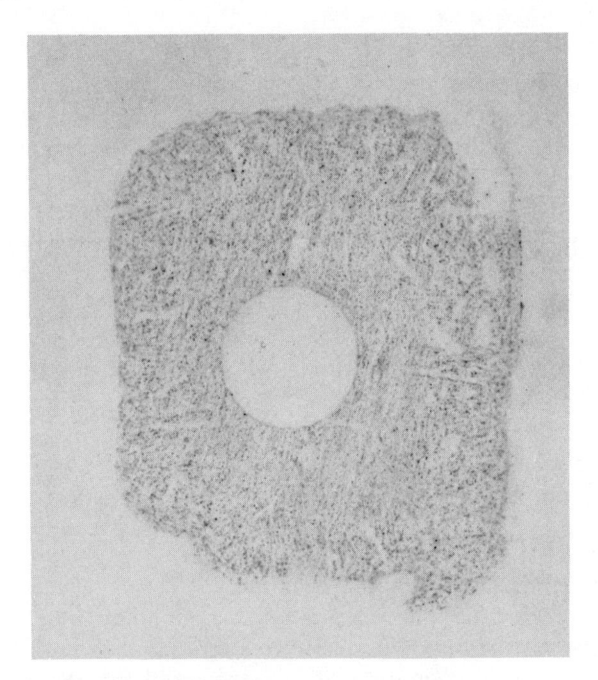

Fig. 11.9 Sulphur print—medium carbon steel casting (3X).

lightly wet with a 10% solution of reagent grade sulphuric acid. The paper is firmly placed on the steel to be tested. Care must be taken to insure uniform contact with the surface. For this reason, the specimen usually has a finer surface finish than in conventional macroetching. The acid reacts with sulphide inclusions in the steel producing H_2S; this reacts with the silver in the photographic paper to form AgS, which is brown. The net result is the formation of sepia patterns on the paper, as shown in Fig. 11.9. Although sulphur printing, like macroetching, is generally qualitative in nature, the intensity and distribution of the discoloration can provide an approximate index of the amount of sulphur present.

A. Pokorny and J. Pokorny[7] have done a remarkable job in the preparation and presentation of cast ingot sections. Their photographs and sulphur prints clearly show that the maximum concentration of sulphur occurs in the interdendritic regions and that the dendrite axes themselves are relatively free of sulphur. The effect of the high sulphur concentration in these regions usually exerts a detrimental effect on the mechanical properties and the corrosion behavior as well, as shown in Chapter 7.

CHEMICAL ANALYSIS

Ordinary quantitative analysis is largely confined to determining macro-segregation. Samples in the form of drillings are taken at specified locations, such as the edge, midradius, and center location of the top and bottom of the bloom or billet. The surface drillings are discarded because of possible contamination and chemical analysis for the desired elements is performed on the balance. The results obtained correspond to the mean composition of the volume tested and do not describe the distribution of elements through that volume.

Similarly, spectrographic analysis can also be used to determine the presence of segregated areas. Specimens are taken from the area under consideration; frequently these are cut from macroetch discs and analyzed in the spectrograph in a conventional manner. As with wet chemical analysis, the results do not yield information on the nature of the micro-segregation, though the volume tested is generally smaller.

MICROPROBE ANALYSIS

Microprobe analysis, or more accurately, electron beam probe micro-analysis, is a technique, that permits routine quantitative determination of all elements with atomic numbers greater than sodium. With special techniques and instrumentation, it is reported[8] that analysis of elements down to lithium is also possible.

The microprobe analysis operates by means of a fine beam of high energy electrons which is focused on the metal surface. The impingement of these electrons produces X-rays whose wavelengths are characteristic of the elements under the beam. The elements present are determined by measuring the wavelength of the excitation produced; the intensity of such excitation corresponds to the concentration of the elements. Specimen preparation is very similar to that required for metallography in that a flat smooth surface is desirable. Surface contamination is obviously a problem, and the use of polishing compounds that will contaminate the metal surface are to be avoided. Diamond polishes are generally satisfactory for this purpose.

A number of variations can be employed, depending on the information required. Analysis of a single point can be produced, as well as a linear traverse. Figure 11.10 shows linear traverses across an inclusion for two elements, Si and Mn, with the inclusion in the background.

AUTORADIOGRAPHY

The autoradiographic technique extends the range of observation for segregation studies. Radioisotopes introduced into the melt permit the detection and definition of segregation, which might be too small to observe by other means.

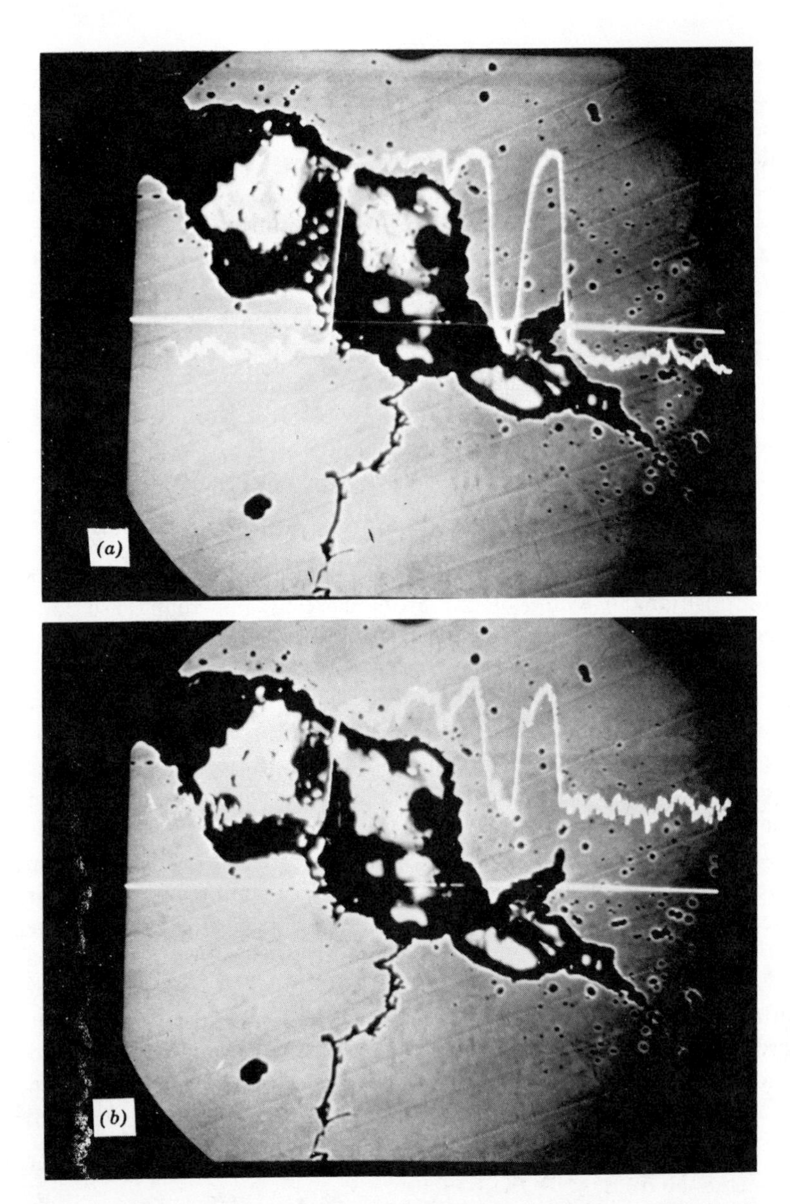

Fig. 11.10 Microprobe trace of an inclusion—linear traverse for Mn and Si. (*a*) Electron probe backscatter photograph-Mn trace. (*b*) Electron probe backscatter photograph-Si trace.

The basis of the technique is the incorporation of a radioactive isotope, a tracer, into the melt to be studied. This effectively tags the element under investigation. A relatively large number of isotopes that can be used are available either from the Atomic Energy Commission or from commercial houses. However, to be useful the isotope should possess a long half-life, and a high specific activity (curies/g). After solidification, the orientation to be examined is selected and the surface is polished, as for microanalysis. A stripping film is placed firmly in contact with the specimen and is thereby exposed. The film may then be processed by ordinary photographic means. The high magnifications attainable with stripping films illuminate details of the segregation that could not otherwise be observed.

Leymonie[9] discussed in detail the application of radioactive tracers in metallurgical investigation. These techniques offer a fruitful means for attacking a number of metallurgical problems.

11.1.7 Effect of Segregation

The net result of segregation is the production of material with a range of compositions. It would be unreasonable to assume that these compositions would exhibit identical properties, and they do not.

The presence of localized regions that deviate from the nominal composition can have far-reaching ramifications in many areas, such as corrosion resistance, forging and welding characteristics, mechanical properties, fracture toughness, and fatigue resistance. In heat-treatable alloys variations in the composition can produce unexpected responses to heat treatments, which result in hard or soft spots, quench cracks, or other defects.

It is not possible to specify the degree of degradation since this depends on several details, including the alloy species segregating. Most metallurgical processes operate on the basis of a nominal composition and the premise that the material is entirely of this composition. However, the mean or average composition in a volume of metal is not the major consideration. Rather, the important factor is the distribution. If the concentration of segregating elements, such as phosphorus and sulphur, can rise manyfold over a short distance, for example, at a grain boundary, the purpose of a specification is defeated and catastrophic failures are invited.

11.2 GAS HOLES AND POROSITY

Gas holes and porosity are caused by the evolution and entrapment of gas in the solidifying metal. This gas may result from several sources: a decrease in solubility upon cooling from the liquid state, reaction of metallic oxides with carbon to form CO and CO_2, and the reaction of liquid metal with mositure in green sand molds.

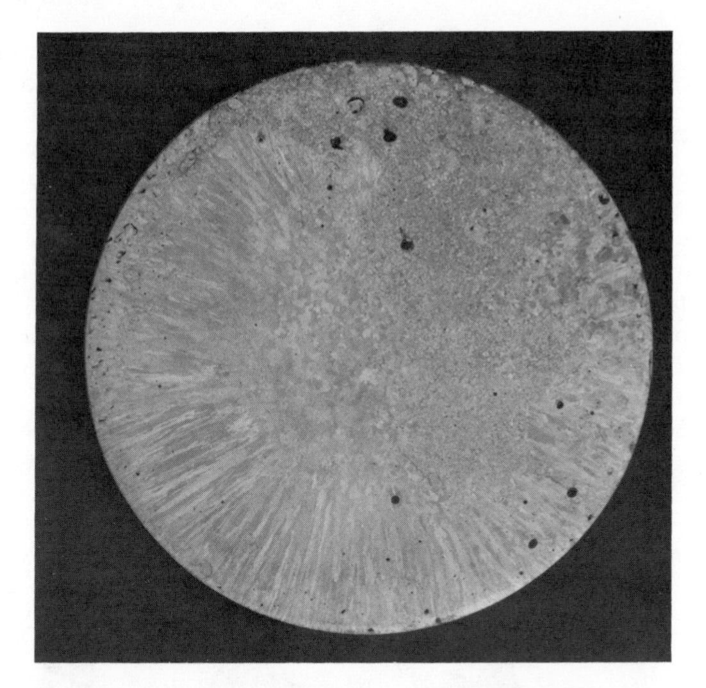

Fig. 11.11 Gas holes as revealed in a macroscopic cross-section (1/4X).

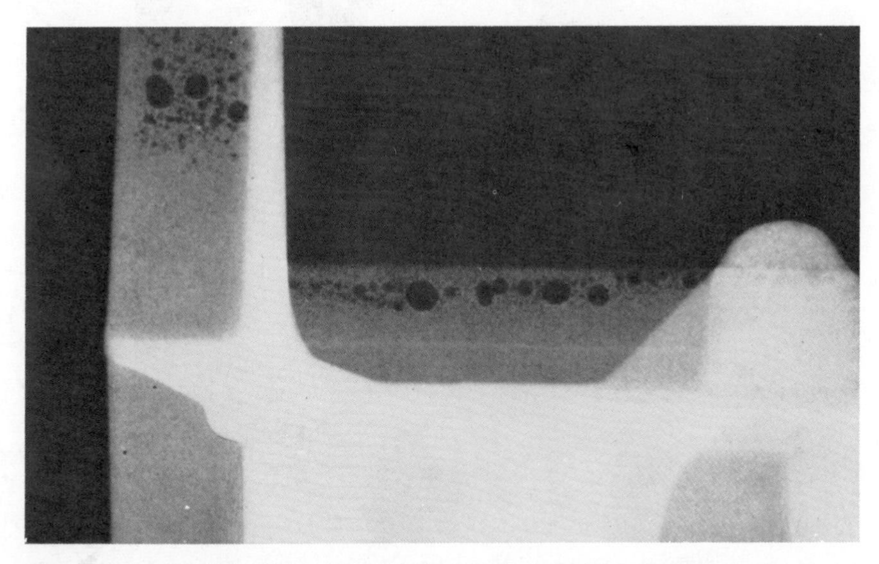

Fig. 11.12 Radiograph illustrating gas holes and gas porosity in the same casting.

In ingot sections, gas holes (or blow holes) appear as round or elongated cavities of smooth contour, as shown in Fig. 11.11. They may occur individually or in massive clusters, depending on the gas concentration. It is better practice to refer to individual cavities as gas holes and to reserve the term gas porosity for those clusters or groupings where individual cavities are barely discernible as discrete entities. The radiograph in Fig. 11.12 illustrates the occurrence of porosity and gas holes in the same casting. These appear as dark rounded indications. When these gas holes appear in an indeterminable number, they are generally referred to as gas porosity.

The primary effects of gas holes and porosity are that they reduce the load carrying capability of the member in which they occur, and act as stress concentrators,[10] thereby increasing the effective stress in that member.

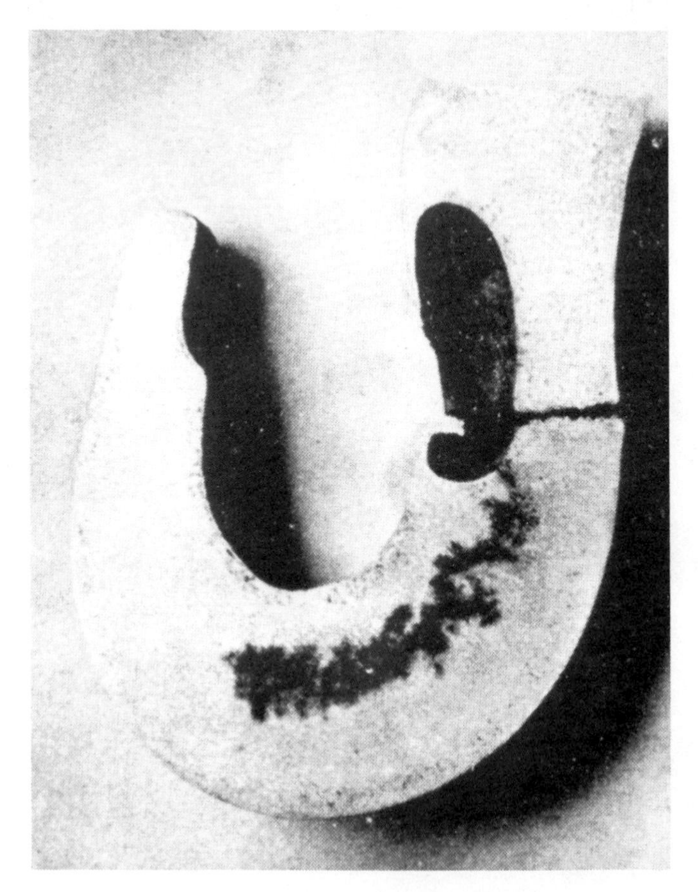

Fig. 11.13 Failure of a cast hook resulting from the presence of a massive blow hole. (Reproduced by permission, Elsevier Scientific Publishing Co.)

An example of failure[11] caused by the presence of a gas void is shown in Fig. 11.13. Note the unfortunate location of the hole on the inner surface, which would coincide with the maximum tensile stresses in the hook upon loading. In some instances, gas holes in a casting may occur in an aligned grouping. Such configurations are particularly dangerous since they tend to act as a single discontinuity, and may act as nuclei for fatigue cracks.

In some instances surface porosity or subsurface porosity exposed by machining or chemical polishing is encountered. Such defects are a particular problem in corrosive environments since they are apt to act as nuclei for subsequent pitting and corrosion fatigue damage. Figure 11.14 illustrates surface pits in the surface of a cast stainless steel surgical plate.

11.3 SHRINKAGE

Solidification of a metal casting begins at the casting surface and proceeds inward. As cooling and solidification proceed, contraction occurs. The contraction experienced is a result of: (*a*) cooling of the liquid, (*b*) change from the liquid to the solid, and (*c*) cooling of the solid.

Fig. 11.14 Surface porosity in a cast stainless steel surgical plate.

Usually, these various volume changes are occurring simultaneously in various parts of the casting. Thus the last metal to solidify, usually near the center of the main casting or in heavier sections, bears the full impact of the contraction, and if the casting is not fed from a molten pool, a cavity will form. The volume of the cavity depends on the mass of metal involved, the casting temperature, and the specific volume changes occurring in the liquid.

The form of shrinkage may vary from massive elongated cavities of irregular shape to small voids at the grain boundaries. Shrinkage defects are most easily discovered and identified by means of radiographic inspection. Since they consist primarily of void space, the shrinkage areas appear as darkened irregular areas, river-like or feathery in appearance. These indications are more or less prominent, depending on the magnitude of the shrinkage, and can range from the severe condition shown in Fig. 11.15 to the mild case shown in Fig. 11.16.

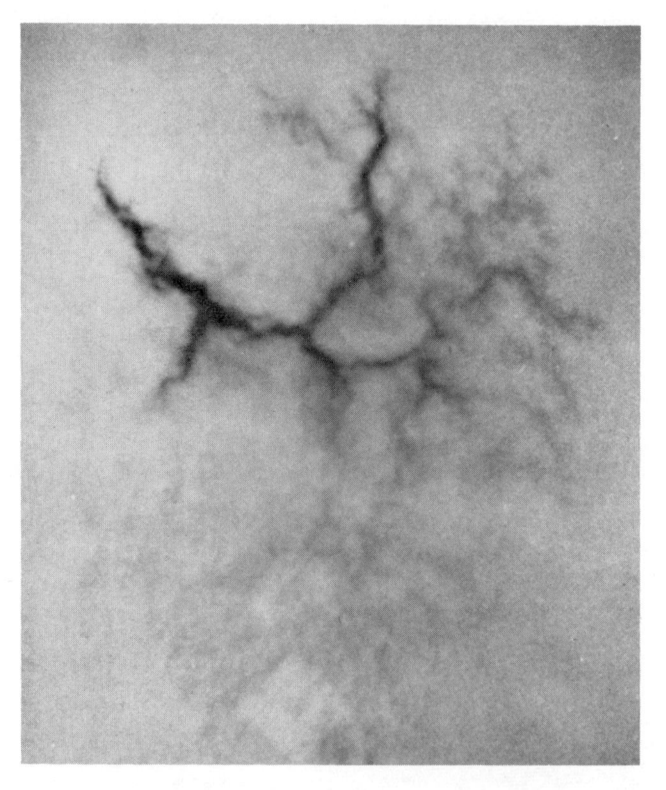

Fig. 11.15 Radiograph showing severe shrinkage in a steel casting (1X).

Fig. 11.16 Radiograph showing mild shrinkage in a steel casting (1X).

11.4 PIPE

Pipe is a particular form of shrinkage which occurs in the top of an ingot because of the contraction of a metal upon solidification from the molten state. As a rule, most metals and alloys contract upon solidification. The exceptions are Bi, Sb, Si, some of their alloys, and certain cast irons. The volume of contraction from the liquid to the solid state is about 4%.[12] It is with this change in volume that pipe, in particular, is associated. Pipe occurs in the last metal to solidify, in the center and top of the ingot. The form of shrinkage cavity is irregular, but is generally shaped like an inverted, elongated cone, as shown in Fig. 11.17. In addition to the main cavity, porous regions may exist around the margins. Regions of secondary pipe and porosity may extend deeper into the ingot.

One of the obvious negative features of pipe is an economic one, the substantial cost of metal lost in cropping the piped regions. Cropping is necessary even when the cross-section of the pipe is small, because surface

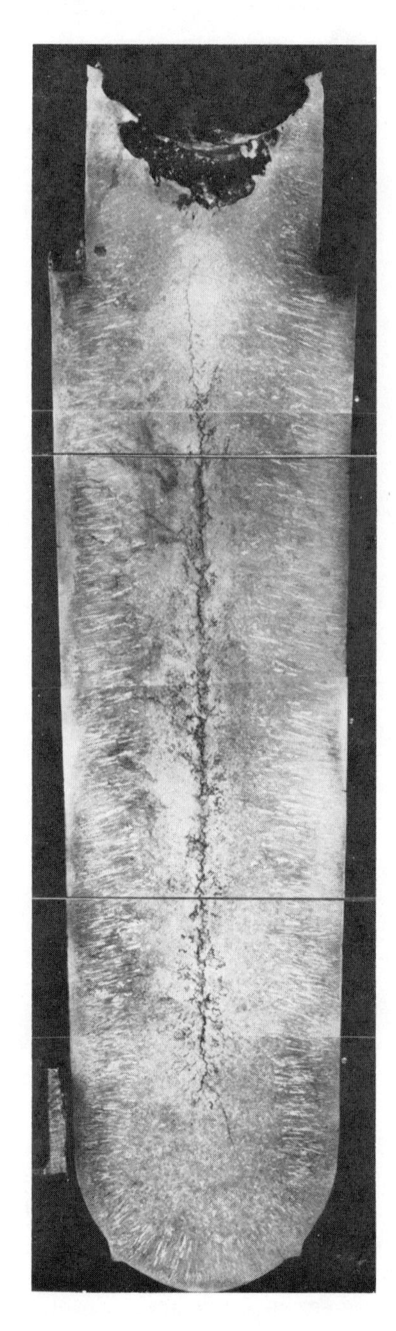

Fig. 11.17 Longitudinal section through an ingot illustrating a piped condition (1/8X).

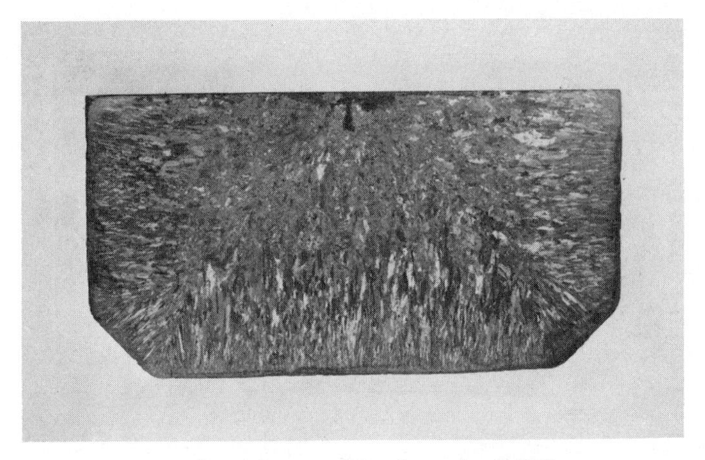

Fig. 11.18 Centerline defects resulting from pipe (1/3X).

areas exposed to air become oxidized and will not weld during rolling or forging.

From a failure analysis standpoint, primary pipe is not a severe problem since it is generally eliminated during mill processing. Secondary pipe and its associated porosity may, however, present problems since they may escape detection and cause centerline defects in bar and wrought products, as shown in Fig. 11.18.

11.5 HOT TEARS

Hot tears are crack-like defects in a casting which occur either internally or externally. The external tears are not important in failure analysis since they connect with the surface and are readily discernible and rejectable. The internal defects, however, are extremely important since they may go undetected unless radiographic or ultrasonic inspection is applied. Internal hot tears do not intersect the surface but may be revealed upon machining. Failure to detect the flaws may result in catastrophic failure since the tears can easily undergo extension under service conditions.

Hot tears are ragged and irregular in appearance and are generally discontinuous, as shown in Fig. 11.19. Often several tears are present in a roughly parallel array, and may exhibit some branching.

Radiographically hot tears are indicated by dark lines that have a wavy or irregular appearance. A radiograph showing a prominent example of internal hot tearing, as disclosed by radiographic inspection, is presented in Fig. 11.20.

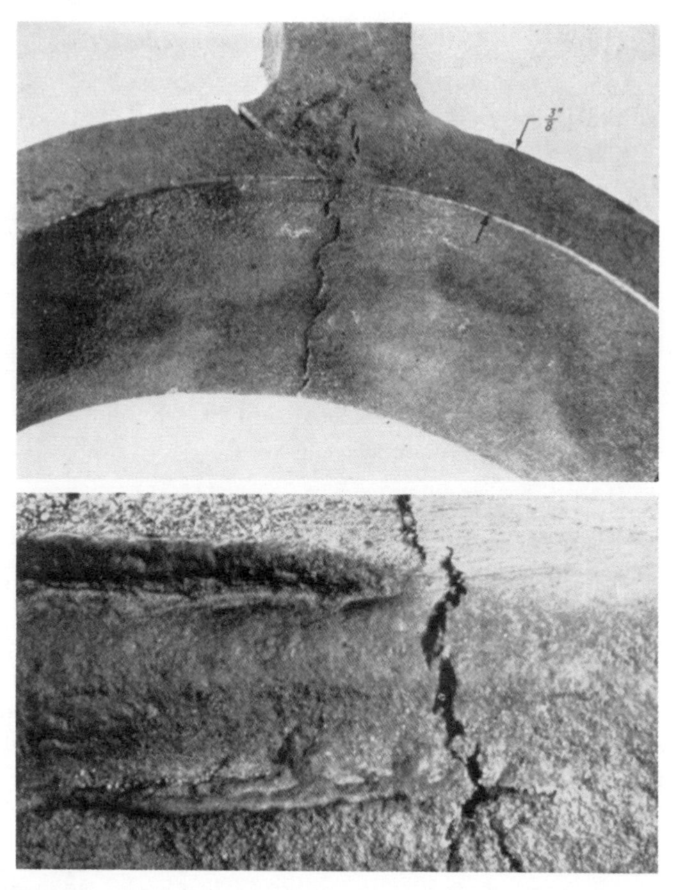

Fig. 11.19 Macroscopic appearance of hot tearing in a steel casting (1/2X) (from Ref. 13).

Hot tearing is caused by contraction stresses imposed on the metal shortly after its solidification. At this time, the metal exhibits low mechanical properties and cracking can occur at stress levels much lower than would be required at room temperature. These stresses result from shrinkage occurring in the solid state. A 30% carbon steel, for example, exhibits a linear contraction of 2.4%.[13] The actual shrinkage occurring in a casting is reduced, however, because of restraints imposed by the mold material and by the configuration of the casting. Briggs and Gazelius[14] have shown that the stress acting on a solidifying casting rises as the amount of free contraction is hindered.

The location of hot tears is usually in areas of design weakness, such as

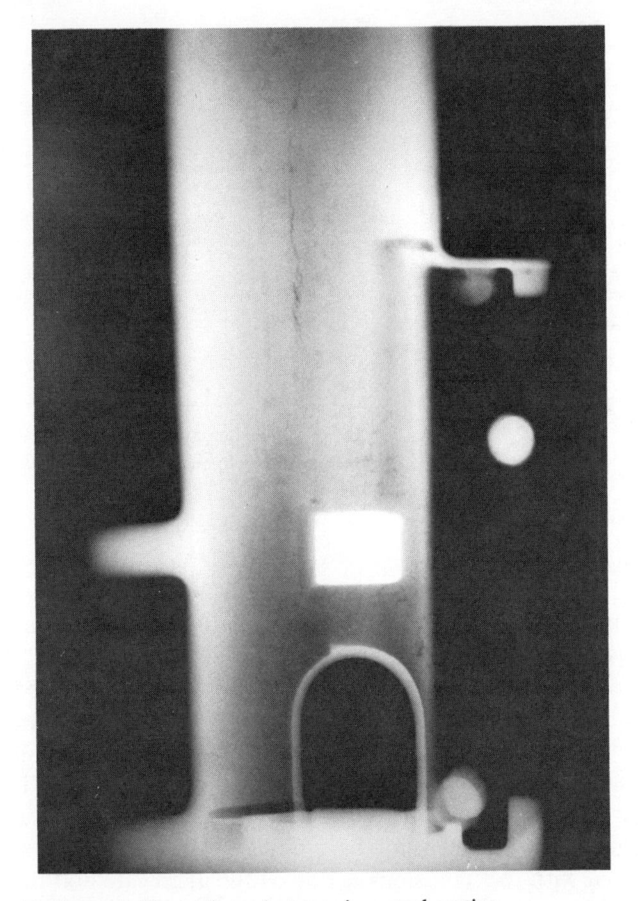

Fig. 11.20 Radiograph illustrating a hot tear in a steel casting.

the juncture between sections of markedly different thickness (Fig. 11.21) or in sections weakened by hot spots. If one section of a casting is contracting more rapidly than another, as a result of marked temperature gradients, solidified sections at a lower temperature can pull away from solidified sections at a higher temperature.

11.6 INCLUSIONS

Since inclusions are originally associated with the metal in the molten state, and subsequently appear in the casting or ingot, they are treated here, rather than in Chapter 10.

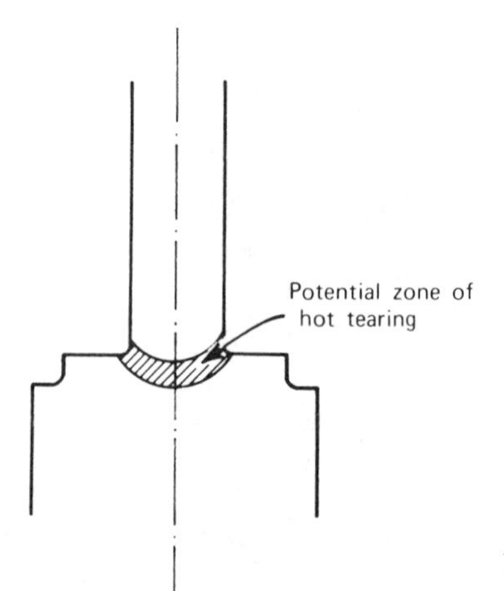

Fig. 11.21 Schematic diagram showing how sharp changes in section can size cause hot tears.

Two distinct kinds of inclusions exist in castings, exogenous and indigenous. Exogeneous inclusions result from the accidental entrapment and subsequent inclusion of foreign bodies, such as particles of refractory from furnace lining, ladle, and molds. These are usually large in size but few in number. Indigenous inclusions occur naturally in the metals, particularly in steels, as a result of changes in temperature or composition. They are due to normal steelmaking practices, such as aluminum deoxidation or silicon deoxidation. These inclusions, by definition nonmetallic, usually possess characteristics markedly different from the bulk material. In fact, the inclusion/metal system may be simplistically considered as a composite material with the inclusions acting as the aggregate and the metal as the matrix. Therefore, it is obvious that there are a number of factors that affect the performance of the whole. Among these are the volume percentage, shape, orientation, and mechanical properties of the inclusions and the direction of the principal stresses with respect to this orientation. Listed in Table 11.1, are a number of inclusion types common to alloy steels. An examination of the accompanying properties reveals that many of these are refractory in nature and tend to be hard and brittle. The sulphides, however, are relatively plastic and deform readily. The obvious inference is that the volume of included material is not the only important factor; the kind of inclusion is equally important.

11.6.1 Assessment of Inclusion Severity

Before the effect of inclusions on mechanical properties can be established, the relative level of the inclusions present must be determined.

Table 11.1 Characteristics of Inclusions Commonly Found in Steel

Inclusion	M. P. (°F)	Crystal System	Moh's Hardness
MnS	2948	Isometric[a]	3.5–4
FeS	2160	Hexagonal	3.5–4.5
Fe_2O_3	2466	Hexagonal	5
$MnSiO_3$	2323	Triclinic	5
MnO	3092	Isometric	5–6
FeO	2480	Isometric	5.5–6.5
SiO_2	—	—	6–7
Al_2O_3	3722	Hexagonal	9

[a] Simple cubic.

There are several methods for accomplishing this; the more common procedures are metallographic in character. Among these are the *J-K* ratings, lineal analysis, and the Fairey inclusion count.

J-K METHOD (COMPARATIVE)

The entire surface of a polished ¼-in.[2] sample is surveyed at a magnification of 100× with a field diameter of 3.15 in., and each field of the sample is compared with representative photomicrographs provided with the specification. The field number shown, for each of four inclusions types (*A*—sulfides, *B*—aluminates, *C*—silicates, and *D*—oxides) which appears most like the field under observation, is for both a thin and a heavy series.

FAIREY INCLUSION COUNT (COMPARATIVE METHOD)

A representative area of the material, in a plane parallel to the direction of applied stress, is polished and microscopically examined. The count number of an inclusion viewed at a magnification of 200× is given by

$$n = \frac{d^2K}{2}$$

where d is the apparent width in millimeters of the inclusion at the above magnification, and K is the stress concentration factor obtained by manipulating the formula for notches

$$K = 1 + \gamma\left(\frac{h}{R}\right)^{1/2} \qquad (K = 3 \text{ for a sphere})$$

where K is the theoretical stress-concentration factor, γ is a constant and equals 2, h is the depth of the inclusion, and R is the radius at the root of the inclusion. One hundred fields of 3 in. diameter at 200× are examined; the total of the count numbers for each individual inclusion gives the Fairey inclusion count number for that sample. To obviate calculating the value of each inclusion, the image is compared to a chart of representative inclusions with their respective count numbers (Fig. 11.22), where a higher number indicates a more detrimental inclusion. This method does not differentiate between specific types of inclusions. This discrepancy is supposedly offset by the geometrical configurations of the comparison chart.

LINEAL ANALYSIS METHOD (DIRECT MEASUREMENT METHOD)

This counting method consists of viewing 25 continuous fields at a magnification of 125×. The grid is moved across the sample transverse to the mechanical working direction, thereby measuring the length of each inclusion until an area equivalent to 25 visual fields has been counted. The total length of inclusions per area is determined by adding the lengths of all the inclusions in a specified range (mm) and dividing by 25.

Other methods have also been devised which utilize ultrasonic or magniflux techniques. These are concerned primarily with the size of the inclusion

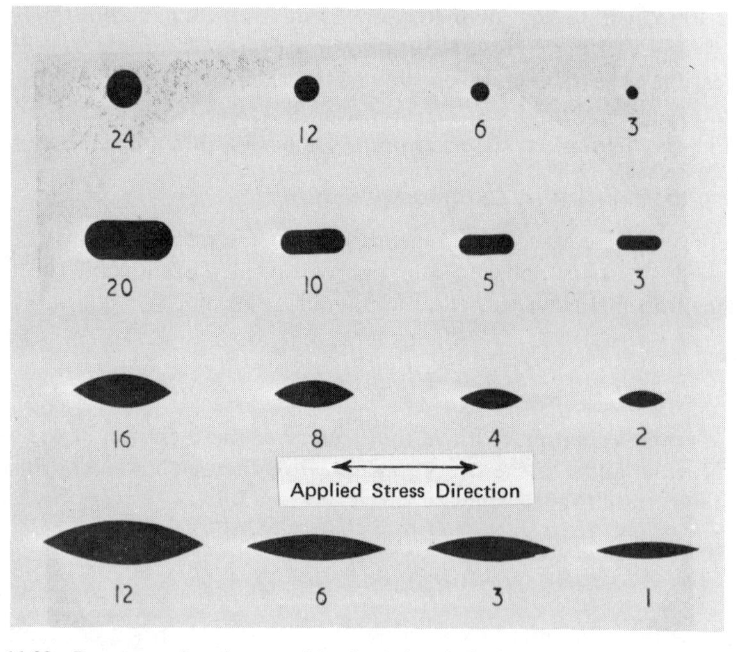

Fig. 11.22 Representative chart used in the Fairey inclusion count.

stringers and do not differentiate the type and orientation of the inclusions. The ultrasonic technique is quite effective, however, for detecting the potentially dangerous exogenous inclusions.

Radiographic methods can also be used. Inclusions appear on a radiograph as light or dark areas, depending on their composition. In general, these are relatively small and well defined.

11.6.2 Influence of Inclusions on Mechanical Properties

A broad review of the influence of inclusions on mechanical properties has been prepared by Thornton.[15] In brief, he states that the effect of inclusion content on strength is not clearly defined. The ultimate tensile strength and yield strength appear to be relatively unaffected[16] by inclusion content over a wide range when tested in air, but they do show some

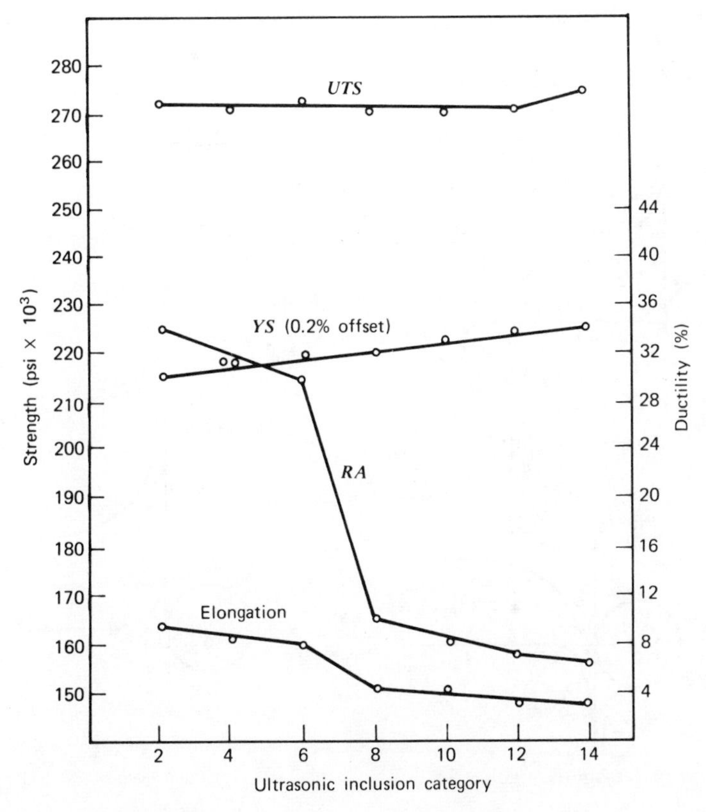

Fig. 11.23 Mechanical properties of 4340 steel as a function of inclusion content (from Ref. 16).

degradation when tested in a corrosive medium. The effect on ductility is more pronounced. Figure 11.23 shows that as the inclusion content increases on an ultrasonic scale, there is a marked decrease in both longitudinal and transverse specimens, of the percentage reduction in area and the percentage of elongation obtained in a tensile test.

The primary effect of the individual inclusion, with respect to the matrix, is a short range increase in the stresses on the matrix.[17] The form of the stress field is shown in Fig. 11.24. Fractographic evidence tends to support this view. Figure 11.25 shows inclusions present predominantly as nuclei for each dimple cavity. In a similar view (Fig. 11.26), taken with the scanning electron microscope, the relationship of the inclusion to the surrounding cavity is shown more clearly.

11.6.3 Influence on Inclusions on Fatigue Properties

The difficulty in assessing the role of inclusions on fatigue is that fatigue life is not well-defined since it is a statistical value and is dependent on

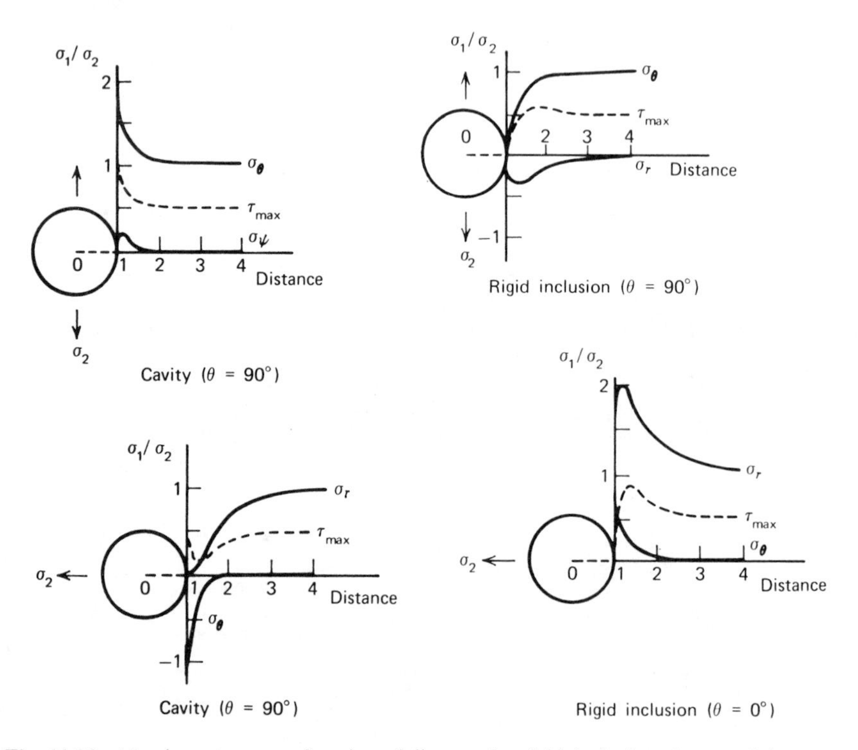

Fig. 11.24 Matrix stresses as a function of distance for rigid inclusions (σ_r = radial stress; $\sigma\theta$, $\sigma\psi$ = tangential stresses, τ_{max} = maximum shear stress) (from Ref. 17).

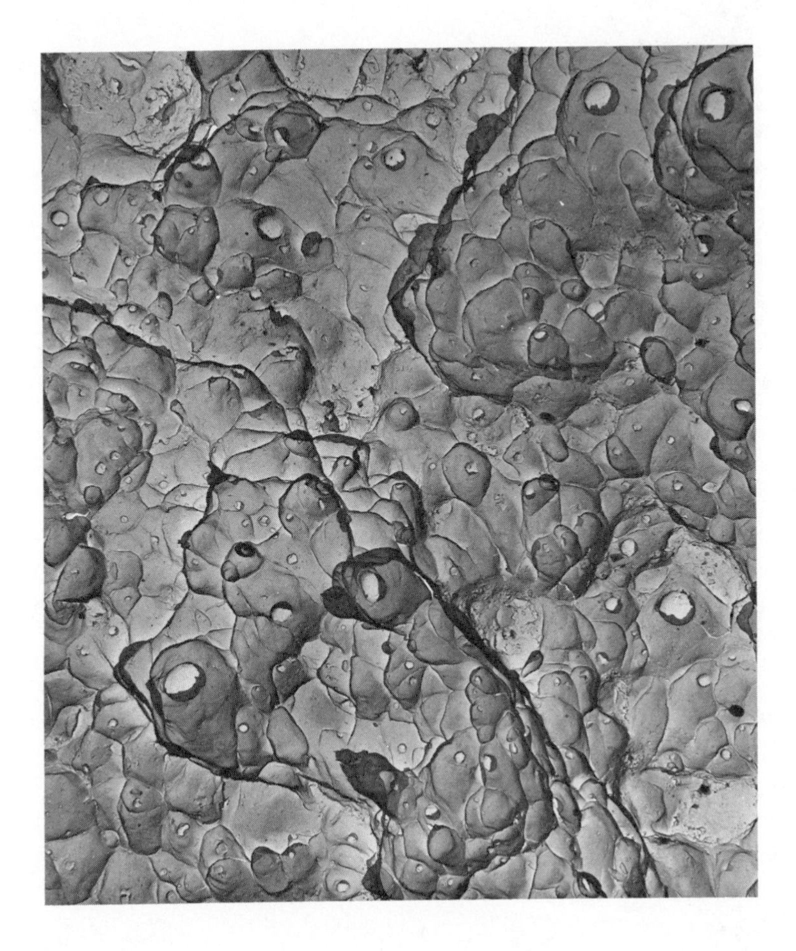

Fig. 11.25 Electron microscope fractograph showing dimples and cavities originating from inclusions (3500 X).

many factors, including the type of test and the stress level. When combined with the deficiencies inherent in establishing inclusion content levels, precise correlations are not possible; however, certain general evidence has been obtained.

Pelloux[18] has shown the effect of constituent particles on fatigue strength in two heats of 7178 aluminum alloy which differed only in impurity content. Particles as small as 100 Å were observed by transmission electron micros-

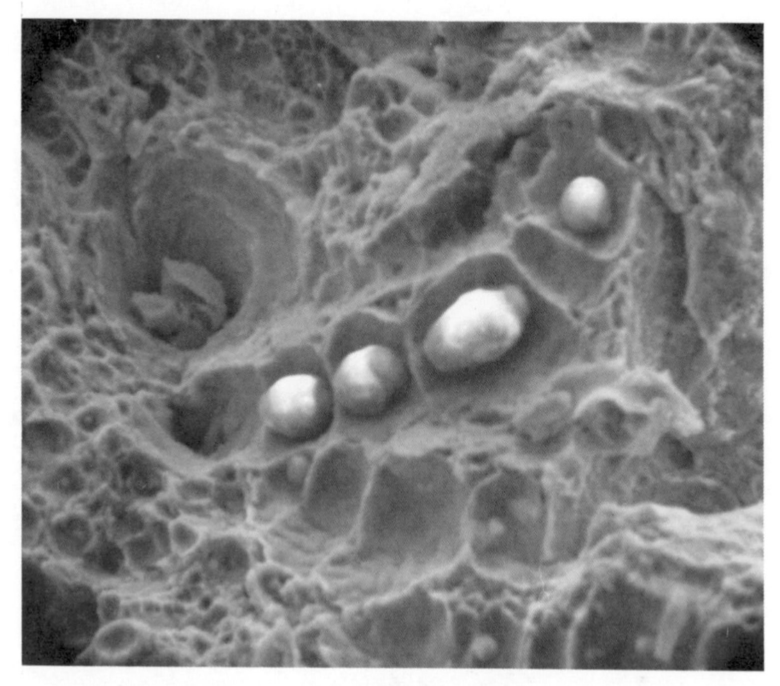

Fig. 11.26 Photograph of fracture surface using scanning electron microscope. Inclusions within the rupture cavities are readily visible (1000X) (from Ref. 15).

copy. The volume fraction of the constituent particles was estimated by a point count technique, which employed a coarse mesh grid applied at a magnification of 500× to unetched metallographic samples of the two heats. The results of this determination are listed as follows:

Alloy	Volume fraction (second phase)	Estimated error
A	0.38%	±0.078%
B	4.78%	±0.36%

Figure 11.27 (*b*, *c*, and *d*) is representative of the fracture features in alloy *B*. Some flat regions where the crack propagated slowly and regularly through the matrix are surrounded by regions where brittle fracture suddenly took place, adjacent to the second phase particles *A*, *B*, and *C*).

These observations suggested that the macroscopic crack growth is the sum of two terms: (*a*) advance of the crack front through the matrix and

Fig. 11.27 Fracture surfaces of two heats of 7178 aluminum alloy with varying cleanliness. Photograph *a* (4000X) is from the cleaner material whereas photographs *b*, *c*, and *d* (2000X) are from the "dirtier" heat (from Ref. 18).

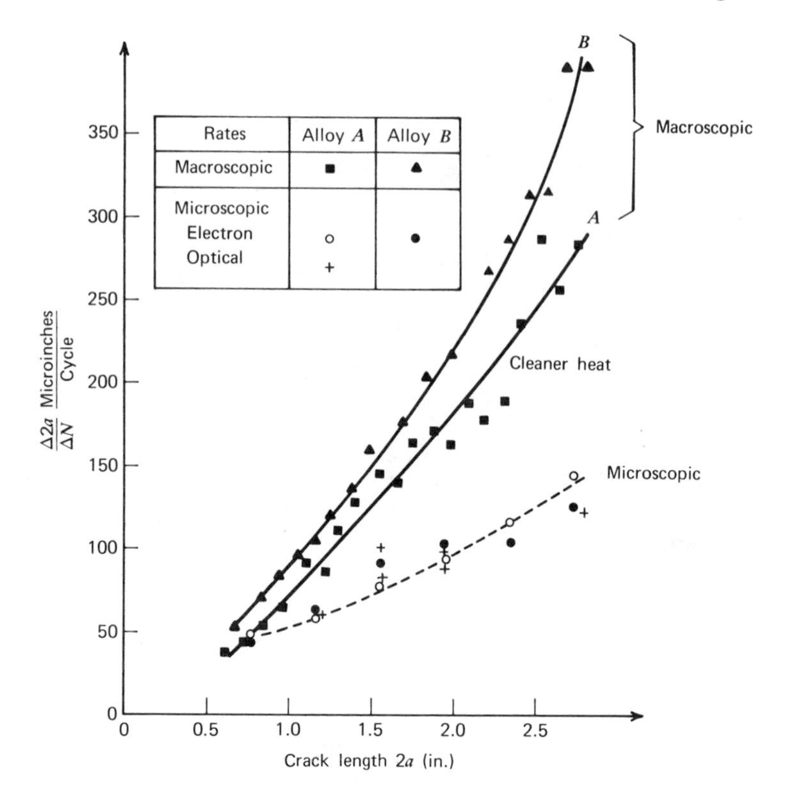

Fig. 11.28 Crack growth rates as a function of crack length for two heats of 7178 aluminum alloy.

(*b*) advance of the crack front due to cracking through and around constituent particles.

The microscopic growth rate (fatigue striations) represents the local rate of crack propagation through the matrix. The larger macroscopic rates, for the two alloys shown in Fig. 11.28, demonstrate the effect of the inclusions in the "dirty" alloy.

In was indicated that the influence of constituent particles on the rate of crack growth could best be understood by comparing the size of the plastic zone at the crack tip to the interparticle spacing.

1. When the plastic zone width is an order of magnitude (or less) smaller than the interparticle spacing, the crack growth is controlled principally by the properties of the matrix.

2. When the plastic zone width is of the order of magnitude of the interparticle spacing, the crack growth is the result of propagation through matrix and particle.

3. When the plastic zone width is an order of magnitude larger than the interparticle spacing, crack growth is controlled by extensive cracking that takes place through and around the second phase particles.

Other studies along this line have also been conducted. Fatigue crack propagation was investigated in hot rolled, normalized steel plate.[19] No attempt was made to establish a quantitative relationship between inclusions and fatigue strength, but the data indicated that the fracture mechanism which had the greatest control over the observed macroscopic growth rate was local fracture at inclusion-matrix interfaces. As described by Pelloux, the macroscopic growth rate was considered to be the summation of several fracture mechanisms, the most important of which was striation formation and local fracture of brittle microstituents (inclusion-matrix fracture or cleavage). Striation spacing was seen to be independent of orientation, suggesting that fracture of inclusions was responsible for the crack growth rate anisotropy. Fastest growth rates were noted in the short transverse orientation, where inclusion matrix fracture occurred parallel to the fracture plane and direction; the slowest growth rates were noted in the crack arrester orientation, where inclusion-matrix fracture occurred normal to the crack growth plane and direction.

Another significant investigation was described by Kiessling,[20] regarding the size of Al_3O_2 inclusions on fatigue life. He found that a correlation existed between fatigue life and inclusion size if the inclusions were larger than a critical size. The reduction in the ratio of fatigue lives (with and without inclusions) was proportional to the cube root of the inclusion diameter. The critical size of the inclusion was somewhat variable since it was related to its distance below the surface. The critical inclusion size necessary to produce an observable effect on fatigue life, increased from 10μ, when located just below the metal surface, to 30μ when located at a depth of 100μ below the surface.

In summary, the crack initiation aspects of inclusions are the most significant, with respect to fatigue damage. The effectiveness of a particular inclusion set in creating localized fracture depends on several features.

1. *The distribution of the inclusions.* Inclusions spaced close together and arranged linearly to act as a single unit raise the local stresses very effectively.

2. *The shape and orientation.* Elongated inclusions aligned parallel to the principal stresses are less damaging than those aligned normal to these stresses. Similarly, angular inclusions are more damaging than rounded ones.

3. *The physical characteristics of the inclusions.* Hard refractory inclusions, such as Ca-aluminate, Al_2O_3, and spinels are more detrimental than sulphide type inclusions.

REFERENCES

1. W. A. Tiller, K. A. Jackson, J. W. Rutter, and B. Chalmers, *Acta Met.*, Vol. 1, 428 (1953).
2. W. C. Winegard, *Trans. Am. Foundryman's Soc.*, Vol. 61, 352 (1953).
3. B. Chalmers, *Principles of Solidification*, Wiley, New York, 1964, p. 180.
4. D. E. Adams, *J. Inst. Met.*, Vol. 75, 805 (1948).
5. G. L. Kehl, *Principles of Metallographic Laboratory Practice*, McGraw-Hill, New York, 1949, pp. 409–448.
7. A. Pokorny and J. Pokorny, *DeFerri Metallographia, Solidification and Deformation of Steels*, W. B. Saunders Co., 1947.
8. A. Franks and K. Lindsey, *The Electron Microprobe*, The Electrochemical Society, Washington, 1966, p. 83.
9. C. Leymonie, *Radioactive Tracers in Physical Metallurgy*, Wiley, New York, 1963.
10. R. E. Peterson, *Stress Concentration Design Factors*, Wiley, New York, 1953.
11. E. P. Polushkin, *Defects and Failures of Metals*, Elsevier, Amsterdam, 1956, p. 63.
12. A. Pokorny and J. Pokorny, *op. cit.*, p. 19.
13. C. W. Briggs and R. Gazelius, *Trans. Am. Foundryman's Assoc.*, Vol. 42, 449–476 (1934).
14. C. W. Briggs and R. Gazelius, *Trans. Am. Foundryman's Assoc.*, Vol. 44, 1–32, (1936).
15. P. Thornton, Journal of Materials Science, Vol. 6, 347–356, 1971.
16. C. J. Carter, R. A. Cellitti, and J. W. Abar, Technical Report, Air Force Materials Laboratory—TR–68–303, January 1969.
17. J. Gurland and J. Plateau, *Trans. ASM*, Vol. 56, 442 (1963).
18. R. M. N. Pelloux, *Trans. ASM*, Vol. 57, 511 (1964).
19. F. A. Heiser, Ph.D. Thesis, Lehigh University, 1969.
20. R. Kiessling, *Non-Metallic Inclusions in Steel*, Part III, Iron and Steel Institute Publication 115, London, 1968, p. 92.

12

Heat Treatment

12.1 INTRODUCTION

Failures due to heat treatment may be related to cracking during heat treatment. Improper microstructure caused by faulty heat treatment can also cause failures during performance. Just as certain service and design conditions, such as low temperature and notches, lead to brittle failure, so too, certain microstructural characteristics, such as a grain boundary network, can cause brittle failures, even under nonembrittling service and design conditions.

A knowledge of physical metallurgy is important for proper heat treatment. Low alloy steels are hard if they are quenched from certain temperatures, whereas aluminum alloys or maraging steels are soft under similar quenching conditions. Some steels crack if they are quenched in water, and others are soft if they are quenched in the same water. A knowledge of constitution diagrams, transformation diagrams, and metallography are important to understand the significance of heat treatment.

Although the various metal systems respond to heating and cooling differently, many problems are common to all of them. For example, the physical metallurgy of the hardening of steel and aluminum are entirely different. Yet, both involve a quenching operation and if it is too severe, the part may crack, whether it is aluminum or steel. Also, if metals are heated in a reactive atmosphere, surface problems may result even though the specific reactions vary with the different metals.

Rather than considering each system separately, heat treating can also be subdivided into those operations which affect the metal throughout and those which affect at the surface. For the sake of nomenclature, the former is called "through heat treatment" and the latter "surface heat treatment".

12.2 THROUGH HEAT TREATMENT

12.2.1 Cracking

Cracking in heat treatment usually occurs during a quenching operation because of localized stresses caused by uneven cooling, or by a flaw in the piece being treated. For example, problems that exist in the initial ingot are very often not eliminated, but are compounded in subsequent operations. Thus surface flaws caused by forging can lead to cracks during heat treatment.

Many heat treatment problems are related to expansion during heating and contraction during cooling. If it were possible to heat and cool uniformly throughout a section, these problems might be minimized. Since this is not generally possible, stress and strain gradients are developed, thereby resulting in possibly distortion or cracking.

The crux of the problem is illustrated in Fig. 12.1, which shows the difference in cooling rate between the center and surface of a part.[1] As the severity of the cooling rate increases, for example, quenching in water versus cooling in still air, the temperature gradient increases. Also, as the section thickness increases, the gradient increases. The thermal gradient sets up the stresses that can cause cracking.

During quenching the cooling rate must be severe enough to obtain the

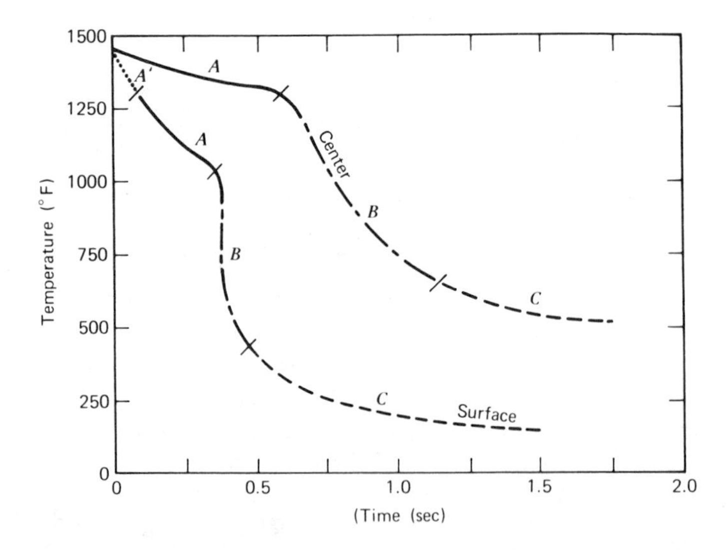

Fig. 12.1 Typical surface and center cooling curves (Reproduced by permission, from Metals Handbook, American Society for Metals, 1964).

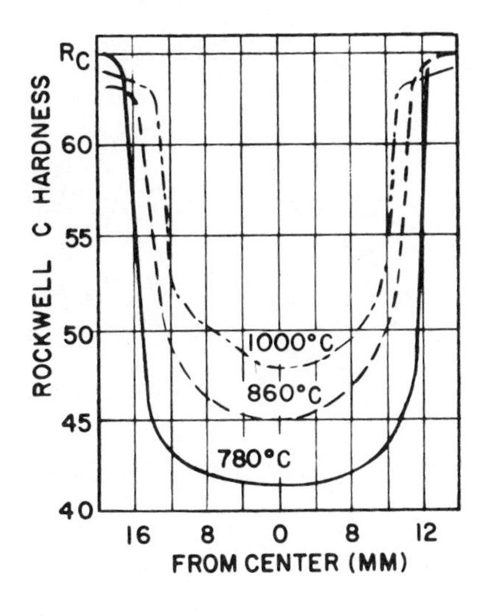

Fig. 12.2 Hardness distribution after quenching as a function of quench temperature (Reproduced by permission, Temple Publications, Inc.).

desired microstructure, but not severe enough to cause cracking. For example, if the alloy content is high in steel, an air quench might be sufficient to develop martensite. Anything more severe could cause cracking. However, if the alloy is lean, a severe water quench may be necessary to develop martensite. Figure 12.2 shows how the thermal gradient illustrated in Fig. 12.1 affects the hardness by using the hardness as an indicator of structure (the lines represent different quenching temperatures).[2] The surface hard-

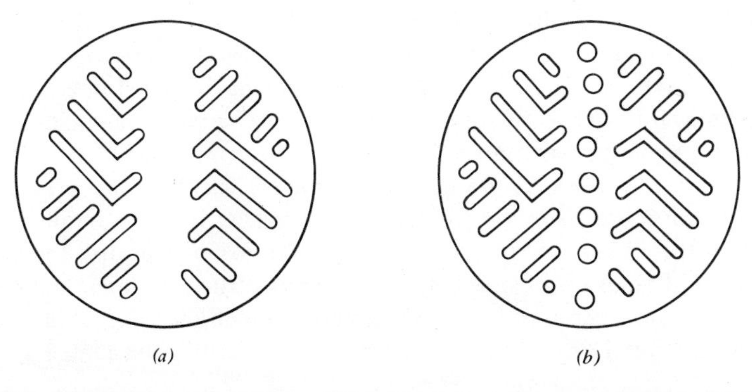

Fig. 12.3 Elimination of a warping problem by removal of excess metal to equalize distribution of metal. In (*a*) the relatively heavy rib caused warpage. Drilling holes (*b*), eliminated the problem (Reprinted through the Courtesy of Carpenter Technology Corporation, Reading, Pennsylvania).

Fig. 12.4 Hardening crack caused by cooling in two directions at right angles (Reprinted by permission, Elsevier Scientific Publishing Company).

ness is higher than the center, which indicates that martensite is present near the surface only.

The stresses may not always be great enough to cause cracking, but may cause distortion. Distortion and cracking during cooling are controlled to a large degree by the design. Figure 12.3 shows a distortion problem caused by poor design in a blanking die, and the actions taken to resolve it. The holes drilled in the center rib equalized the distribution of metal, minimized the cooling stresses, and eliminated warpage.[3] Figure 12.4 shows an unusual cracking situation caused by cooling in two directions at right angles to each other.[4] Although both problems were related to design, they were caused by stresses set up by the uneven thermal distribution.

Most metals may crack during quenching; steels, however, are the most likely to crack, since, in addition to the normal expansion-contraction associated with change in temperature, there is a factor associated with dimensional change when austenite transforms to martensite.

Since it is the thermal gradient that causes cracking, the greater this gradient, the more likely that cracking will occur. It is an axiom in heat treatment that the minimum temperature which will accomplish the desired result should be used. In steel, this is the temperature at which austenite is formed. Too high an austenitizing temperature can cause cracking during quenching (Fig. 12.5), which might not occur if a lower temperature is used.[5]

Carelessness can often cause problems. Figure 12.6 shows cracks in a rotary slitter knife quenched in a grooved quenching press.[6] Oil pumped through the grooves quenched those sections, but the sections of the workpiece in contact with the platens were not quenched, resulting in an increased thermal gradient. A similar phenomenon can be seen if the workpiece is held by tongs during quenching. Where contact is made by the tongs, quenching is obstructed. Even if cracking did not result, the final product would contain soft spots.

Since untempered martensite contains high residual stresses, it is often advisable to temper a quenched piece of steel immediately. In some operations, the workpiece is not allowed to reach room temperature. If the part is quenched cold, without tempering, cracking may result (Fig. 12.7).[7] Conversely, if the part is not quenched sufficiently before tempering, cracking may also result (Fig. 12.8).[8] The latter is possible since, if the quench is incomplete, austenite still remains. Since many components are water

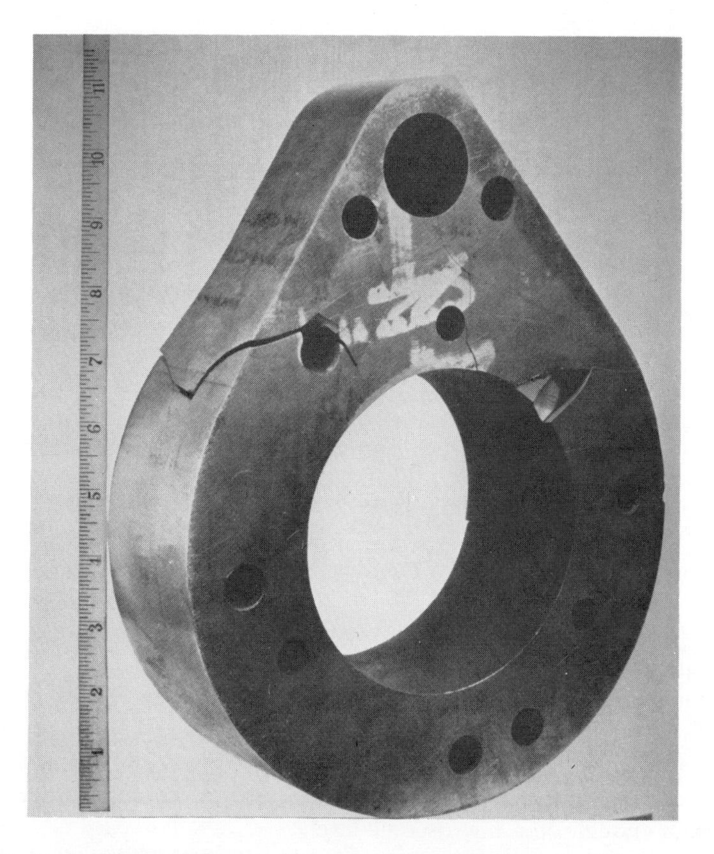

Fig. 12.5 Tool steel cam that cracked because of quenching from too high a temperature (Courtesy, Bethlehem Steel Corp.)

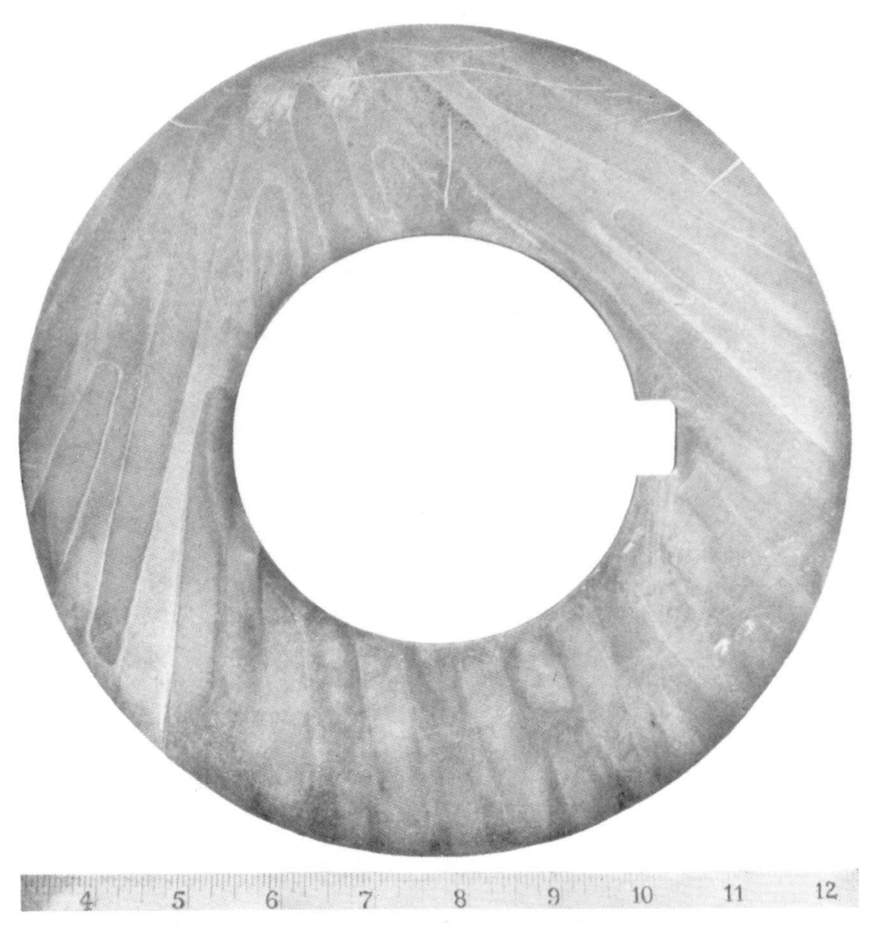

Fig. 12.6 Cracking due to stresses caused by uneven quenching (Courtesy, Bethlehem Steel Corp.).

quenched from tempering, with austenite still present after tempering, the water quenching can be too severe, as in the illustration.

Quench cracks are usually open cracks, easily seen by visual inspection. However, this is not always true. In fact, quench cracks can be internal. Incipient quench cracks in steel are shown in Fig. 12.9. Apparently, the overall stresses were not great enough to cause macroscopic cracking, but were high enough on a microscopic level.[9]

A failure caused by a quench crack is easily identifiable. If a chevron pattern exists, it may be possible to trace it to its origin. Quench cracks

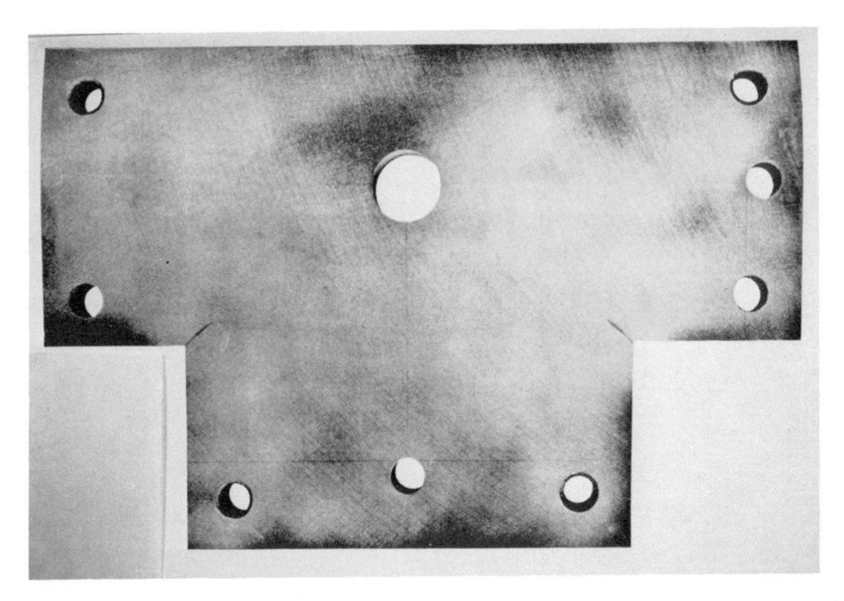

Fig. 12.7 Quenching in an oil hardening steel. The steel was not tempered immediately after quenching. The cracking occurred during the waiting period (Courtesy, Bethlehem Steel Corp.).

assume a characteristic shape that is often macroscopically smooth and elliptical (Fig. 12.10). In addition, since quenching is followed by tempering, the surface of the crack is discolored (Fig. 12.11).[10]

Steel experiences the majority of thermal cracking problems. Not all steels are prone to quench cracking, the problem is more prevalent as the alloy content increases. However, any metal that is quenched is liable to crack.

12.2.2 Microstructure

The purpose of heat treating is to develop a microstructure with properties adequate for the application. When the quench rate is too slow in solution treatment-age hardening alloys, such as aluminum, second phase particles precipitate undesirably during the quench. In steel, mixed microstructure, which is undesirable for performance if ductility and toughness are significant, could result. Mixed structures can also result if the component is not fully austenitized. In this case, after quenching, the resultant structure is a combination of martensite, pearlite, and ferrite. Figure 12.12 shows this microstructure, and the resultant failure of a die that exhibited it.[11]

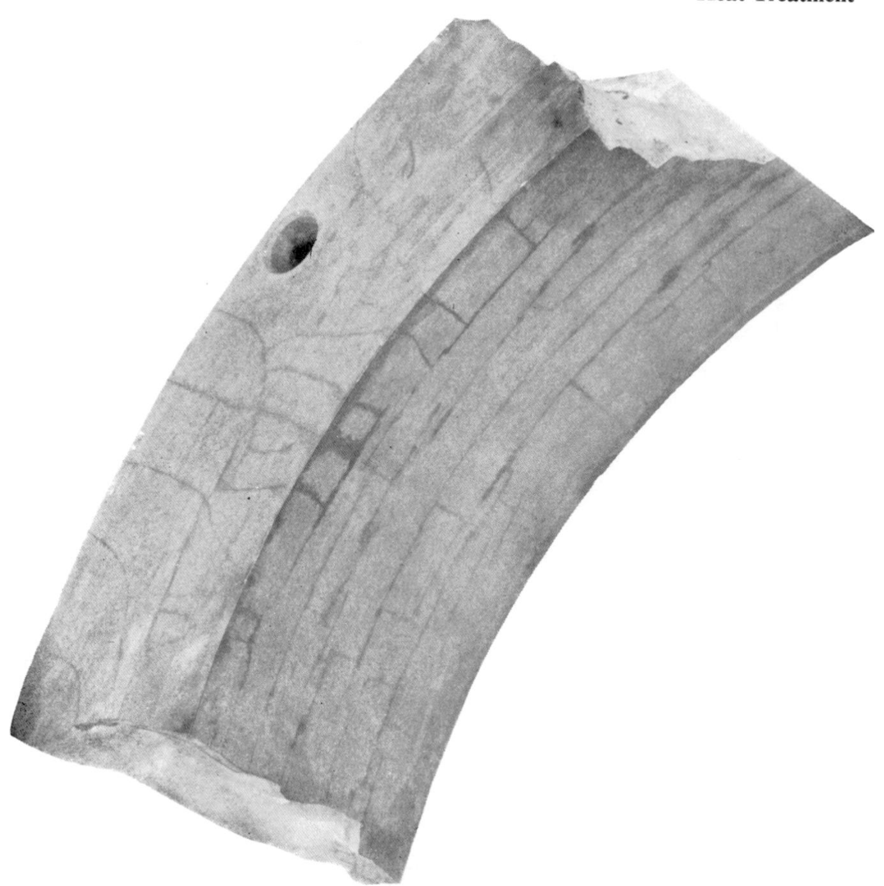

Fig. 12.8 Cracking in an oil hardening steel caused by tempering too soon after quenching. The quenching was stopped before all the austenite had transformed to martensite. After tempering, the steel was water quenched, causing the cracking. If all the austenite were previously transformed, the quenching operation would have been safe (Courtesy, Bethlehem Steel Corp.).

Although quenching practice exerts a major influence, subsequent treatments also influence microstructure. Spheroidizing is sometimes used to develop rounded second phase particles. Generally, as the particles become rounder, the material becomes tougher and more ductile.

Figure 12.13 shows the result of poor spheroidizing practice. Note the relative shape of the carbide particles in the properly and improperly spheroidized microstructures.[12]

Both the shape of the second phase particles and their location are

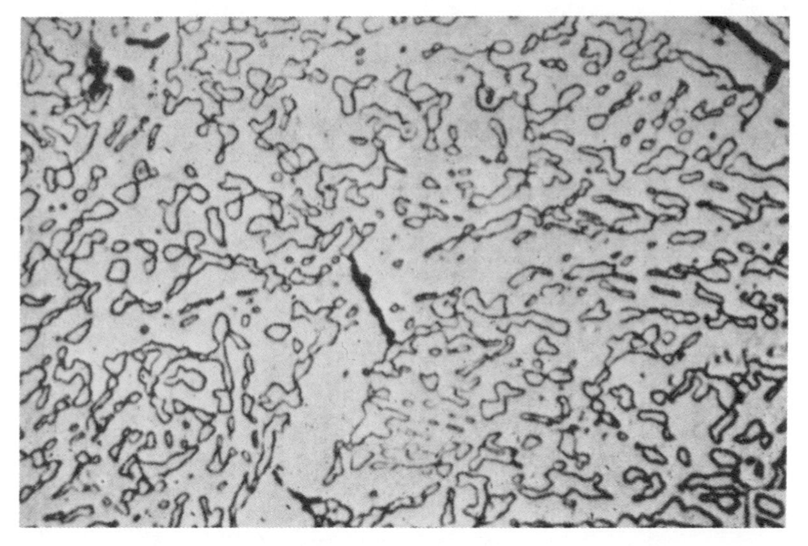

Fig. 12.9 Incipient quench cracks in steel showing that quench cracks need not be external (Reproduced by permission, from Transactions ASM, American Society for Metals, 1934).

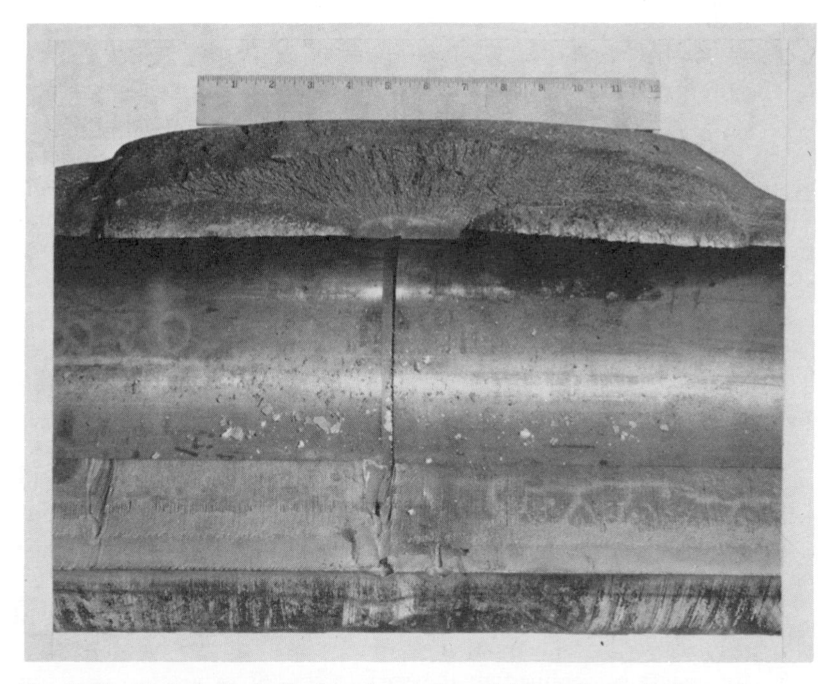

Fig. 12.10 Typical elliptically shaped quench crack in a hollow cylinder.

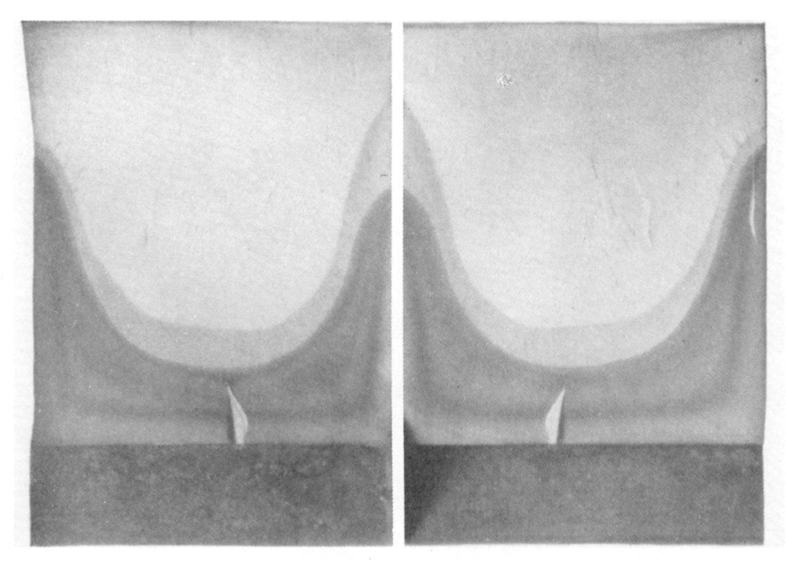

Fig. 12.11 Typical shape and discoloration associated with quench cracks. The discoloration shows that the crack was open during tempering (Courtesy, Bethlehem Steel Corp.).

important. In the discussion on fractography, examples of intergranular and transgranular fracture were shown. It was pointed out that transgranular fractures could be ductile or brittle, but intergranular fractures were brittle. The latter often result from continuous films of second phase material in grain boundaries. They occur in many alloy systems, and usually result from improper heat treatment. Generally, it is possible to show microscopically the existence of the grain boundary precipitate in a failure, as in Fig. 12.14 where a carbide network that caused the failure of the steel die is clearly visible,[13] and in Fig. 12.15, which shows a carbide network in 304 stainless steel.[14]

However, the cause of the intergranular fracture is not always visible. This is particularly true for temper embrittled steels. Temper embrittlement is a phenomenon that occurs when steel is tempered in the range of 850 to 1000°F. It was once thought that with proper etching it would be possible to show the grain boundary precipitate. However, this is not certain. Presently, a shift in the FATT in impact testing is used to determine temper embrittlement, as described in Chapter 2.

Embrittlement during heat treatment can also be caused by reaction with the atmosphere, or by inadequate control of temperature. The atmosphere effects are generally confined to the surface, but if exposure exists long

Fig. 12.12 (*a*) Oil hardening die that failed during service. Only the outer portion had been hardened because the austenitizing temperature was too low. The hardened and unhardened microsctructures are shown in (*b*) and (*c*) (Courtesy, Bethlehem Steel Corp.).

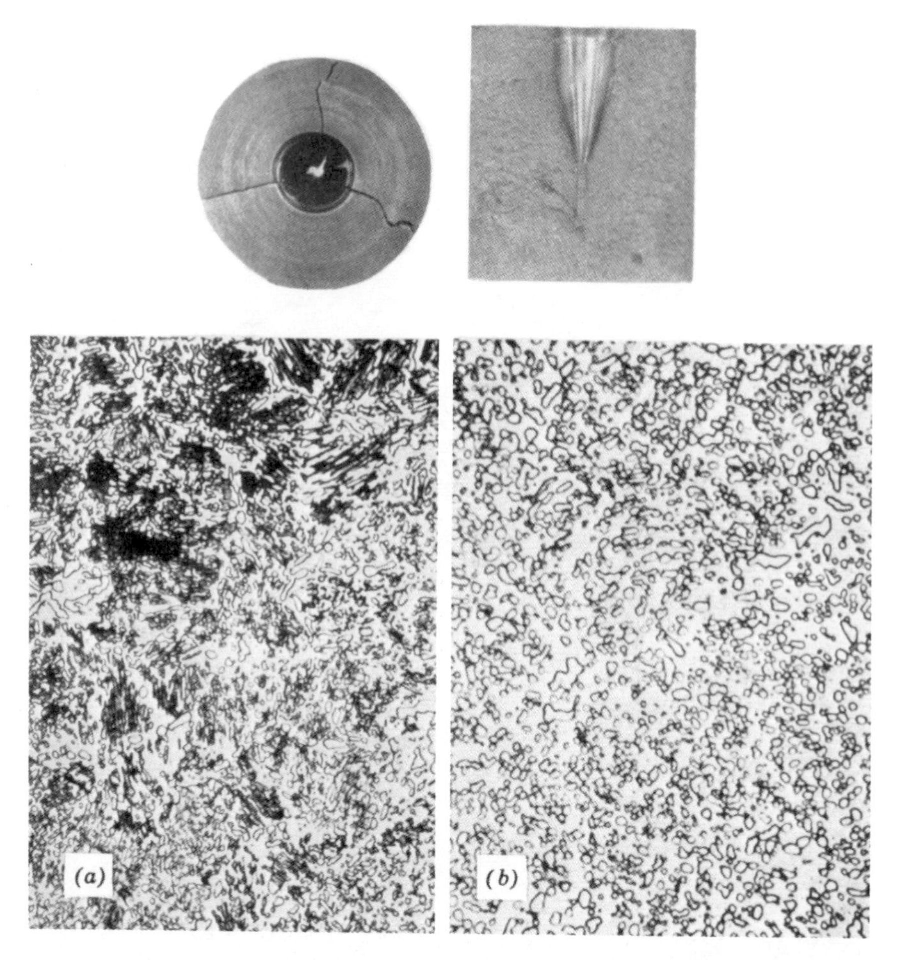

Fig. 12.13 Comparison of (*a*) improperly annealed (lamellar) structure and (*b*) properly annealed (spheroidized) structure. The lamellar structures are brittle (Courtesy, Bethlehem Steel Corp.).

enough for diffusion to occur, the effects may be more penetrating. Steel and titanium are subject to hydrogen embrittlement. In addition, steel can become embrittled by nitrogen,[15] as illustrated by the nitrides in Fig. 12.16.

Furnace temperature control is important in all heat treating operations, but particularly in those near the melting point of the material. Because of segregation and the necessity to heat treat close to the liquidus temperature, that is, near the melting temperature for the gross alloy content, it is possible that the melting point in highly segregated regions may be ex-

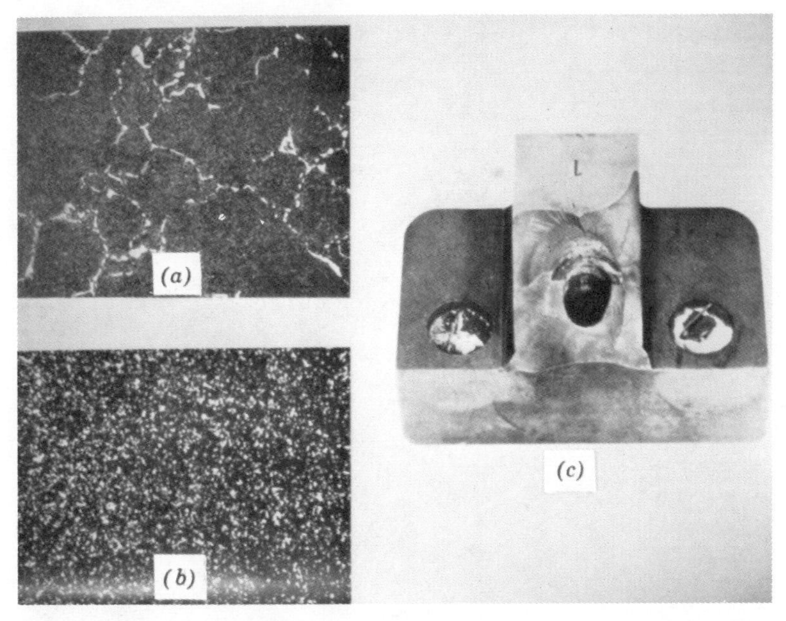

Fig. 12.14 (*a*) Carbide network in improperly annealed tool steel compared to (*b*) properly annealed structure. (*c*) The brittle carbide network caused failure in the steel die (Courtesy, Bethlehem Steel Corp.).

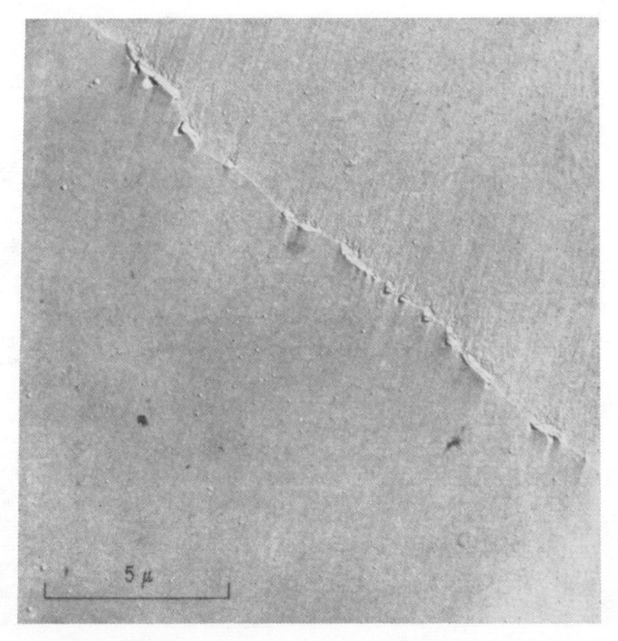

Fig. 12.15 Electron micrograph of a grain boundary precipitate in 304 stainless steel (Reproduced by permission, from Transactions, ASM, American Society for Metals, 1966).

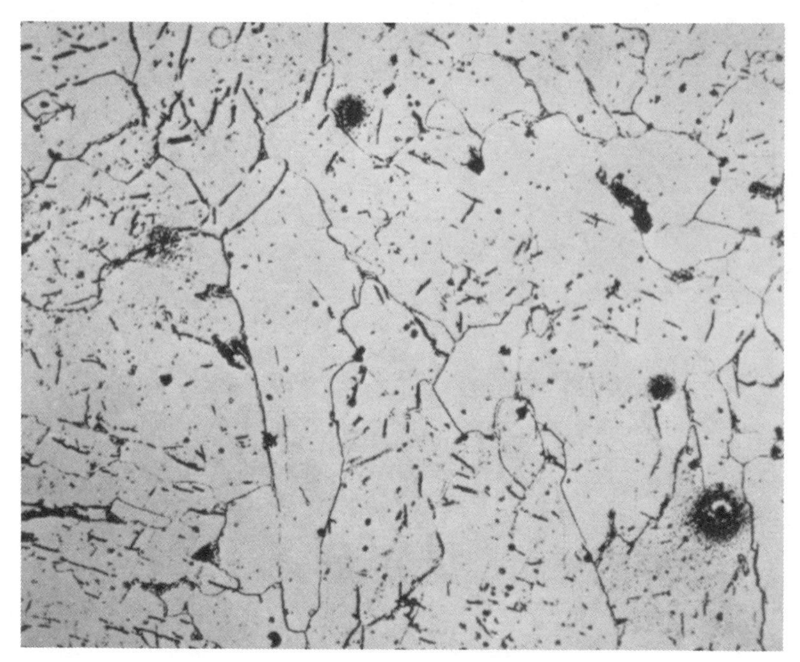

Fig. 12.16 Nitride network in overheated low carbon steel (Reproduced by permission, Elsevier Scientific Publishing Co.).

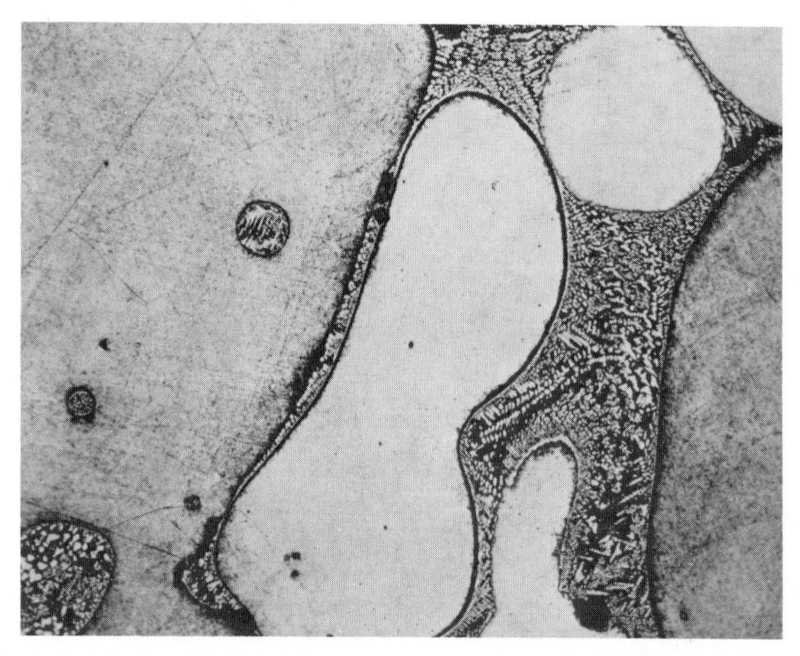

Fig. 12.17 Burning in cast aluminum caused by a combination of segregation and overheating.

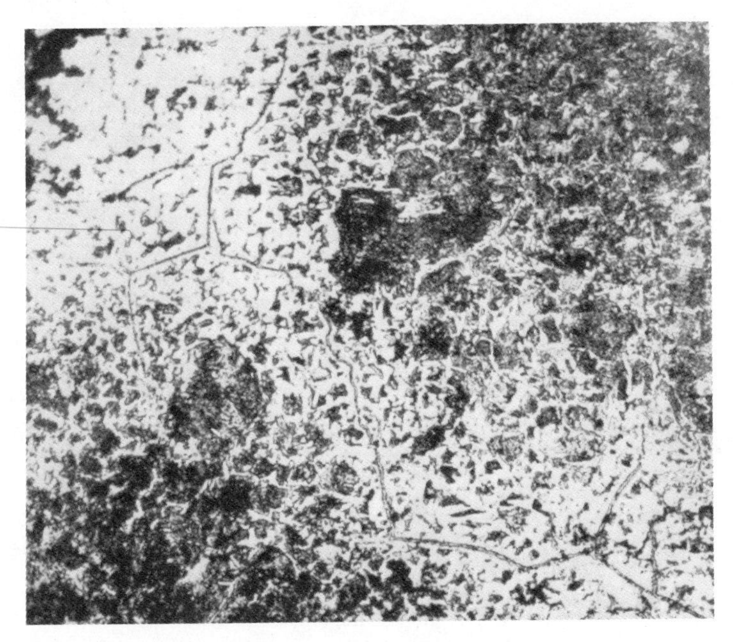

Fig. 12.18 Burned tool steel showing penetration of oxide and decarburization (Reproduced by permission, Elsevier Scientific Publishing Co.).

ceeded. Thus, a liquid-solid combination exists at the heat treatment temperature. Since segregation is usually most severe at grain boundaries, melting occurs there. The final result is a weak grain boundary layer. This can occur in any number of alloy systems, for example, cast aluminum (Fig. 12.17).

A phenomenon that combines the effects of overheating and surface reaction is burning in steel. This is often seen in preheating for forging. At temperatures in the vicinity of 2400°F and in an oxidizing atmosphere, oxidation may occur so rapidly that the temperature of the metal is increased. This results in the structure in Fig. 12.18, which shows the penetration of oxide and decarburization along the grain boundaries in the steel.[16]

Much emphasis has been placed on the effect of microstructure in decreasing ductility and toughness; there are other facets. Grain boundary precipitates may enhance intergranular corrosion, such as in 304 stainless steel (Fig. 12.19).[17] The mixed structure of bainite and martensite in steel is more susceptible to hydrogen embrittlement than either is alone. Improper stress relief may result in cracking or in stress corrosion during performance.

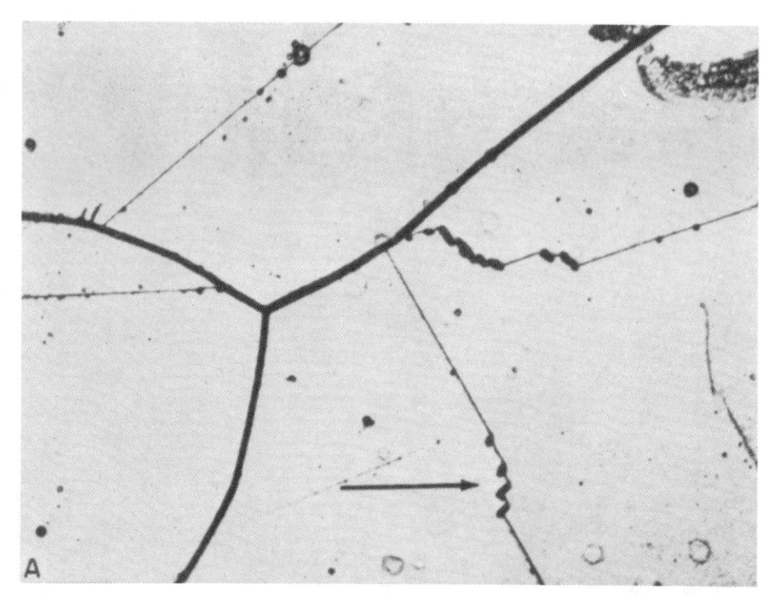

Fig. 12.19 Intergranular attack of 304 stainless steel in boiling nitric acid (Reproduced by permission, from Transaction ASM, American Society for Metals, 1966).

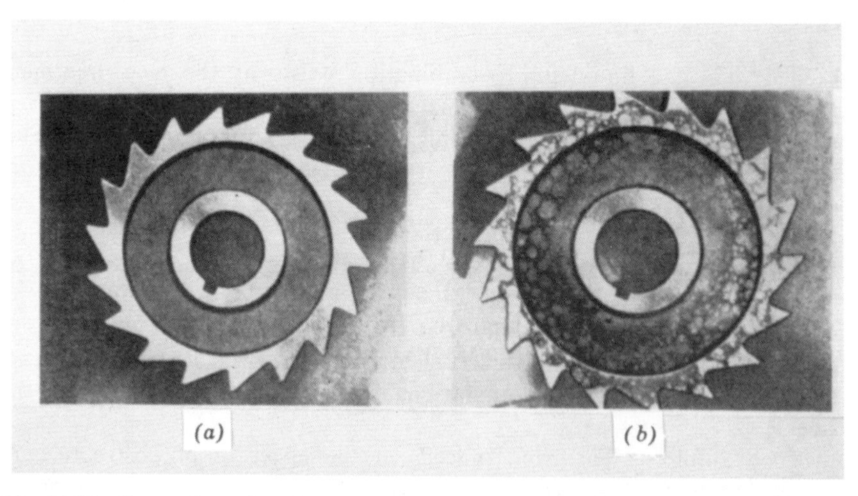

Fig. 12.20 Comparison of steel hardened in (*a*) a reducing atmosphere and (*b*) an oxidizing atmosphere. The light areas in (*b*) are soft spots (Reprinted through the Courtesy of Carpenter Technology Corporation, Reading, Pensylvania).

Hard-and-fast rules in heat treatment are rare. The structure that performs satisfactorily depends on the application and the alloy system. Grain boundary precipitates are usually deleterious. However, it is not always possible to relate a failure to its microstructure. The properties and performance may vary greatly, and it may not be possible to note any difference in microstructure.

12.3 SURFACE TREATMENT

The surface chemistry and, thereby, the surface microstructure and mehcanical properties are sometimes changed to utilize the overall properties of the core with the special properties of the case. For example, steel is nitrided or carburized for wear resistance without losing the toughness in the core. These surface alterations can also occur accidently, for example, by heating in an oxidizing atmosphere, the carbon may be depleted from the surface of a piece of steel. In effect, surface treated metals are composites of two different metals. It is necessary to consider the properties of both.

An example of the effect of atmosphere is shown in Fig. 12.20, which compares the results of hardening in a reducing atmosphere and hardening in an oxidizing atmosphere. The oxidizing atmosphere depleted carbon from certain areas, which resulted in the surface defects, soft spots, that are shown.[18]

Most surface treatments entail the reaction of the metal at the surface with an atmosphere, gas or liquid. One of those used most often is carburizing. In steel, as the carbon content increases, the strength and wear resistance increase, but the toughness and ductility decrease. Therefore, by using a low carbon core with a high carbon case, it is possible to have a tough, ductile material that is wear resistant. Care must be taken in heat treating after carburizing, since this controls the properties. Figure 12.21 illustrates the grain boundary carbides that resulted from box cooling from the carburizing temperature rather than quenching.[19]

Because of the differential chemistry, problems can arise from differing thermal expansion and contraction. Tensile stresses can be developed at the case-core boundary. If they are sufficiently high, cracking may result. These cracks can be particularly dangerous since they occur internally without breaking through to the surface. This potential for cracking is common to most surface treatments.

The depth of case is important for proper functioning. If the case is too deep, its properties will predominate. It may crack brittly and cause failure. However, if the case is too thin, it may not be sufficiently strong to carry

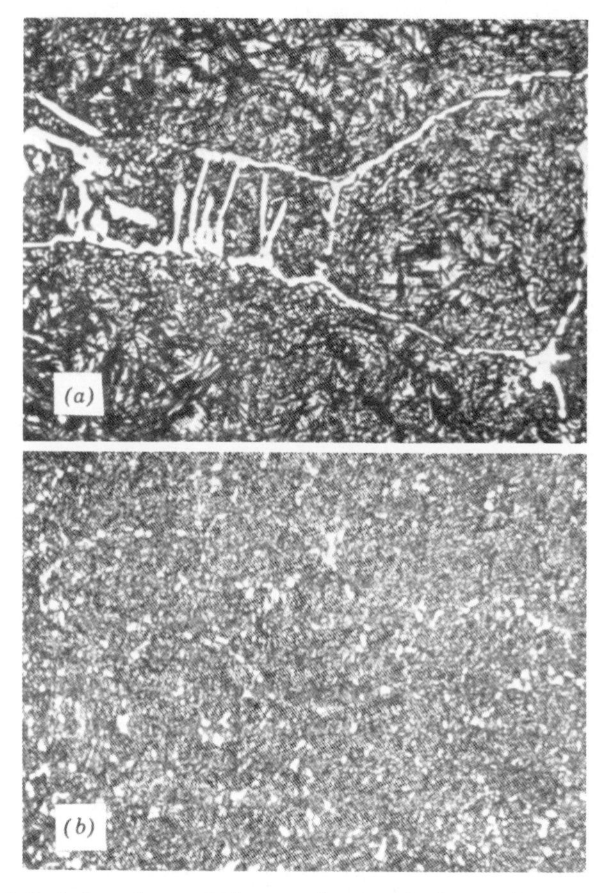

Fig. 12.21 (*a*) Grain boundary carbide network caused by box cooling a carburized steel from the carburizing temperature. (*b*) Direct quenching avoids the network (Reproduced by permission from Metal Progress 1962).

the load. This can result in crushed cases. Cases are often used to carry local heavy contact loads.

Carburizing is not always performed deliberately. When it occurs in service, failure often results, since the condition has not been considered in the design. For example, parts subjected to impact loading are seldom carburized. A hard carburized case on these parts can lead to failure.

In addition, an unexpected increase in carbon can confuse the heat treater. Tempering operations are controlled by hardness, that is, a heat treater tempers to a certain hardness. Thus tempering to achieve hardness on a carburized section results in an undesirable softening in the core.

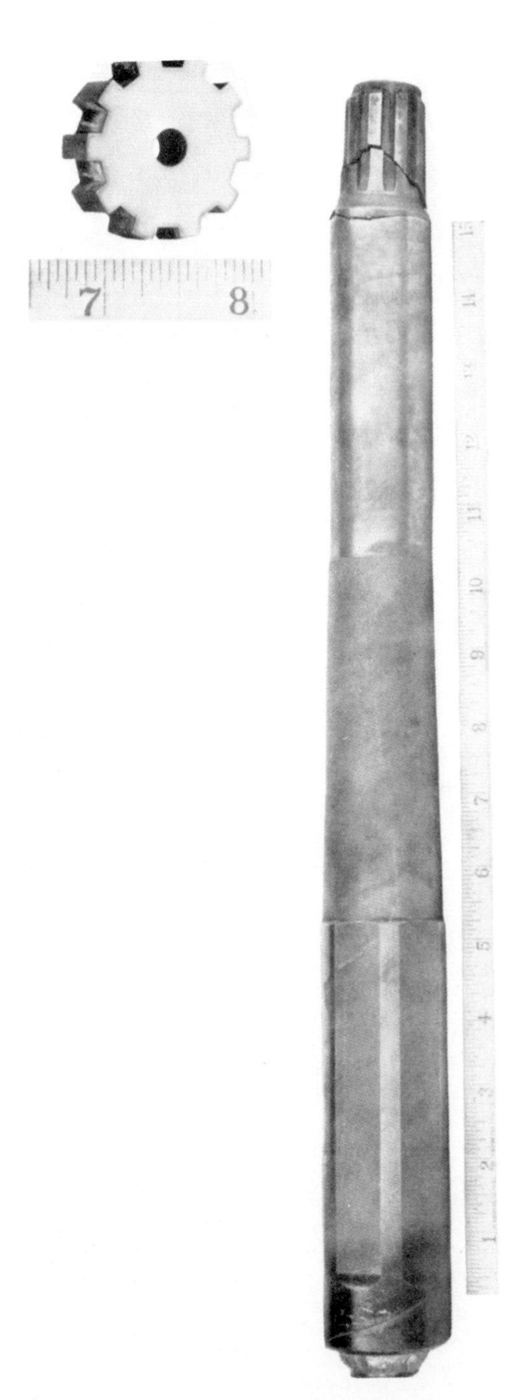

Fig. 12.22 Crank shaft that failed in service due to a brittle, carburized case accidently put on during heat treating (Courtesy, Bethlehem Steel Corp.).

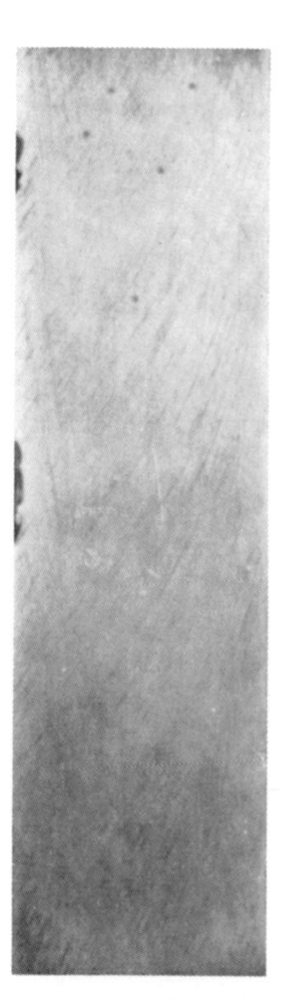

Fig. 12.23 Rail ending blade that chipped in service because of carburization (Courtesy, Bethlehem Steel Corp.).

Two failures caused by accidental carburizing are shown in Figs. 12.22 and 12.23.[20,21] In Fig. 12.22, a crank shaft wrench cracked from a carburized case introduced during heat treating. Complete fracture, such as this, may not necessarily occur. Because the surface layer is so brittle, it may simply chip off, as in the rail ending blade in Fig. 12.23. When the carburizing is not anticipated, the necessary or adequate heat treating operations to render it useful are not employed and the case is harder and more brittle than usual.

At the other extreme from carburizing is decarburizing. Decarburizing is never deliberately employed; if the surface atmosphere is not adequately controlled, however, it can occur. Decarburized cases are relatively weak and can reduce fatigue life by initiating cracks. The basic problem again is that the decarburization is not anticipated by the design. Figure 12.24 shows the change in structure that results. Since the carbon is depleted, the surface chemistry approaches pure iron, that is, ferrite.[22]

Decarburized layers can often be seen adjacent to cracks, which indicates

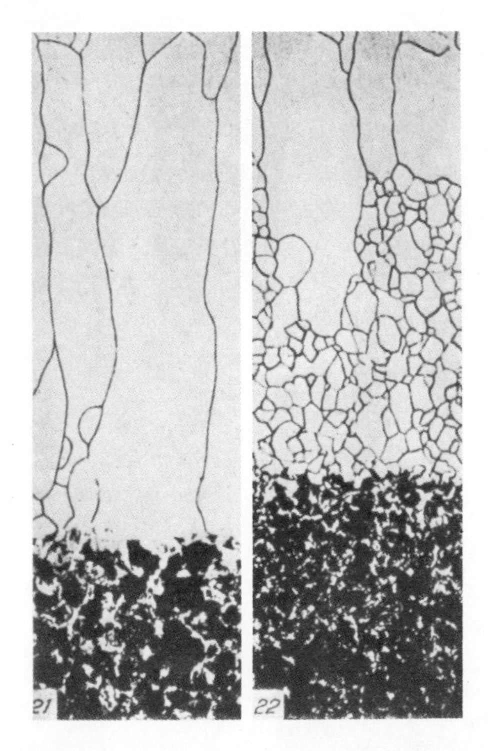

Fig. 12.24 Columnar grains of ferrite caused by depletion of carbon (decarburization) (Reproduced by permission, from Transactions ASM, American Society for Metals, 1935).

that the cracks have been open to the atmosphere for some length of time. Decarburization can also cause failures, as in Fig. 12.25, which shows the failure of several teeth in a thread chaser.[23] Probably, many applications can tolerate a certain degree of decarburizing. However, if it becomes excessive, as in the illustration, its inferior properties control the functioning.

Surface hardening and improved fatigue life can also be effected by introducing nitrogen into the surface. If the alloy content is proper, nitrides will be developed. Since not all alloying elements form nitrides equally, the lack of nitride forming elements can result in a soft case.

A white layer is often formed on the nitrided surface (Fig. 12.26). It is usually advisable to machine it off, since it sometimes initiates cracks.[24] The surface condition affects the nitriding results and subsequent performance (Figs. 12.27–12.29).[25,26] As a result of nitriding, there is an increase in dimension. Therefore, subsequent machining may be necessary. If too much metal is removed, the case may be removed and excessive wear may result (Fig. 12.27). Nitriding over a decarburized layer can result in spalling of the case (Fig. 12.28), as can nitriding on an excessively cold worked surface (Fig. 12.29).

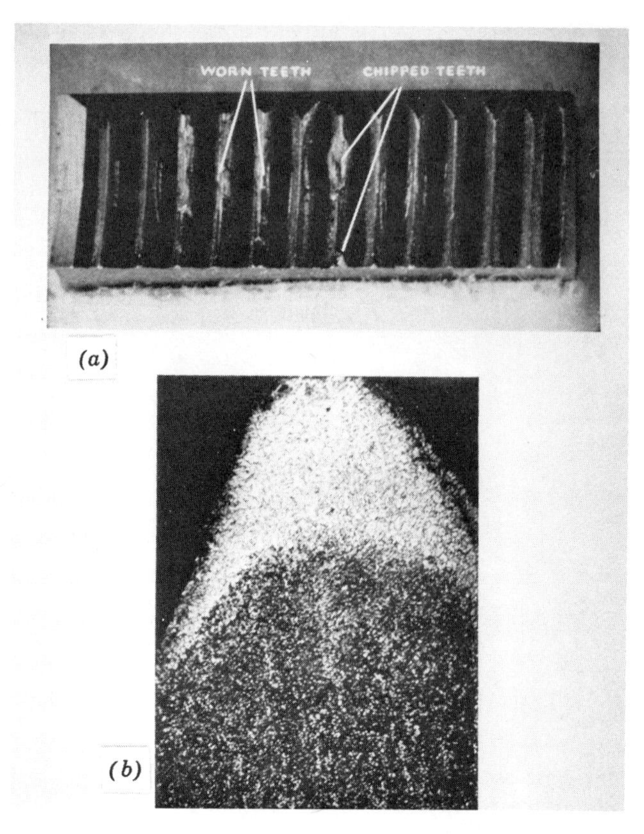

Fig. 12.25 (*a*) Thread chaser that failed in service because of (*b*) heavy decarburization on the teeth (Courtesy, Bethlehem Steel Corp.).

Not all surface treatments require a change in surface chemistry. Local rapid heating at the surface can develop a layer of austenite in steel, which transforms to martensite upon quenching. Two such operations are flame hardening and induction hardening. The choice of surface treatments depends on the application. If a hard case with soft core is required, carburizing can be used. However, if a relatively hard case on a strong core is required, and the design is adequate, induction or flame hardening may be used.

Gear teeth are often hardened this way. The application of heat must be closely controlled. If heat is applied for too long, an excesssive case depth may result. Figure 12.30 shows a properly hardened and an improperly hardened gear. Note the cracking on the improperly hardened gear.[27,28]

Fig. 12.26 White layer formed during nitriding (Reproduced by permission, from Metals Handbook, American Society for Metals, 1964).

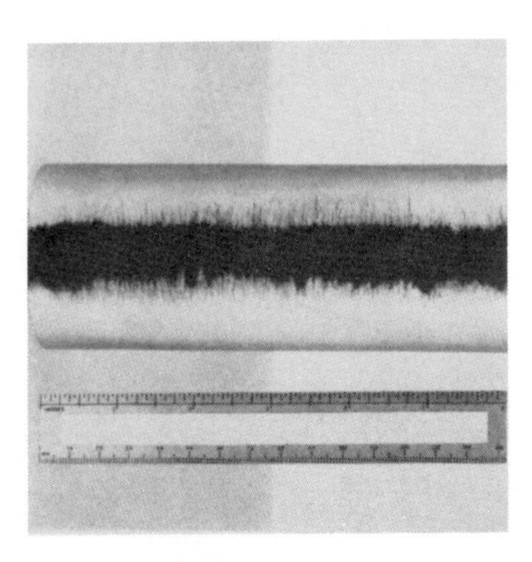

Fig. 12.27 Wear caused by excessive removal of material after nitriding (Reproduced by permission, Temple Publications, Inc.).

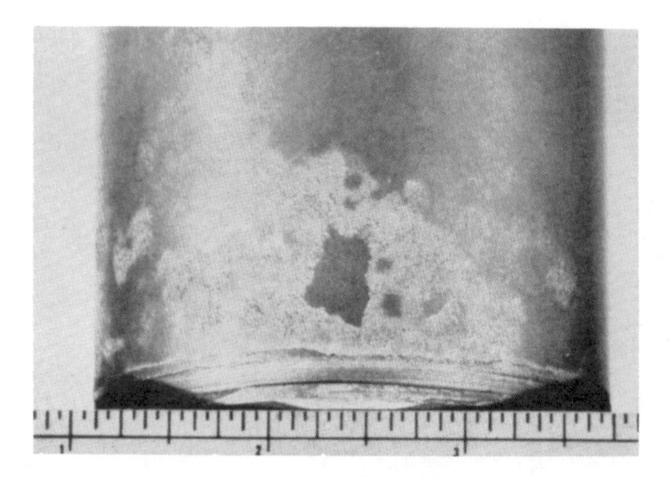

Fig. 12.28 Spalling caused by failure to remove decarburized layer prior to nitriding (Reproduced by permission Temple Publications, Inc.).

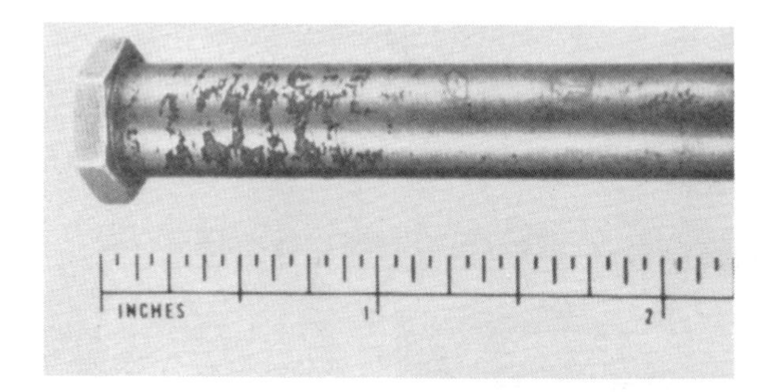

Fig. 12.29 Spalling caused by excessive cold working prior to nitriding (Reproduced by permission, Temple Publications, Inc.).

The emphasis has again been on steel. However, other metals can also be surface treated. Any metal that reacts with the atmosphere can have its surface altered. The main method for showing surface layers is metallography. The change in microstructure can easily be seen. It may not always be possible to prove a cause-effect relation; secondary cracking should be examined. Often, partial cracks reveal more than complete fractures.

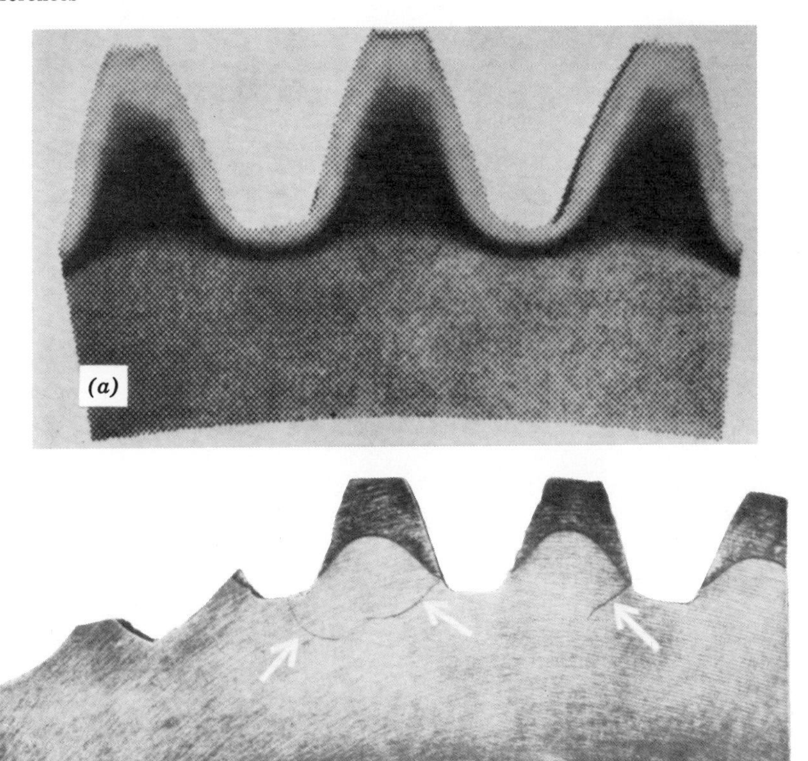

Fig. 12.30 (*a*) Properly hardened gear (Reproduced by permission, from Metals Handbook, American Society for Metals, 1964) and (*b*) improperly hardened gear. Note the cracks in the improperly hardened gear (Reproduced by permission, Republic Steel Corporation).

REFERENCES

1. *Metals Handbook*, Vol. 2, 8th ed., ASM, 1964, p. 16.
2. *Met. Treat.*, April–May, 1961, p. 13.
3. F. R. Palmer and G. V. Luerssen, "Tool Steel Simplified," The Carpenter-Steel Co., 1948, p. 393.
4. E. Polushkin, "Defects and Failures of Metal," Elsevier Publishing Co., 1956, p. 205.
5. *The Tool Steel Trouble Shooter*, Bethlehem Steel Company, 1952, p. 60.
6. *The Tool Steel Trouble Shooter*, *op. cit.*, p. 72.
7. *The Tool Steel Trouble Shooter*, *op. cit.*, p. 53.
8. *The Tool Steel Trouble Shooter*, *op. cit.*, p. 74.
9. E. S. Davenport, E. L. Roff and E. C. Bain, Trans. ASM, Vol. 22, April 1934, p. 305.

10. *The Tool Steel Trouble Shooter*, *op. cit.*, p. 56.
11. *The Tool Steel Trouble Shooter*, *op. cit.*, pp. 66–67.
12. *The Tool Steel Trouble Shooter*, *op. cit.*, p. 83.
13. *The Tool Steel Trouble Shooter*, *op. cit.*, p. 84.
14. K. Aust, J. Armijo, and J. Westbrook, *Trans. ASM.*, Vol. 59, 549 (1966).
15. E. Polushkin, *op. cit.*, p. 230.
16. E. Polushkin, *op. cit.*, p. 200.
17. K. Aust, J. Armijo, and J. Westbrook, *op. cit.*, p. 546.
18. F. R. Palmer and G. V. Luerssen, *op. cit.*, p. 450
19. O. McMullan, *Met. Prog.*, April 1962, p. 70.
20. *The Tool Steel Trouble Shooter*, *op. cit.* p. 79.
21. *The Tool Steel Trouble Shooter*, *op. cit.*, p. 77.
22. D. H. Rowland and C. Upthegrove, Trans. ASM, Vol. 24, March 1935, p. 112
23. *The Tool Steel Trouble Shooter*, *op. cit.*, p. 82.
24. *Metals Handbook*, Vol. 2, ASM, 8th ed., 1964, p. 151.
25. P. Lisk, *Met. Treat.*, February–March 1964, p. 10.
26. P. Lisk, *op. cit.*, p. 12.
27. *Metals Handbook*, Vol. 2, ASM, 8th ed., 1964, p. 181.
28. *Heat Treatment of Steels*, Republic Steel Corp., Booklet 1302, 1961, p. 22.

13

Welding

13.1 WELDING PROCESSES

There are numerous welding processes involving various techniques and procedures. They are markedly different in their process details and in the equipment required to achieve the joint. Savage,[1] however, classified them into two basic categories based on the response of the metal in the weld joint.

1. *The fusion welding processes.* The joining operation involves melting and epitaxial solidification, and any external forces applied to the system play no active role in producing coalescence.
2. *The pressure welding processes.* Externally applied forces play an important role in the bonding operation, whether consummated at room or elevated temperature.

Examples of the first category are shown in Fig. 13.1 and examples of the second are shown in Fig. 13.2. Note that the pressure welding processes have been further subdivided into "solid-state" and "pressure-fusion" groups.

A brief description of some of these processes is included for background purposes. A complete and detailed description of all the processes, however, is outside the scope of this work. Adequate and detailed descriptions are available elsewhere.[2,3]

13.1.1 Fusion Welding

GAS TUNGSTEN ARC WELDING

This is a fusion welding method in which the heat is generated from an electric arc between a nonconsumable tungsten electrode and the work. The heated material in the vicinity of the arc is protected from atmospheric contamination by a gas-shielding medium, which is usually argon, helium, or both.

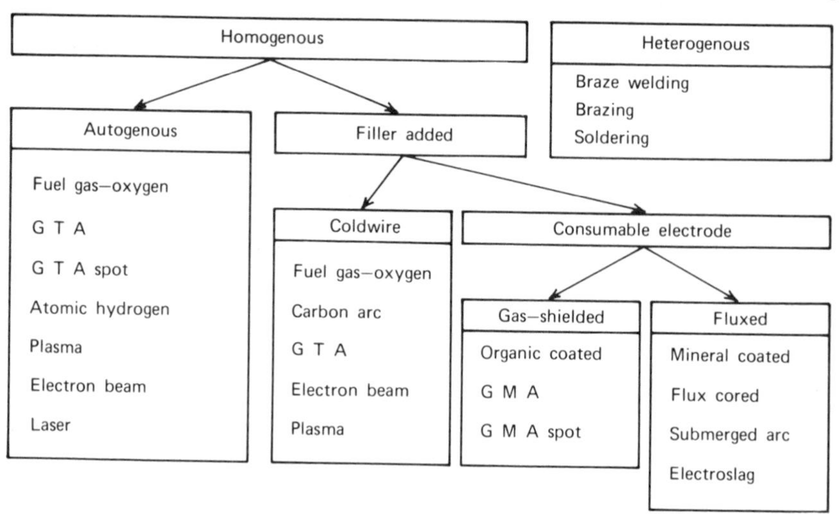

Fig. 13.1 Fusion welding processes (from Ref. 1).

The gas shield should be well-controlled to insure adequate protection of the metal in its elevated temperature condition. Unlike hydrogen in atomic hydrogen welding, an inert gas does not afford any reducing effect. It simply occupies the volume from which it shuts out oxygen and nitrogen in the air. Its transparency results in submerged arc welding under conditions of good visibility. For this reason, arc action and flow of the metal can be kept under observation at all times. Proper maintenance of the shielding protection necessitates the following precautions: (1) the gas cap should be of proper size and held as close as possible to the work; (2) the flow of gas should be adequate to cover the work during the movement of the electrode; (3) the welding must be conducted in a relatively still atmosphere, free from drafts; (4) the shield should be maintained for a proper cooling interval when the arc is broken.

Helium and argon are equally effective for providing adequate inert shielding. Helium, however, produces a hotter arc and is more effective in supporting higher speeds, faster melting, sharper arc action, and deeper penetration. Argon gives smoother action, and, therefore, is usually preferred on light gauge work where appearance as-welded is important. A mixture of the two gases is best for the majority of applications.

The process may utilize filler material when required, or flanges and excess material at the weld interface may be melted down. The weld deposits are usually clean, smooth, free of spatter, and slag free, which eliminates slag removal between subsequent passes.

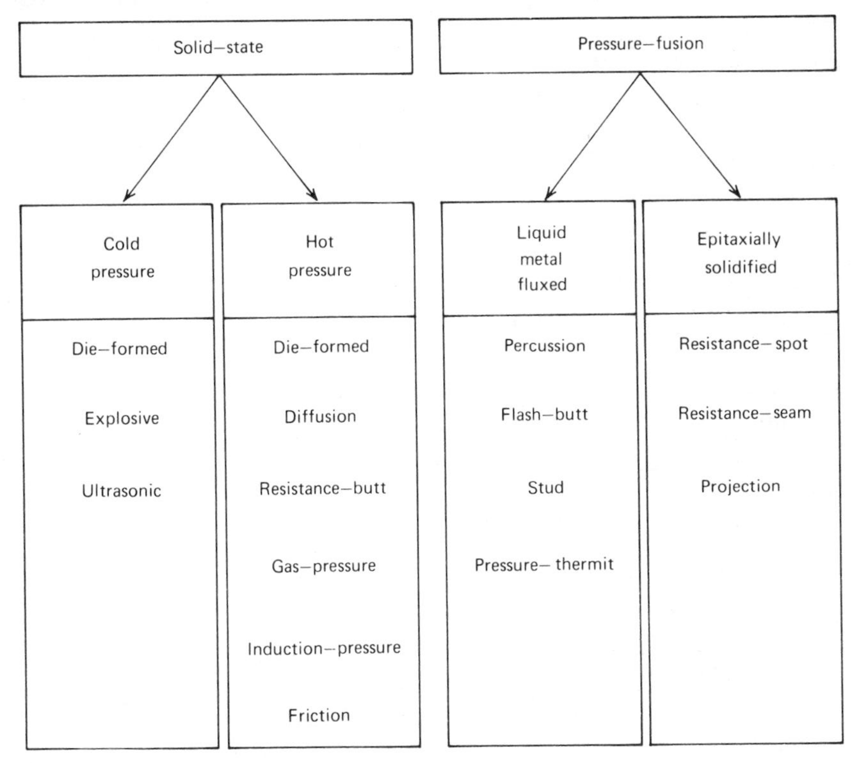

Fig. 13.2 Pressure welding processes (from Ref. 1).

GAS METAL ARC WELDING

This process is similar to gas tungsten arc welding except that welding heat is generated from an electric arc between a consumable bare-wire electrode and the work. The gas medium is usually argon, helium, or a mixture of both. Occasionally, 1 to 5% oxygen is added to improve the bead contour or penetration. The original gas metal arc process consisted of a continuous operation requiring high current densities to achieve a smooth transfer of molten metal. Recently, two modifications of this original spray-type arc have been developed for various applications. These are the shorting arc and pulsed-spray arc techniques. The shorting arc technique is essentially the same as the basic spray arc method, except that the welding cycle consists of alternating periods of arcing and short-circuiting. Metal transfer occurs during the short-circuit phase of the cycle in the form of a molten, but relatively cold droplet. The process is characterized by relatively low heat inputs, controlled penetration, and good out-of-position capability.

Another variation of the gas metal arc process is the pulsed-spray arc technique. In this process, the current is pulsed between high and low current densities, which causes a semicontinuous arc. Metal transfer occurs during the cyclic pulses of higher current. The process permits welding with minimal splatter, uniform penetration, and good out-of-position capability.

PLASMA ARC WELDING

This is a weld process which may be considered a modification of the gas tungsten arc process, in that a nonconsumable electrode in a gaseous environment is used. The gas flow is constricted and delivered under high pressure, raising its temperature and, thereby, creating a columnar arc of great stability. The columnar arc can achieve high penetration and weld speeds because of the high temperatures and velocities of the plasma. To protect the weld zone, however, additional gas shielding is usually required.

SHIELDED METAL ARC WELDING

This is a fusion welding process in which the heat is obtained from an arc between a manually held electrode and the workpiece. The consumable, flux coated electrode is melted and transferred across the arc, and provides the filler metal. The flux covering on the electrodes provides a slag and gaseous shield that protect the molten metal of the weld. The covering may also be used as a source of alloying elements to supplement the composition of the rod, or to compensate for losses of alloying elements from the weld.

SUBMERGED ARC WELDING

The process of welding with a submerged arc is similar to the shielded arc process. Weld heat is derived from an arc between an uncoated consumable wire electrode and the workpiece. The weld zone is protected by a covering of granular flux that is deposited, in advance, in the groove to be filled. The arc melts a portion of the flux and base metal, as well as the electrode, thereby providing an effective slag blanket.

GAS WELDING

This method utilizes the heat from a gas flame, usually oxygen-acetylene, to achieve fusion. Filler metal is added in the form of wire or rod, but an autogenous weld can be made by melting down excess metal in the weld zone. Compared to arc welding methods, gas welding is slower and more difficult to control. Although the flame temperature is lower, the heat produced is less concentrated and is spread over a relatively wide area surrounding the weld.

Protection of the metal at high temperature from contact with the surrounding atmosphere is of vital importance to weld quality. This is also true of other methods, particularly when stainless steels are involved. Protection can be insured by two basic methods. The first is through the use of a flux. The other is to maintain a protective shield of natural gases from the products of combustion of the flame.

Flux properly applied to all exposed hot metal affords more continuous protection than does the gas shield. It should never be omitted. The protective gas shield is subject to variations, interruptions, and disturbance by air currents. In addition, the neutral character of the gas shield must be maintained. Therefore, oxygen and acetylene must be proportioned during mixing to avoid an excess of one or the other. Too much oxygen leads to weld porosity, and too much acetylene promotes a carburizing atmosphere.

13.1.2 Resistance Welding

Resistance welding includes a group of methods that depend on the combined use of heat (generated by the resistance to an electrical current flowing across an intended joint location) and pressure to produce a homogeneous bond. Resistance spot and seam welds are examples of this type of process.

SPOT WELDING

To achieve a spot weld, the pieces of metal to be joined are placed between a set of electrodes, as shown in Fig. 13.3. An electrical current of specific magnitude and duration is sent from one electrode through the workpieces to the other electrode. The contact between the workpieces undergoes resistance heating and achieves the maximum temperature rise. The contact between the electrode and the workpiece reaches lower temperatures because of a number of factors, among which are decreased resistance due to the contact pressure and higher heat dissipation caused by the construction material. As the temperature of the metal adjacent to the interface of the workpieces rises, it approaches the melting point. Fusion on both sides of the interface occurs and this combined with external pressure transmitted through the electrodes results in coalescence. An ellipsoid nugget is formed, with its maximum height in line with the center line of the electrodes. The growth of this nugget is controlled by stopping the current. There are a number of variables in the process which affect the production of quality welds. The amount and length of the pressure, the amount and length of the current flow, and the thickness of

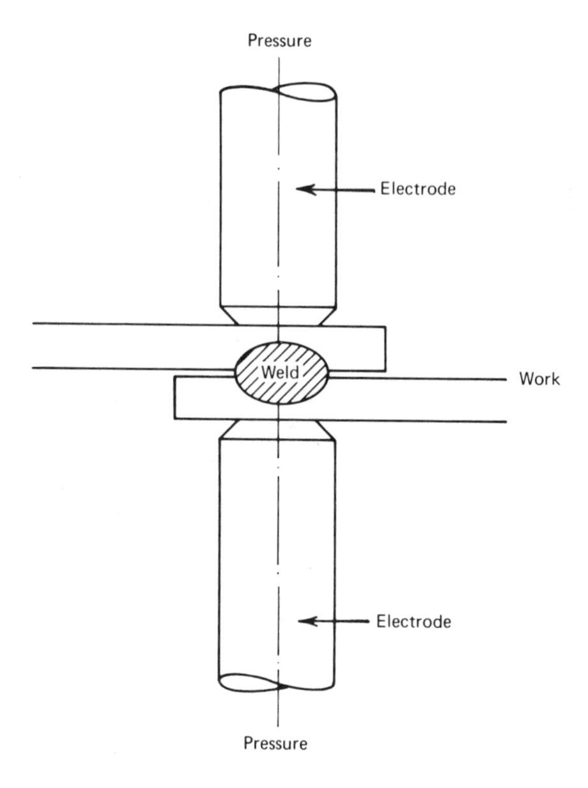

Fig. 13.3 Schematic sketch of the spot welding process.

the metal to be joined are all factors that must be adjusted to yield an optimum weld.

SEAM WELDING

Seam welding is a resistance welding technique in which a series of overlapping spot welds are made along a line by a set of disk-shaped electrodes. The degree of overlap can vary from a wide to a closely spaced series of weld spots that are liquid and gas tight. As with spot welding, the pressure, current amplitude, and current duration are significant factors in the production of sound welds. Because the individual welds are closely spaced, a portion of the current is shunted through the previously completed weld, thereby decreasing the heat input. Therefore, increased currents are necessary to achieve seam welds, as compared to spot welds.

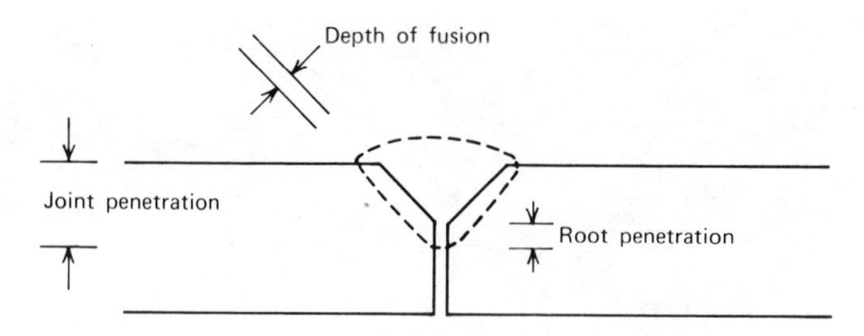

Fig. 13.6 Schematic diagram illustrating the concepts of fusion and penetration.

Fig. 13.7 Cross-section showing incomplete fusion in a steel weld.

and travel speed are probably the most important. Low heat inputs and high speeds produce only superficial heating and melting, and result in shallow penetration.

13.2.5 Undercutting

Undercutting is a condition where a groove in the margin of the weld (the toe) is created by the melting of the base metal. This condition is

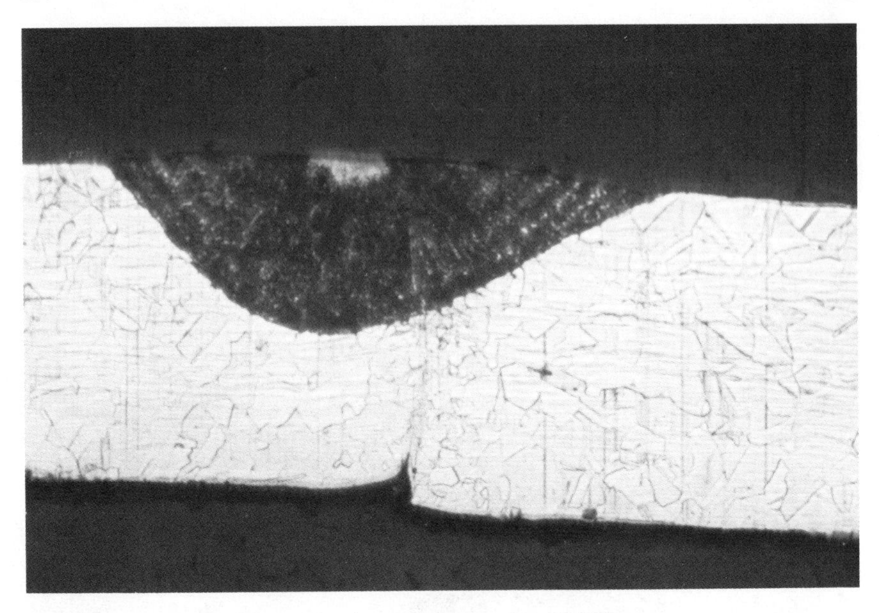

Fig. 13.8 Incomplete penetration in a stainless steel weld (5X).

shown in Fig. 13.9. Examining this figure shows that the rift created by undercutting has all the characteristics of a notch. The creation of this notch in the very area where the transition from weld metal to base metal occurs has very serious consequences from the fatigue standpoint. The resultant concentration of stresses at this point favors early crack initiation and considerably shortens fatigue life. Linnert[7] states that undercutting, though often ignored, has the worst record for causing mechanical failures in weldments.

13.2.6 Lamellar Tearing

Lamellar tearing is a form of cracking that occurs in welded corner or "tee" joints, as shown in Fig. 13.10. When these joints are severely loaded, high tensile stresses are created normal to the rolling plane of the plate. Because of the anisotropy often found in steel plate, cracking occurs in a plane parallel to the rolling plane. Metallurgical investigations have shown that the cracking or tearing originates with nonmetallic inclusions of the silicate or sulphide types. Figure 13.11 shows a transverse section of a crack origin, with an electron fractograph of the inclusion surface at that point. The stepped nature of lamellar cracks,[8] as shown in Fig. 13.12, indicates that multiple cracks originate on parallel planes at approximately the same time and propagate for a short distance before linking up.

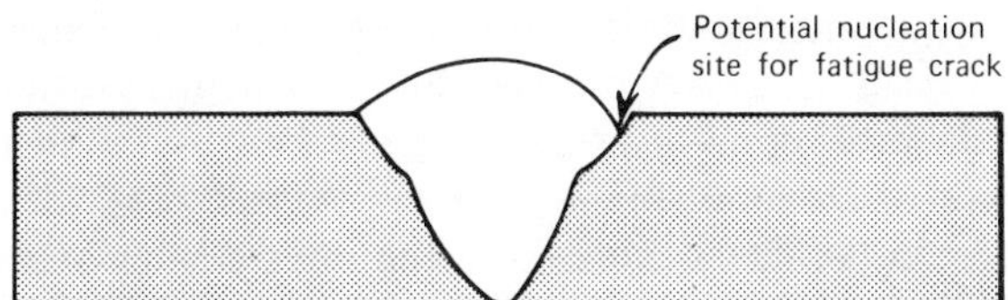

Fig. 13.9 Top view of an undercut weld together with a diagram showing the notch effect created by such under cutting.

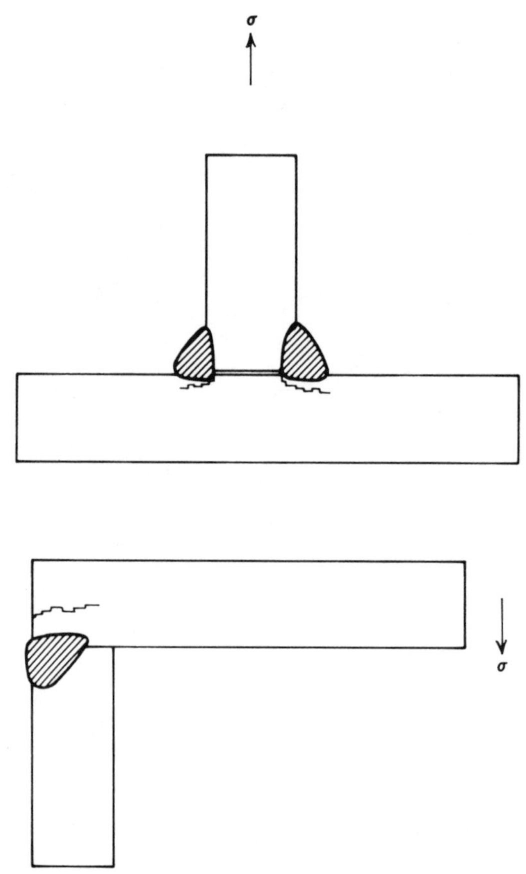

Fig. 13.10 Diagrams illustrating lamellar tearing in a "tee" and corner weld.

13.2.7 Hot Cracking

The primary cause of hot cracking in welds are contraction stresses imposed on the weld metal soon after its solidification. At this stage, the weld metal has poor mechanical properties and may be overstressed relatively easily. Segregation effects can also exert an influence on hot cracking. As a result of segregation, residual elements, particularly sulfur, can reach considerable concentrations in the last portion of metal to freeze. With high travel speeds and high heats, an elongated teardrop pool may occur with the result that rejection into the pool creates an accumulation of residual elements along the center line of the weld. These residual elements tend to form relatively weak, low melting compounds which give rise to cracks under stress.

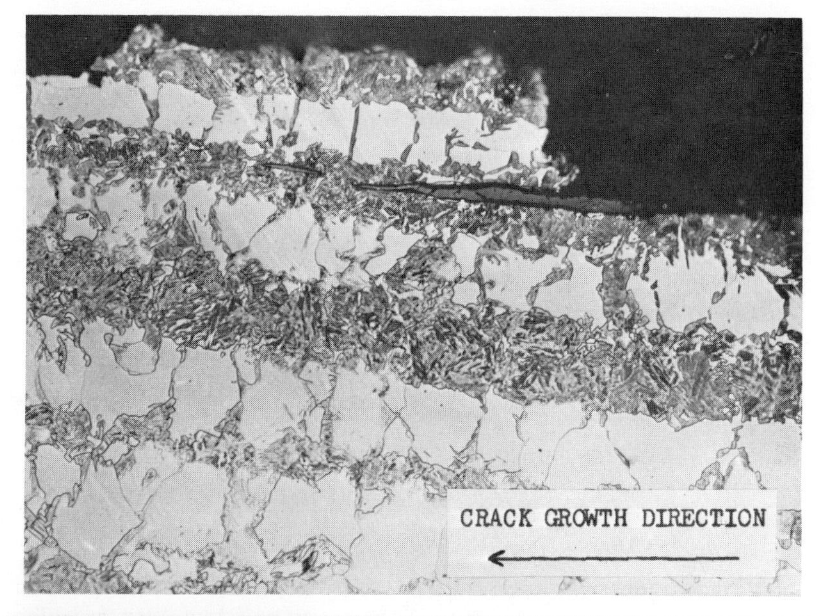

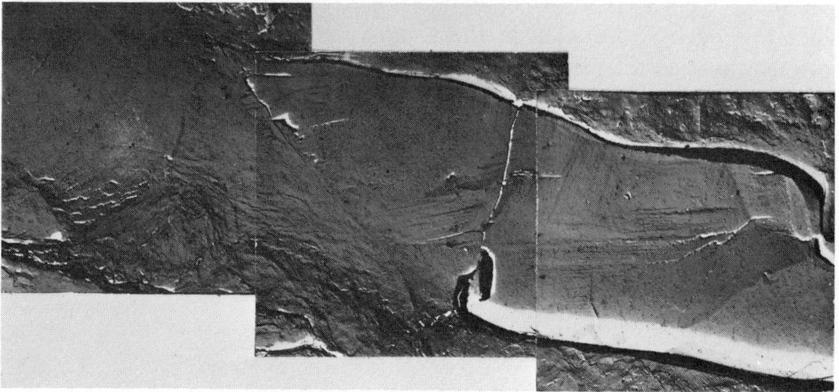

Fig. 13.11 Cracking in a banded structure originating at an inclusion.

Hot cracks are typically longitudinally oriented, as shown in Fig. 13.13, or star-shaped crater cracks. Although these cracks are not overly dangerous in themselves, they may act as nuclei for the formation of fatigue cracks. On a metallographic basis, hot cracks are generally intergranular and are often heavily oxidized along the side wall. The presence of this high temperature oxide is a definite indication of hot cracking.

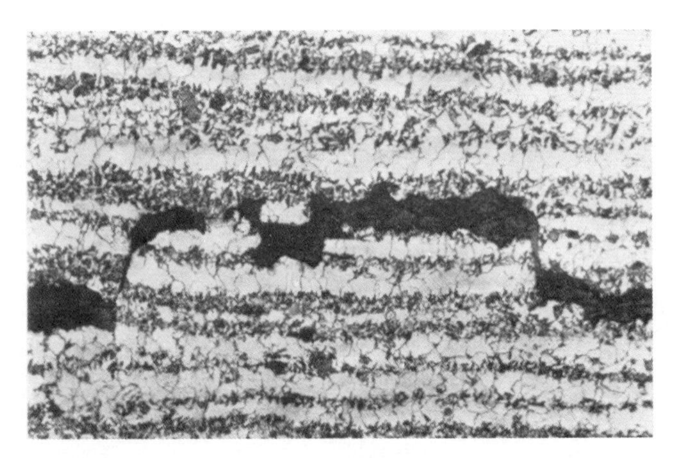

Fig. 13.12 Stepped nature of a lamellar tear (250X).

13.2.8 Hydrogen Damage

Welds made under conditions where high hydrogen concentrations result can suffer from two forms of damage, porosity and hydrogen cracking. Once hydrogen is occluded in the solidifying weld metal, it can react in various ways. As the metal approaches the freezing point, hydrogen that is in excess of the solubility limit is rejected and forms gas holes and fine porosity. As the metal cools further, some hydrogen diffuses from the weld surface and a concentration gradient is established from the weld interior to the weld surface. This action can be shown by submersing a weld recently made under high hydrogen conditions in water. Numerous small bubbles will be seen forming and escaping from the weld surface. Nevertheless, because diffusion, even of hydrogen, is relatively slow at room temperature, residual hydrogen does remain in the weld.

CRACK APPEARANCE

Cracks that occur as a result of hydrogen damage are primarily longitudinal underbead cracks, as shown in Fig. 13.14. These form in the heat affected zone and adjacent to the weld interface. They are intergranular in nature and often nucleate at inclusion sites in the heat affected zone (HAZ). In the partially melted zone, cold cracks can be nucleated in liquated regions in the grain boundaries.

FACTORS AFFECTING DAMAGE

Szekeres,[9] in a study of two heats of alloy steel welded under high and low hydrogen conditions, has shown that the physical manifestation and

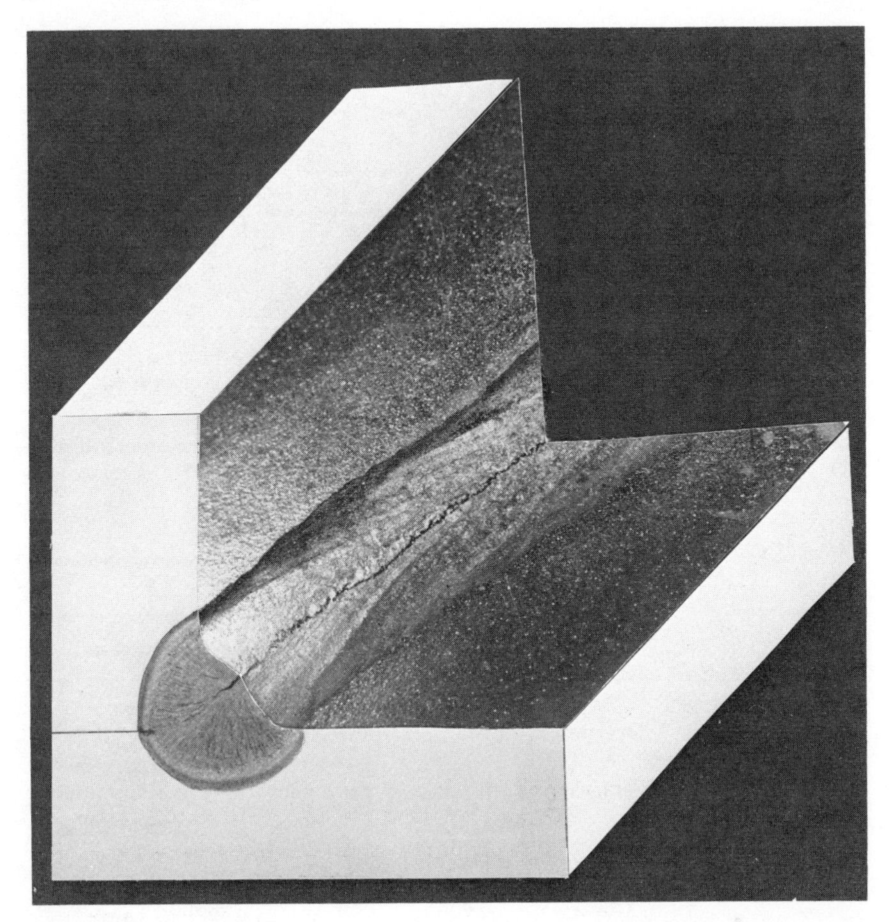

Fig. 13.13 Composite photograph showing longitudinal view and transverse section of a hot crack (2X).

the magnitude of damage depends on the level of hydrogen, the direction of stresses relative to banding, and the alloy content. His results are summarized in Table 13.1. Hydrogen cracks and gas bubbles were found to be associated with elongated sulphide inclusions and appeared to have nucleated at these sites. This factor would account for the decrease in cracking observed with the applied stress parallel to the banding, since the elongated inclusion would be less effective as a stress raiser in this orientation.

Many of the characteristics of hydrogen embrittlement described previously are also true of hydrogen damage in welds. Delayed cracking frequently occurs days or weeks after welding. This cracking can occur

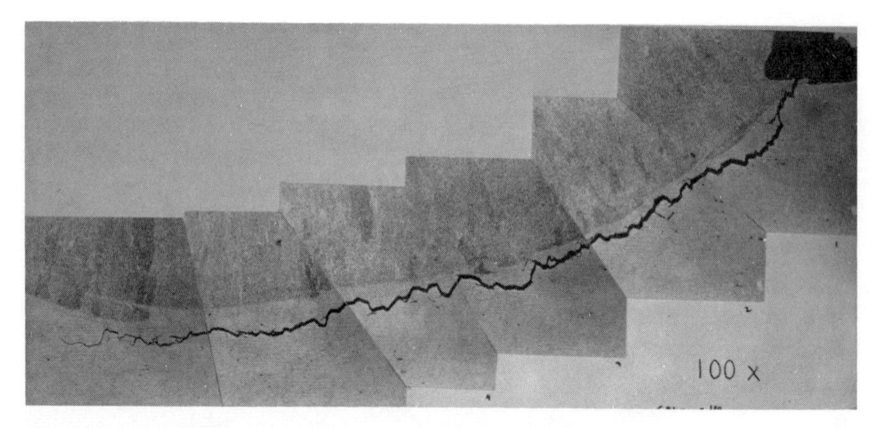

Fig. 13.14 Underbead cracking showing the intergranular nature resulting from hydrogen damage (Courtesy of W. F. Savage).

Table 13.1 Summary of General Observations of Cold-Cracking-Test Specimens of HY-80 Steel[a]

Heat Number	Relative Hydrogen Content	Banding-to-Stress Orientation	Amount of Bubble Evolution	Relative Cracking Observed[b]: Fusion Boundary[c]	Relative Cracking Observed[b]: True Heat-Affected Zone
A	High	Transverse	Profuse	Very many	Very Many (long)
		Parallel	Profuse	Few	Few (short)
	Low	Transverse	None	Few	None
		Parallel	None	None	None
B	High	Transverse	Profuse	Several	Many (medium length)
		Parallel	Profuse	None	None
	Low	Transverse	None	None	None
		Parallel	None	None	None

[a] From Ref. 9.

[b] Approximate rating of the incidence of cracking: few—1 to 3, several—4 to 6, many—7 to 12, very many—over 12.

[c] Includes cracks associated with the unmixed zone and/or the partially-melted zone.

over a wide range of applied stress, but is usually initiated more rapidly with higher stresses. Similarly, the stress required to cause cracking decreases as the hydrogen content increases. Consequently, failure can occur at relatively low stresses if sufficient hydrogen is present.

Microstructure can also exert a major influence. Hydrogen cracking is observed most often when the microstructure is martensitic. This increased susceptibility of martensites to hydrogen cracking is believed to be due to the presence of large, short order, transformation stresses. These stresses tend to produce small microrifts in the martensite matrix which can act as hydrogen sinks. As the carbon content of the alloy increases, the martensites formed are more stress laden and, consequently, more susceptible to hydrogen damage.

SOURCES OF HYDROGEN

Hydrogen may become available for weld embrittlement from a variety of sources. The most common source is moisture, either present on the base metal surface or held in the electrode covering. Moisture on the prepared weld surface can be the result of condensation or actual rainfall, and residual moisture can remain even though it is not visible. Failure to remove this moisture through preheating can have similar effects. Hydrogen-susceptible alloys are usually welded with low-hydrogen electrodes of the EXX15, EXX16, or EXX18 types, rather than electrodes covered with ordinary cellulosic materials, such as wood flour. These cellulosic materials can absorb large quantities of moisture in addition to the hydrogen generated as a combustion product when these materials are burned.

Even low-hydrogen electrodes must be used prudently, however, if hydrogen damage is to be avoided. Although these electrodes are manufactured to low hydrogen concentrations, they must be protected from moisture in shipping and handling. Some precautions should also be taken on the fabrication site. Prebaking followed by the limited issuance of electrodes is effective in minimizing hydrogen contamination. Fluxes used in the submerged arc process should also be protected from moisture pickup by the use of similar procedures.

13.2.9 Corrosion

There are a few areas where the corrosion of welds becomes a major problem. The most significant of these is the selective attack of welds in austenitic stainless steels.

INTERGRANULAR ATTACK

If a weld is made in the austenitic stainless steels, there is a zone of metal, at some unspecified distance within the material surrounding the

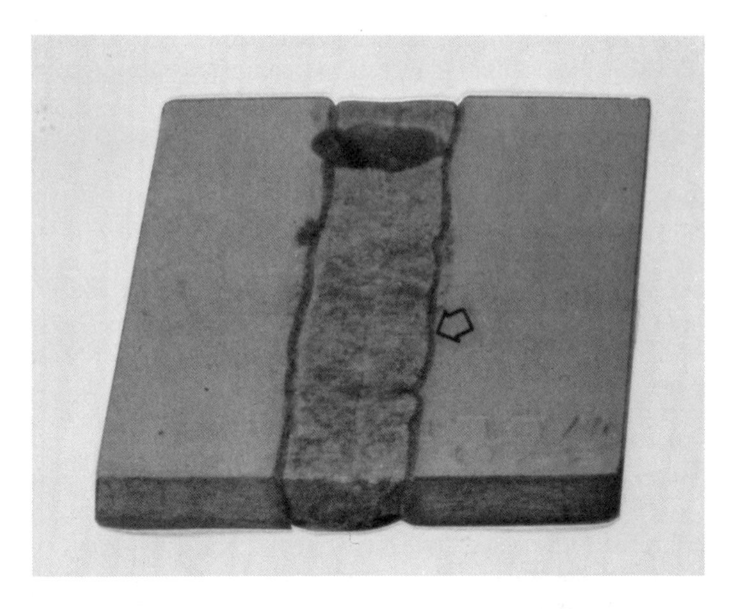

Fig. 13.15 Weld decay resulting from the intergranular attack of a sensitized stainless steel.

weld that experiences temperatures in the sensitizing range (900 to 1500°F) of substantial duration. The sensitization treatment causes a precipitation of carbides at the grain boundaries in this zone. In a suitable environment, weld decay occurs in a band in the parent metal adjoining the weld, as shown in Fig. 13.15. The degree of sensitization that occurs depends on the welding temperatures, thickness of the plate, ambient temperatures, and welding speeds. In general, thicker materials are subjected to higher thermal inputs, cool more slowly, and many consequently suffer greater sensitization than thinner materials.

Sensitization is dependent not only on the time and temperature of the heating cycle, but also on the amount of carbon present, the ratio of nickel and chromium to carbon content, and the presence of residual stresses.

Metal in which stresses are present become sensitized somewhat more rapidly than annealed or stress-free metal. These stresses may result from the use of severely cold-rolled material, from mechanical restraints imposed on the system during welding, or from contraction stresses resulting from forced fits. Such stresses may reduce the lower limit of the sensitizing range, thereby expanding the temperature range over which carbide precipitation may occur.

Except for the role of stresses, most of the factors affecting the sensitization of stainless steels, as well as the mechanism, have been described in Chapter 7.

The removal of precipitated carbides from austenitic stainless steels to restore maximum corrosion resistance can be accomplished by resorting to the following two successive steps.

1. Bring the metal to a temperature above the 900 to 1500°F range (about 1850 to 2000°F) and hold it there long enough to redissolve the precipitate and thus return its chromium and carbon back into their original condition, in solid solution.
2. Rapidly cool the metal down through and below the 900 to 1500°F range so that chromium and carbon do not have time to recombine. This assures the retention of chromium and carbon in solid solution.

KNIFE-LINE ATTACK

Knife-line attack is a form of intergranular corrosion that affects welds in stabilized austenitic stainless steels (type 321 or 347) under certain conditions. During the welding process, a thin band adjacent to the fusion zone experiences quite high temperatures. Research[10] has shown that in the range of 2250°F to the melting point (≅2600°F), the columbium and titanium carbides, which make types 347 and 321 resistant to intergranular attack, go into solution. Upon rapid cooling, these elements remain in solution and do not combine with the available carbon to form columbium or titanium carbides. If the welded component is then subjected to temperatures in the range of 950 to 1450°F, chromium carbides, rather than columbium carbides, precipitate preferentially, resulting in chromium depletion and sensitization. If these components are then exposed to environments that cause intergranular corrosion, a thin band of metal adjoining the fusion zone will be attacked. An example of knife-line attack is presented in Fig. 13.16.

It should be remembered that the sensitization is the direct result of a

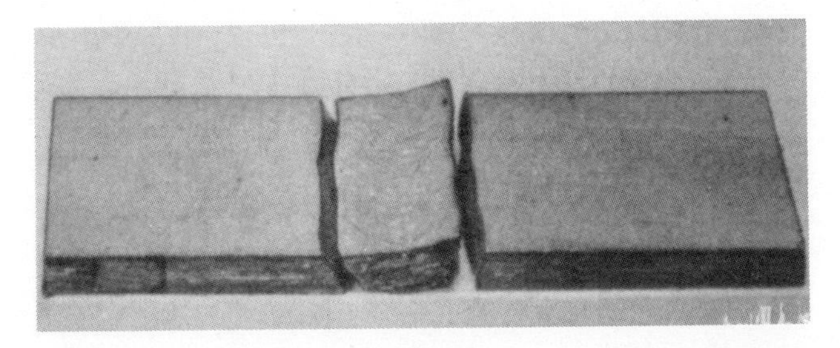

Fig. 13.16 Knife-line attack on a stabilized stainless steel weld.

heating cycle after welding and that knife-line attack will not result unless this occurs. Sensitization of stabilized stainless steels may be eliminated by thermally soaking the weldment at approximately 1950°F. At this temperature, chromium carbides undergo dissolution and columbium or titanium carbides precipitate.

13.2.10 Excessive Reinforcement

This defect arises when the weld bead is considerably higher than the base metal. Savage[1] states that excessive reinforcement reduces the fatigue life of the joint in proportion to the acuity of the contact angle, as shown in Fig. 13.17. This statement seems to be quite valid since, in effect, the

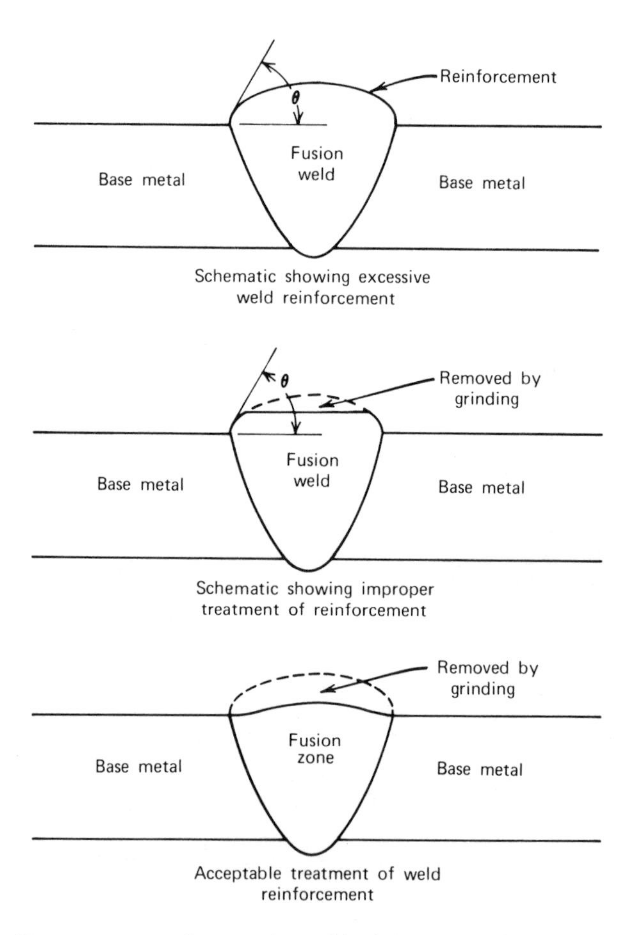

Fig. 13.17 Proper treatment for excessive weld reinforcement (from Ref. 1).

angle between the weld bead reinforcement and the base metal forms a notch. Depending on its root angle, this notch can then be a very detrimental stress raiser. Savage further adds that reducing the weld bead height without changing the contact angle may satisfy the inspectors, since maximum height is the factor specified in the ASME Boiler Code, but it will not improve the fatigue life.

13.2.11 Resistance Welding Defects

Spot and seam welds can exhibit a variety of defects associated with the processes. These defects can range from internal cracks to oversize weld nuggets. Figures 13.18 and 13.19 show the types of defects that can arise for spot welds and the weld parameters to which they are related. A similar set of data is presented in Figures 13.20 and 13.21, which show the kind of defects associated with seam welds and the prevalent causes of their occurrence.

13.3 HEAT TREATMENT

Since a weld is in essence simply a casting, it is subject to and responsive to heat treatment, as is any other casting of the same alloy. Heat treatments applied prior to the welding operation are referred to as preheating, those applied subsequent to the welding are called postheating. Linnert[11] has outlined the major reasons why heat treatment is applied to welded structures.

"1. To avoid cold cracking in the heat-affected zones of hardenable steels.

"2. To increase the toughness of the weld joint and improve its ability to withstand adverse service conditions involving impact loading or low temperatures.

"3. To alleviate the effects of hydrogen which enters the weld metal and the base metal heat-affected zone.

"4. To reduce residual stresses (internal stresses from shrinkage, phase transformation, and reaction to restraint) to a desired low level.

"5. To minimize shrinkage and distortion.

"6. To produce particular mechanical or physical properties in the steel of which the weldment is constructed."

13.3.1 Preheating

Preheating refers to warming or heating the structure to be welded before the welding operation. The effects of the preheating operation are varied since the process temperature can range from barely warm to over 1000°F. However, several basic effects do occur. Preheating lowers the cooling rate subsequent to welding in proportion to the preheat temperature. The

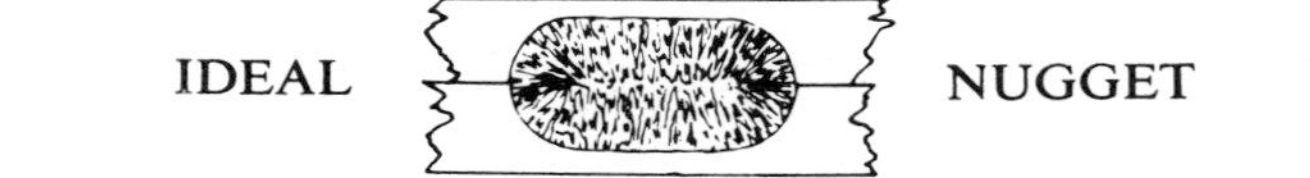

IDEAL NUGGET

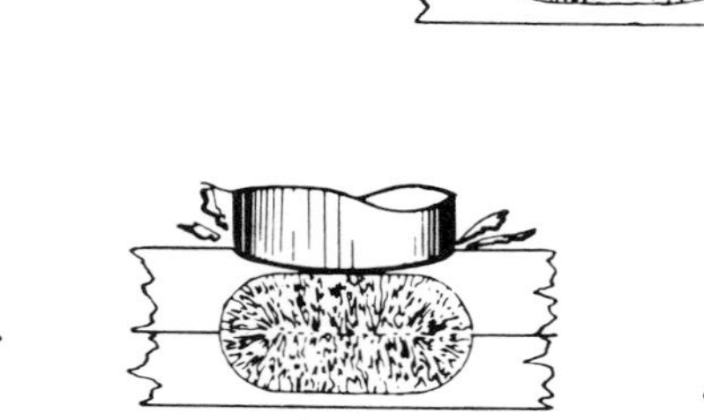

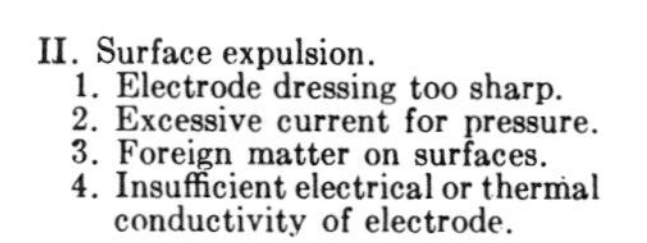

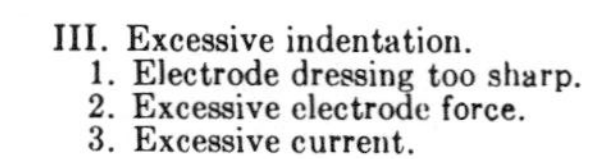

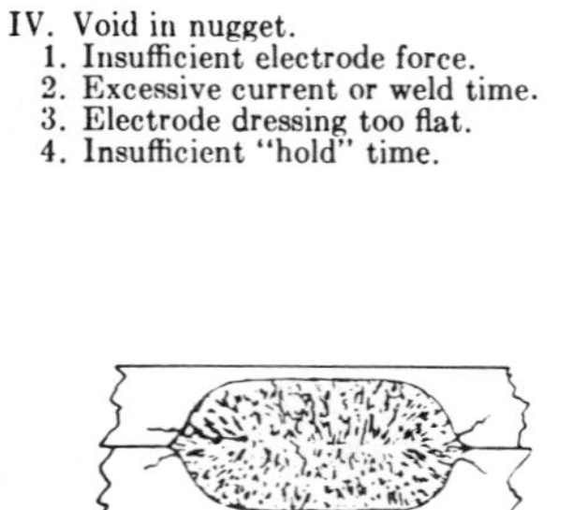

I. Excessive expulsion at interface.
1. Insufficient electrode force.
2. Excessive current or weld time.
3. Electrode dressing too sharp.
4. Foreign matter on interface.

II. Surface expulsion.
1. Electrode dressing too sharp.
2. Excessive current for pressure.
3. Foreign matter on surfaces.
4. Insufficient electrical or thermal conductivity of electrode.

III. Excessive indentation.
1. Electrode dressing too sharp.
2. Excessive electrode force.
3. Excessive current.

IV. Void in nugget.
1. Insufficient electrode force.
2. Excessive current or weld time.
3. Electrode dressing too flat.
4. Insufficient "hold" time.

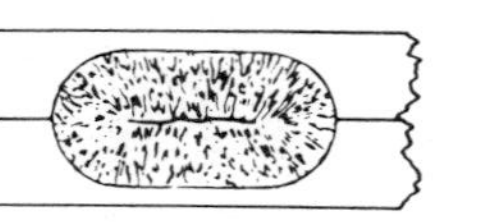

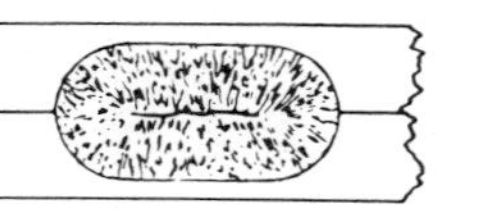

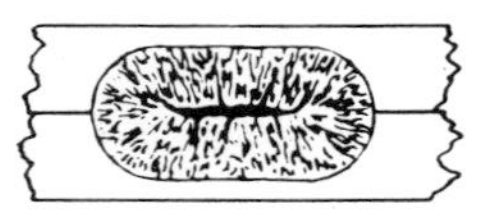

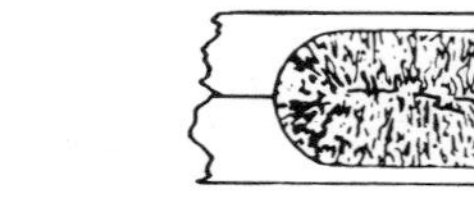

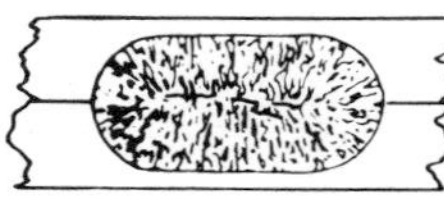

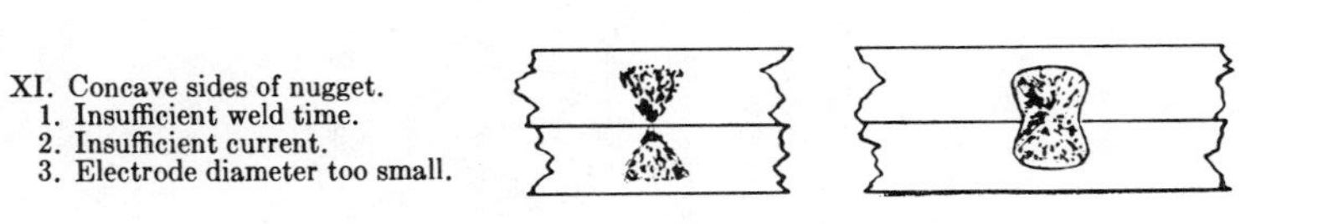

Fig. 13.18 Typical defects found in spot welding materials of equal thickness. (Courtesy of ITE Corporation).

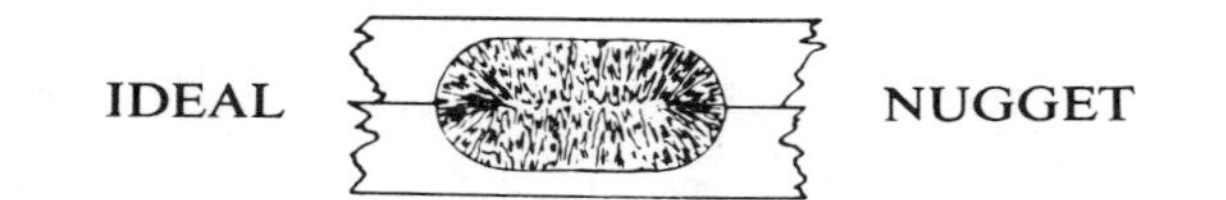

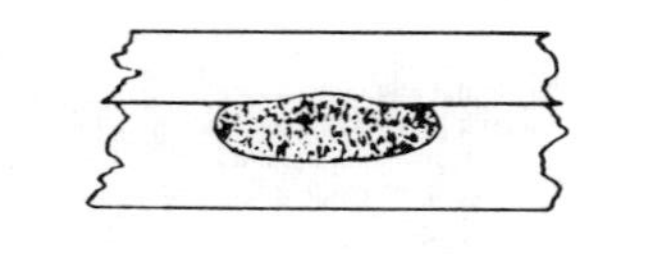

I. Lack of penetration in thin sheet.
   1. Electrode dressing too flat on thin sheet.
   2. Excessive electrode force.
   3. Excessive electrical or thermal conductivity of electrode on thin sheet.
   4. Insufficient electrical or thermal conductivity of electrode on heavy sheet.

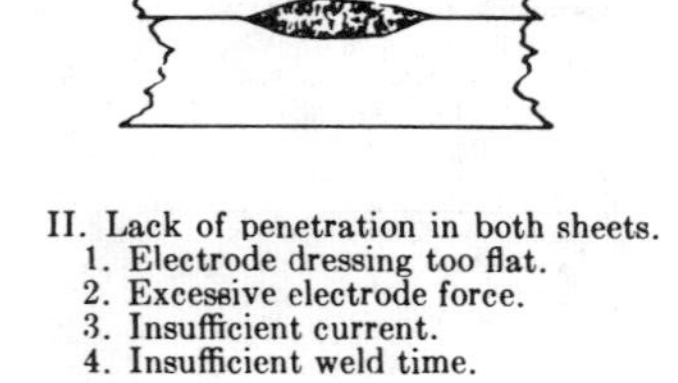

II. Lack of penetration in both sheets.
   1. Electrode dressing too flat.
   2. Excessive electrode force.
   3. Insufficient current.
   4. Insufficient weld time.

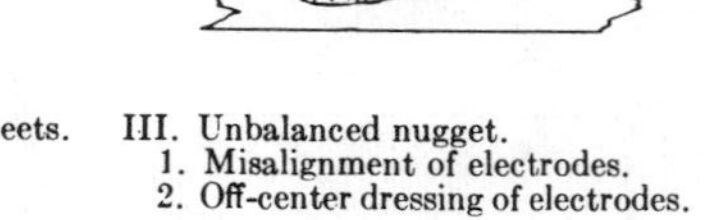

III. Unbalanced nugget.
   1. Misalignment of electrodes.
   2. Off-center dressing of electrodes.

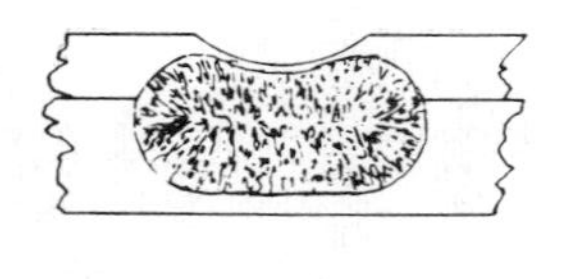

IV. Excessive indentation.
   1. Electrode dressing too sharp at indentation.
   2. Excessive electrode force.
   3. Excessive current or weld time.
   4. Electrode diameter too small at indentation.

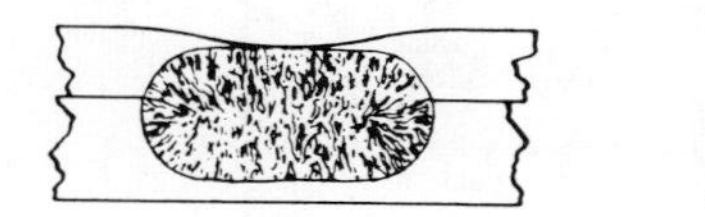

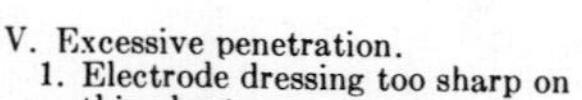

V. Excessive penetration.
1. Electrode dressing too sharp on thin sheet.
2. Insufficient electrode force.
3. Excessive current or weld time.
4. Insufficient electrical or thermal conductivity of electrodes.

VI. Void in nugget.
1. Insufficient electrode force.
2. Excessive current.
3. Excessive weld time.
4. Electrode dressing too flat.
5. Insufficient "hold" time.

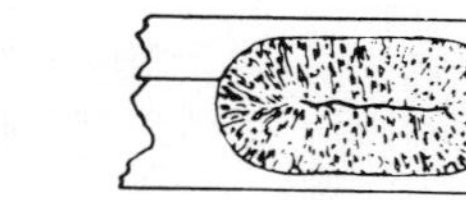

VII. Cracks in Nugget.
A. Horizontal cracks at center.
1. Insufficient electrode force.
2. Insufficient "hold" time.
3. Electrode dressing too flat.
4. Excessive current or weld time.

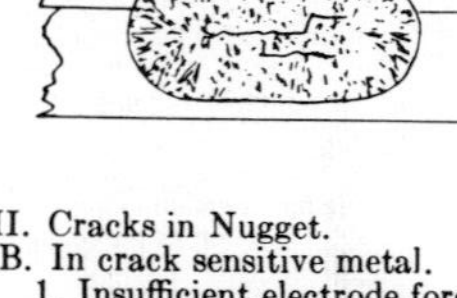

VII. Cracks in Nugget.
B. In crack sensitive metal.
1. Insufficient electrode force.
2. Electrode dressing too flat.
3. Insufficient "hold" time.
4. Insufficient weld time.

VIII. Cracks in parent metal.
1. Electrode diameter too small for nugget diameter.
2. Electrode dressing too sharp.
3. Insufficient electrode force.
4. Excessive current or weld time.
5. Insufficient "hold" time.

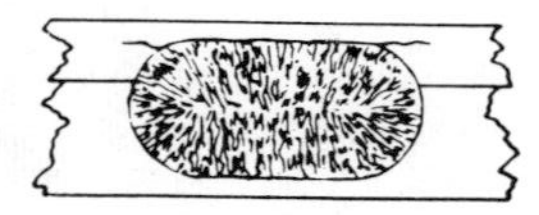

Fig. 13.19 Typical defects found in spot welding materials of unequal thickness. (Courtesy of ITE Corporation).

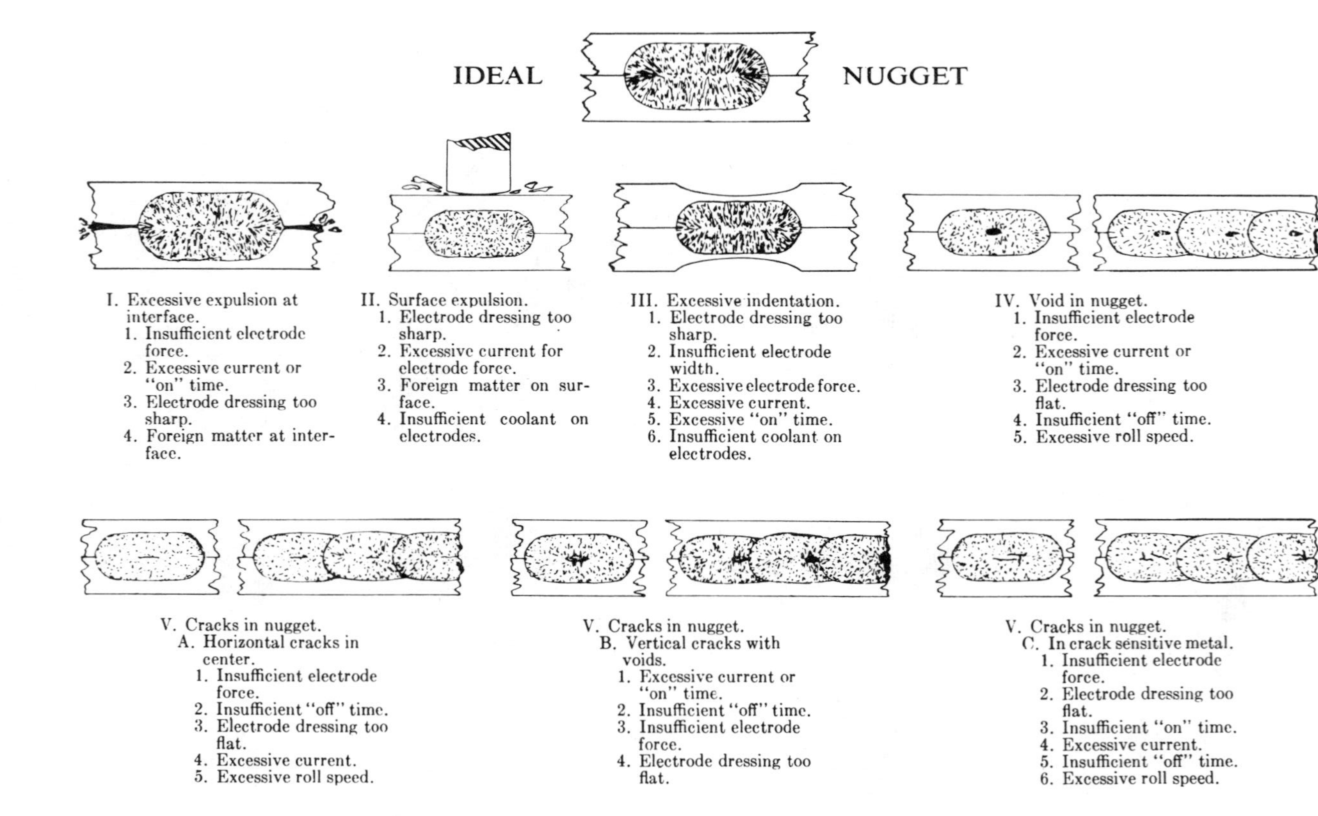
IDEAL NUGGET
I. Excessive expulsion at interface.
1. Insufficient electrode force.
2. Excessive current or "on" time.
3. Electrode dressing too sharp.
4. Foreign matter at interface.
II. Surface expulsion.
1. Electrode dressing too sharp.
2. Excessive current for electrode force.
3. Foreign matter on surface.
4. Insufficient coolant on electrodes.
III. Excessive indentation.
1. Electrode dressing too sharp.
2. Insufficient electrode width.
3. Excessive electrode force.
4. Excessive current.
5. Excessive "on" time.
6. Insufficient coolant on electrodes.
IV. Void in nugget.
1. Insufficient electrode force.
2. Excessive current or "on" time.
3. Electrode dressing too flat.
4. Insufficient "off" time.
5. Excessive roll speed.
V. Cracks in nugget.
A. Horizontal cracks in center.
1. Insufficient electrode force.
2. Insufficient "off" time.
3. Electrode dressing too flat.
4. Excessive current.
5. Excessive roll speed.
V. Cracks in nugget.
B. Vertical cracks with voids.
1. Excessive current or "on" time.
2. Insufficient "off" time.
3. Insufficient electrode force.
4. Electrode dressing too flat.
V. Cracks in nugget.
C. In crack sensitive metal.
1. Insufficient electrode force.
2. Electrode dressing too flat.
3. Insufficient "on" time.
4. Excessive current.
5. Insufficient "off" time.
6. Excessive roll speed.

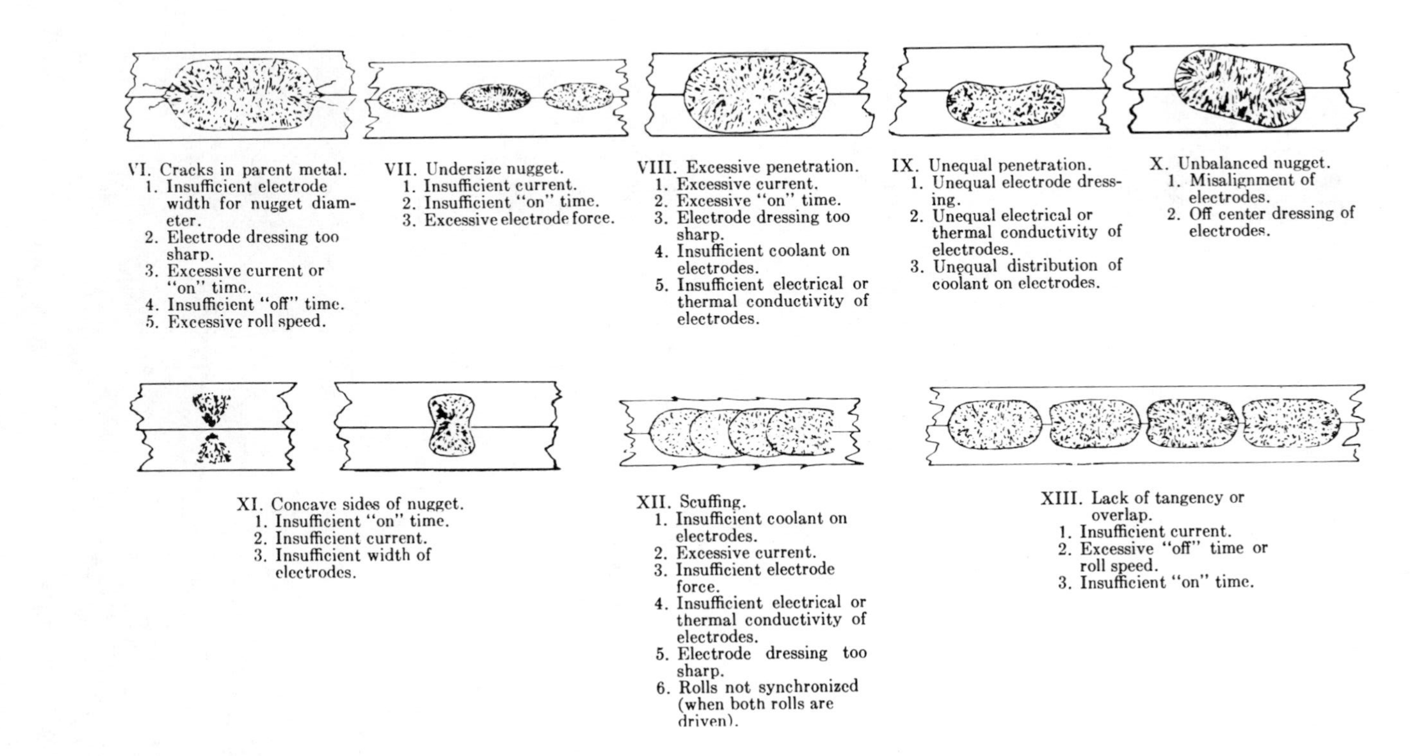

Fig. 13.20 Typical defects found in seam welding materials of equal thickness. (Courtesy of ITE Corporation).

IDEAL  NUGGET

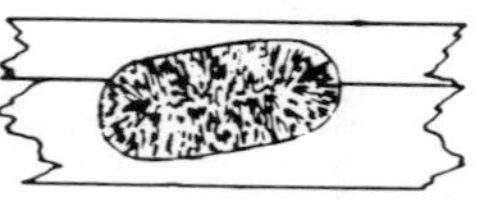

I. Lack of penetration in thin sheet.
1. Electrode dressing too flat on thin sheet.
2. Excessive electrode force.
3. Excessive electrical or thermal conductivity of electrode on thin sheet.
4. Insufficient electrical or thermal conductivity of electrode on heavy sheet.

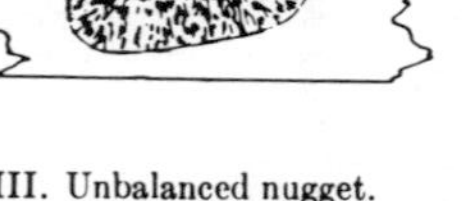

II. Lack of penetration in both sheets.
1. Electrode dressing too flat.
2. Excessive electrode force.
3. Insufficient current.
4. Insufficient "on" time.

III. Unbalanced nugget.
1. Misalignment of electrodes.
2. Off-center dressing of electrode faces.

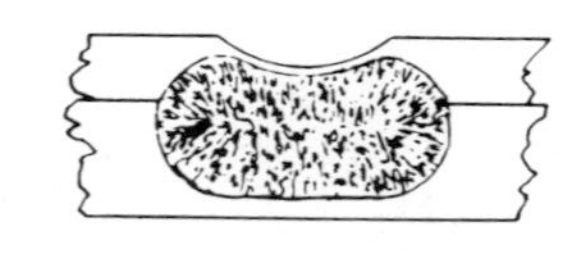

IV. Excessive indentation.
1. Electrode dressing too sharp at indentation.
2. Excessive electrode force.
3. Excessive current or "on" time.
4. Insufficient width of electrode at indentation.

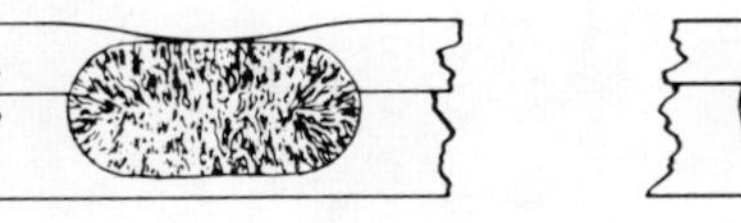

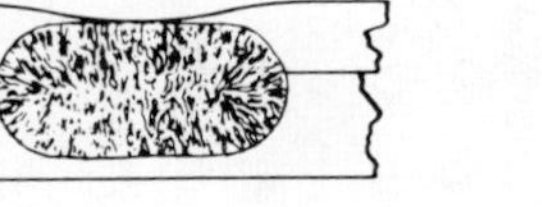

V. Excessive penetration.
1. Electrode dressing too sharp on thin sheet.
2. Insufficient electrode force.
3. Excessive current or "on" time.
4. Insufficient electrical or thermal conductivity of electrode.
5. Insufficient coolant.

VI. Void in nugget.
1. Insufficient electrode force.
2. Excessive current.
3. Excessive "on" time or roll speed.
4. Electrode dressing too flat.
5. Insufficient "off" time.

VII. Cracks in nugget.
A. Horizontal cracks at center.
1. Insufficient electrode force.
2. Insufficient "off" time.
3. Electrode dressing too flat.
4. Excessive current or "on" time.

VII. Cracks in nugget.
B. In crack sensitive metal.
1. Insufficient electrode force.
2. Electrode dressing too flat.
3. Insufficient "off" time.
4. Insufficient "on" time.

VIII. Cracks in parent metal.
1. Insufficient electrode width for nugget diameter.
2. Electrode dressing too sharp.
3. Excessive roll speed.
4. Excessive current or "on" time.
5. Insufficient "off" time.
6. Insufficient electrode force.

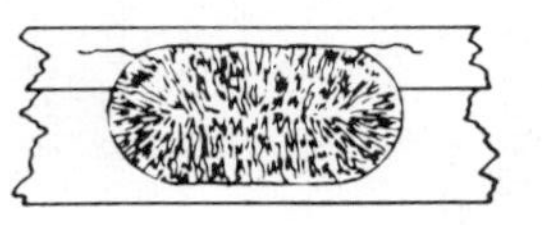

Fig. 13.21 Typical defects found in seam welding materials of unequal thickness (Courtesy of ITE Corporation).

higher the preheat temperature is, the slower the rate of cooling after welding. This reduced cooling rate minimizes the formation of cold cracks, reduces the magnitude of residual stresses, and reduces the hardness of the heat affected zone. All are factors which could act as defects under the proper circumstances.

In addition to minimizing the formation of immediate detrimental mechanical and physical defects, preheating can play a significant role in alleviating delayed cracking caused by hydrogen. Preheating of the weldment above 212°F drives off any adsorbed moisture that might be present. It also raises the ambient air temperature immediately adjacent to the weld, which prevents condensation resulting from changes in the ambient temperatures. Preheating further decreases any tendency toward hydrogen embrittlement by virtue of the decreased hydrogen solubility at elevated temperatures and increased diffusion rates. These factors effectively reduce the overall concentration of hydrogen in the weld.

13.3.2 Postheating

Postheating involves those heat treatments that are given after the completion of welding. They are generally employed to achieve a desired level of mechanical properties or toughness, to relieve residual stresses that might cause premature mechanical cracking or increase the stress corrosion susceptibility, to minimize any propensity towards delayed hydrogen cracking, and to achieve dimensional stability.

The time and temperature schedule that must be employed for postheating varies depending on the desired aim and the materials welded. Stress relief treatments are conducted below the critical temperature of the base metal, usually from 900 to 1300°F. The result desired is the elimination of residual stresses, but the treatment must be controlled to avoid grain growth and other deleterious effects.

The specific postheat schedule used to develop optimum mechanical properties is also dependent on the chemical composition of the alloy welded and its thickness. Table 13.2[12] shows the heat treatment recommended by the International Institute of Welding for a series of alloy steels. Data on postweld heating for specific alloys is also available from other sources, such as the producer of the alloy, or the publications of the American Welding Society (e.g., AWS D10.8–61, *Welding of Chromium-Molybdenum Steel Piping*. These heat treating schedules should be adhered to; deviations in the heat treatment can result in properties and structures markedly different from those desired.

Table 13.2 Recommendations of International Institute of Welding for Heat Treatment of Welded Alloy Steels

A	Type of steel	1.2 Mn	0.5 Mo	0.5 Mo 0.25 V	1 Cr 0.5 Mo	2.25 Cr 1 Mo	3 Cr 0.5 Mo 0.5 V	5 Cr 0.5 Mo	7 Cr 0.5 Mo	9 Cr 1 Mo
	A. S. T. M. Designation	A299	A204		A301 Grade B	A387 Grade D		A357	A213 Grade T7	A231 Grade T9
B	Typical analysis (%) C	~0.20	max. 0.20	max. 0.15	max. 0.15	max. 0.15	~0.20	max. 0.15	max. .015	max. 0.15
	Mn	1.2	0.6	0.6	0.6	0.6	0.6	0.6	0.5	0.5
	Si	0.3	0.3	0.3	0.3	0.3	0.3	0.3	0.7	0.7
	Cr					2.25	3	4.6	6.8	8.10
	Ni									
	V			0.2			0.5			
	Mo		0.5	0.5	0.5	1	0.5	0.5	0.5	1
C	Examples of application (°F)	Pipe work and pressure vessels for temperatures up to					Pressure vessels containing hydrogen up to 900 °F.	Pipe work and vessels for temperatures up to		
		885°	925°	975°	1025°	1065°		1200°	1250°	1300°
		Turbine parts and chemical plant						Chemical plant		Turbine parts
D	Section	Up to 5 in.				Up to 3 in.			Up to 3 in.	
E	Stress relief temperature (°F)	975 to 1200	1100 to 1300	1200 to 1300	1150 to 1350	1200 to 1375	1200 to 1375	1250 to 1375	1300 to 1375	1350 to 1425
	Time (min/in.) (not less than 30)	Up to 125				Up to 100		Up to 75		
	Heating (°F/min)	6	6	6	6	4.5	4.5	3	3	Slow
	Cooling (°F/min)	3	3	3	3	3	3	3	3	Slow
					Preferably immediately after welding before cooling down					
F	Technical benefits from treatment E (besides stress relief and dimensional stability)		Tempering of heat affected zone and weld deposit							
G	Preheating for thicker sections (°F)	400 to 750	200 to 400	400 to 575	400 to 575	400 to 575	400 to 575	400 to 575	400 to 575	400 to 660
H	Other heat treatment		Normalization is often specified. Heating to 1650 to 1750 °F. followed by annealing treatment E				1830 to 1885 °F., followed by annealing treatment E			
I	Scope of treatment H		Grain refining							

REFERENCES

1. W. F. Savage, *Symposium on Weld Imperfections*, *Palo Alto*, *1966*, A. R. Pfluger and R. E. Lewis, Eds., Addison Wesley, 1968.
2. *Welding*, *Brazing and Soldering Handbook*, Technical Brief 69–10264, Technology Utilization Division, NASA, Washington, D. C.
3. *Stainless Steel Fabrication*, Allegheny Ludlum Steel Corporation, Pittsburgh, 1958, pp. 31–97.
4. *Welding Handbook*, 4th ed., Sec. 1, American Welding Society, N. Y., 1957, pp. 8.0–8.4.
5. *Welding Inspection*, American Welding Society, New York, 1968.
6. H. Thielsch, *Materials Evaluation*, February 1969, p. 25–33.
7. G. E. Linnert, *Welding Metallurgy*, Vol. 2, American Welding Society, New York, 1967, p. 218.
8. J. C. M. Farrar and R. E. Dolby, *Metal Construction and British Welding Journal*, February 1961, pp. 32–39.
9. E. S. Szekeres, Ph.D. Thesis, Rensselaer Polytechnic Institute, February 1968.
10. M. L. Holzworth, F. H. Beck, and M. G. Fontana, *Corrosion*, Vol. 7, 441–449 (1951).
11. G. E. Linnert, *op. cit.*, p. 141.
12. *Mat. Prog.*, August 1956, p. 96

Index